MACMILLAN INTEGRATED SCIENCE Book 2

MACMILLAN INTEGRATED SCIENCE Book 2

John Allen
Jenny Crocker-Michell
Maggie Hannon
Richard Page-Jones
Tony Thornley

Editor: Mick Michell

MACMILLAN
EDUCATION

First published 1987
Reprinted 1987, 1988

Published by
MACMILLAN EDUCATION LTD
Houndmills, Basingstoke, Hampshire RG21 2XS
and London
Companies and representatives
throughout the world

Printed in Hong Kong

British Library Cataloguing in Publication Data
Macmillan integrated science.
Bk. 2
1. Science
I. Michell, Michael
500 Q161.2
ISBN 0-333-36226-8

Contents

Introduction for students and teachers

Although there are many good traditional textbooks on the market which concern themselves with the separate parts of science such as Chemistry and Biology we feel that those of you who have been wise enough to continue to study science as broadly as possible have been poorly served. The textbooks that exist are either designed to back up one particular course or are mixtures of bits taken from the separate Biology, Chemistry and Physics syllabuses. We set out with the idea of trying to write as modern a book as possible which would combine the best features of a traditional type of textbook, but which would also make your study of science broader and more interesting. In doing this our main aims are:

1. to provide you with a textbook which covers all that is required for GCSE, O-level and CSE science syllabuses;
2. to write for you, as far as we can, in an interesting and up-to-date way which includes many relevant examples;
3. to cut out any unnecessary scientific 'jargon' and to explain things in simple language so that you understand the difficult, as well as the simple, parts of science;
4. to help you to understand what a scientist does and what science is – and to emphasise that science is not just about remembering a lot of facts;
5. to help you to learn how best to approach your study of science so as to make it as interesting and relevant as possible;
6. and, most importantly of all, to help you to enjoy the study of science and to want to continue finding out about it.

The two books are divided into ten sections which we call *units*. Each unit contains a number of *topics*.

In each topic there are:

summaries of what you need to know at the end of each topic;
questions at the end of each unit to help you test your understanding;
assignments at regular intervals for you to do in school or at home. These are designed to extend your understanding or to reinforce it;
investigations which are aimed at getting you to look at science in a practical way.

Whilst we believe that the experiments can be carried out in safety and comply with the latest informed opinion, advice does change with time and it is the professional responsibility of the teacher to judge what is safe in his/her own circumstances.

Eye protection should be worn wherever chemicals are handled. A reminder about this is included in those experiments where the hazard is particularly great.

Mick Michell

Acknowledgements

The authors and publishers wish to thank the following who have kindly given permission for the use of copyright material:

Edward Arnold (Publishers) Ltd for table from *Illustrated Human Biology* by C.M. Wheeler
The Associated Examining Board for a question from Schools Council Integrated Science Project Examination Project Paper 1 (098/1), 1983
Daily Telegraph for extracts from 'Fish stocks hit by acid rain' by Austin Hatton, 30.10.83
Imperial Chemical Industries for adaption of article from *Steam*, Issue 2
Observer for 'Pines fight for life' by Robin McKie, 11.11.84 and extracts from 'The town America wrote off for cash' by Joyce Egginton, 27.2.83, 'Erosion sweeps Himalayan mountains down to the sea' by Geoffrey Lean, 29.5.86. and 'Rich harvest for organic farmers' by Paul Lashmar, 20.10.85.
Radio Times for 'Good Health' by Roger Woddis
Times Newspapers Ltd for 'The genius baby begins its climb' by Majorie Wallace, 21.8.83

The authors and publishers wish to acknowledge the following photograph sources:

Anglo-Chilean Society, page 186; Barnabys Picture Library, page 224 top right; British Aerospace, pages 70 right, 180; Jim Brownbill, pages 1, 62 right, 81 bottom, 83, 96, 118; from Rowlinson and Jenkin: *Human Biology*, diagram on growth curves and lifespans reproduced by kind permission of Cambridge University Press, page 132; Camera Press, pages 69 top right, 69 bottom right, 73, 78 top right, 79 top left, 217 left, 274; J. Allan Cash Ltd, pages 69 left, 79 bottom left, 136 bottom left, 136 bottom right, 140 bottom left, 224 bottom right, 246 bottom left, 269, 272 left; Central Electricity Generating Board, page 180; Clean Air Society, page 192; Bruce Coleman Ltd, pages 95 bottom, 156; Gene Cox, page 147 (a) and (b); Family Planning Information Service, page 134; Farmers and Stockbreeders, page 163 left; Food and Agriculture Organisation, pages 131, 261, 263, 264, 267; by kind permission of the Geological Museum (crown copyright), pages 5, 171; GeoScience Features, pages 5, 171, 181; by kind permission of Her Majesty's Stationery Office (crown copyright), pages 3, 68; ICI Agriculture Division, page 30; JBP Associates, page 194; Rodney Jennings, pages 28, 36, 39, 52, 57 right, 81 top, 106, 154 top, 154 middle, 180, 183 bottom, 187, 204, 225, 226, 228, 246 top, 251, 270 left; Mansell Collection, page 184; Nature Photographers Ltd, pages 94 left, 124 left, 140 top left, 176; the *Observer*, page 130; Photo Source, pages 46 right, 221 left, 244, 249, 270 right, 272 right; by kind permission of the Commissioner of the Police of the Metropolis, page 152; Popperfoto, pages 31 bottom, 53, 57 left, 62 left; Rolls Royce plc, page 70 top left and bottom left; RoSPA, page 44; M. Rowland, page 154 bottom; Science Photo Library, page 147 (c); by kind permission of Shell from the Shell Book of How Cars Work, page 78 left; Shell Photographic Service, pages 248, 250; Spear and Jackson Tools Ltd, page 182 top; Stanton and Staveley Ltd, page 183 top; Topham Picture Library, pages 10, 67, 78 bottom right, 217 left, 224 top left, 224 bottom left, 277, 281; Transport and Road Research Laboratory (crown copyright), page 63; Vauxhall Motors, page 180; C. James Webb, page 140 bottom right; World Health Organisation, page 278; World Wildlife Fund/Tony Eckersley, page 268; Zoological Society of London, pages 1, 163 right.

Illustrations by Illustra Design.

The authors and publishers would like to thank Dr T. P. Burrows, Chairman of the ASE Laboratory Safeguards Sub-committee, who offered advice on safety.

Every effort has been made to trace all the copyright holders but if any have been inadvertently overlooked the publishers will be pleased to make the necessary arrangement at the first opportunity.

6. Classification

CLASSIFICATION SYSTEMS

Football teams are sorted into divisions with each division containing teams of a similar standard. The 'For Sale' page of a newspaper has sections for things like houses, cars and pets. It has a separate heading for each type of item. This page is often called the 'classified' section of the paper because the advertisements are put into groups, or *classified*, as they come in.

A CLASSIFICATION SYSTEM is a set of labels that we can use to put things into groups. The labels refer to properties, so things in the same group have similar properties; those in another group have a different set of properties. Whenever we have a great deal of information we can store it

Fig. 6.1 What do these pictures all have in common?

more easily if we classify it. If we find a good way to do this, then we can

find information quickly;
add new information easily;
use the pattern of groups to make predictions and help in the hunt for new information.

You will have used at least one of the classification schemes shown in Fig. 6.1. If you go to the supermarket to buy cornflakes you probably look for the sign saying 'Cereals'; if you want to see lions at the zoo, you follow the signs to the Big Cats.

People carrying out scientific investigations have always tried to classify the data they collect. Not only can a classification scheme help us to store information where we can find it again, but a good scheme can highlight an unexpected pattern and help us to make useful predictions. Do Investigation 6.1 now.

In Investigation 6.1, each time you grouped the elements, you used a different property. You were given three properties, but you could have used others if you had the information. You might have decided to group the elements by the number of letters in their name, when they were discovered or how expensive they are. Some of these may be useful classifications, others may not.

The 'trick' in making a good classification is to make a wise choice of properties as the labels for the groups. If we choose well, then all the information fits easily into groups, and new information can be added as easily.

Classifications can change. The information can be re-grouped to suit the needs of the person using it, or as new information comes along. For example, a record shop displays its stock under labels like 'Classical', 'Solo Artists' or 'Groups'. You might decide to store your own collection in alphabetical order by title if that method makes it easier for you to find a record. It is the same with scientific information. The way it is grouped can also change. This unit deals with some of the most well-known scientific classifications.

As you study them, remember that today's version could change when new information is discovered, or when someone invents a better system. People who use the classification schemes a great deal often disagree about the best versions, as you will see later. Sometimes, though, we rely upon a classification scheme staying the same. If all doctors did not use the same scheme for classifying blood types, it could be dangerous for many patients.

CLASSIFYING BLOOD TYPES

In Unit 2 you learned that blood contains red cells, white cells and platelets in a liquid called

Investigation 6.1 Putting elements in groups

The table below contains some information about a number of elements. Use it to put the elements into the following groups:

(a) solids, liquids, gases;
(b) metals, non-metals;
(c) ion charge (+1, +2, +3, −1, −2, −3 etc.).

1 Did you put the elements into the same groups each time?
2 Can you find a better way to group the same elements?

Element	State at room temperature	Conducts electricity?	Appearance	Charge on ion
Bromine	liquid	no	red-brown	−1
Calcium	solid	yes	silvery	+2
Carbon	solid	yes	black, slippery	−
Chlorine	gas	no	greeny yellow	−1
Magnesium	solid	yes	silvery	+2
Mercury	liquid	yes	silvery	+2
Oxygen	gas	no	colourless	−2
Sodium	solid	yes	silvery	+1
Sulphur	solid	no	yellow	−2

plasma. At the beginning of this century it was found that mixing samples of blood from different people could have worrying results. Sometimes the red cells in the mixture clumped together. To make it safe for one person to be given another's blood in a transfusion, doctors needed to classify blood types into those that were safe to mix and those that made harmful clumps of red cells.

It is possible to divide people into four groups on the basis of their blood type. These blood groups are called A, B, AB and O. If you know your own blood group, use Table 6.1 to find out which types of blood your body will accept in a transfusion. One way of testing blood to find out to which group it belongs is to mix it with both type A and type B blood. By noting what happens, and using the sort of key shown in Table 6.2, your blood group can be decided. Look at the key and you may be able to see the link between the names of the groups.

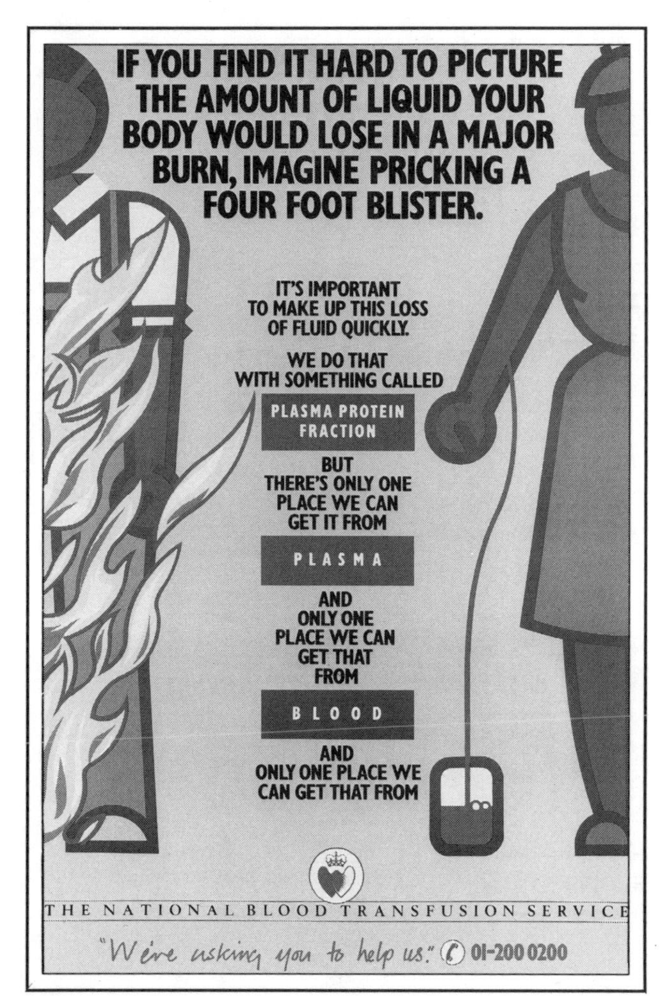

Fig. 6.2

Table 6.1 Blood transfusions which are safe

Group	Can give to	Can receive
A	A & AB	A & O
B	B & AB	B & O
AB	AB	all groups
O	all	O

Table 6.2 Key to blood groups

If your red cells ...	Your blood group is ...
clump with group A	B
" " group B	A
" " groups A & B	AB
don't clump with group A or B	O

Once you are 18 years old, you may decide to become a blood donor, and help to provide blood for use in operations and treatment of accident victims. Table 6.1. shows how the different blood types can be mixed safely.

The blood that donors give is used for more than just transfusions. Many important substances can be taken from blood. One of these is called 'Factor 8', which is used to help people suffering from haemophilia. As you will see in Unit 8, this is an inherited disease which prevents the blood from clotting properly. A haemophiliac will bleed heavily from a very small wound. Treatment with 'Factor 8' makes it possible for a haemophiliac to have injections or operations when necessary. These operations would otherwise cause them to lose a huge quantity of blood.

The Rhesus system

There are, in fact, many classification systems for blood types. The ABO system is probably the best known, but there is another, separate system, developed in 1940. About that time it was discovered that most people had blood containing a substance named the RHESUS FACTOR (because it was also found in the blood of Rhesus monkeys). People with the Rhesus factor in their blood are labelled *Rhesus positive (Rh+)* and those without it, *Rhesus negative (Rh−)*.

A problem arises if a Rh− woman is expecting a Rh+ baby. As the baby is born, the contractions of the womb may allow some of the baby's

Rh+ blood cells to pass into its mother's bloodstream. The woman's blood will develop substances called *Rh+ antibodies*, which will destroy any Rh+ blood cells. If she is pregnant again, and the baby is also Rh+, then these Rh+ antibodies will pass from her blood into the baby's bloodstream. Once there, they will destroy the baby's red blood cells, which will cause it to be stillborn.

By testing the blood of mother and father it is possible to discover whether this situation is likely to arise. You will learn more about the way a baby inherits characteristics from its parents in Unit 8, but Fig. 6.3 summarises the possibilities for a woman with Rhesus negative blood.

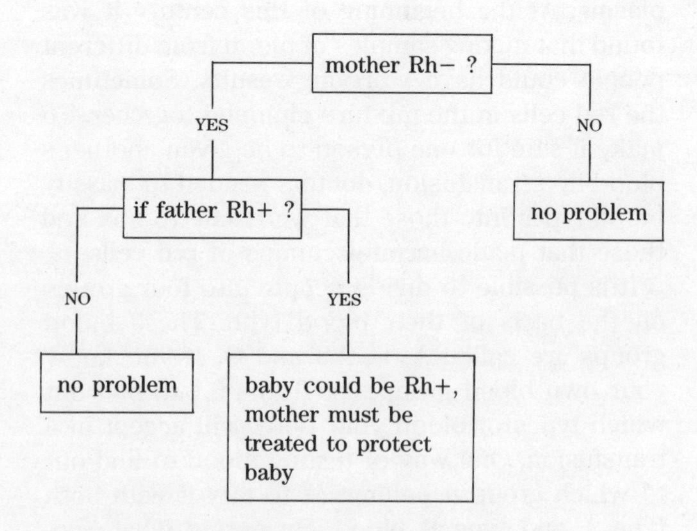

Fig. 6.3 Key to Rhesus factor problems

FINDING THE RIGHT KEY

The method chosen to present classified information can be very important. One useful way is to use the information to make a key. In Fig. 6.4 you can see two different layouts of the same key.

In Book 1 you learned that rocks can be metamorphic, igneous or sedimentary, and Fig. 6.4

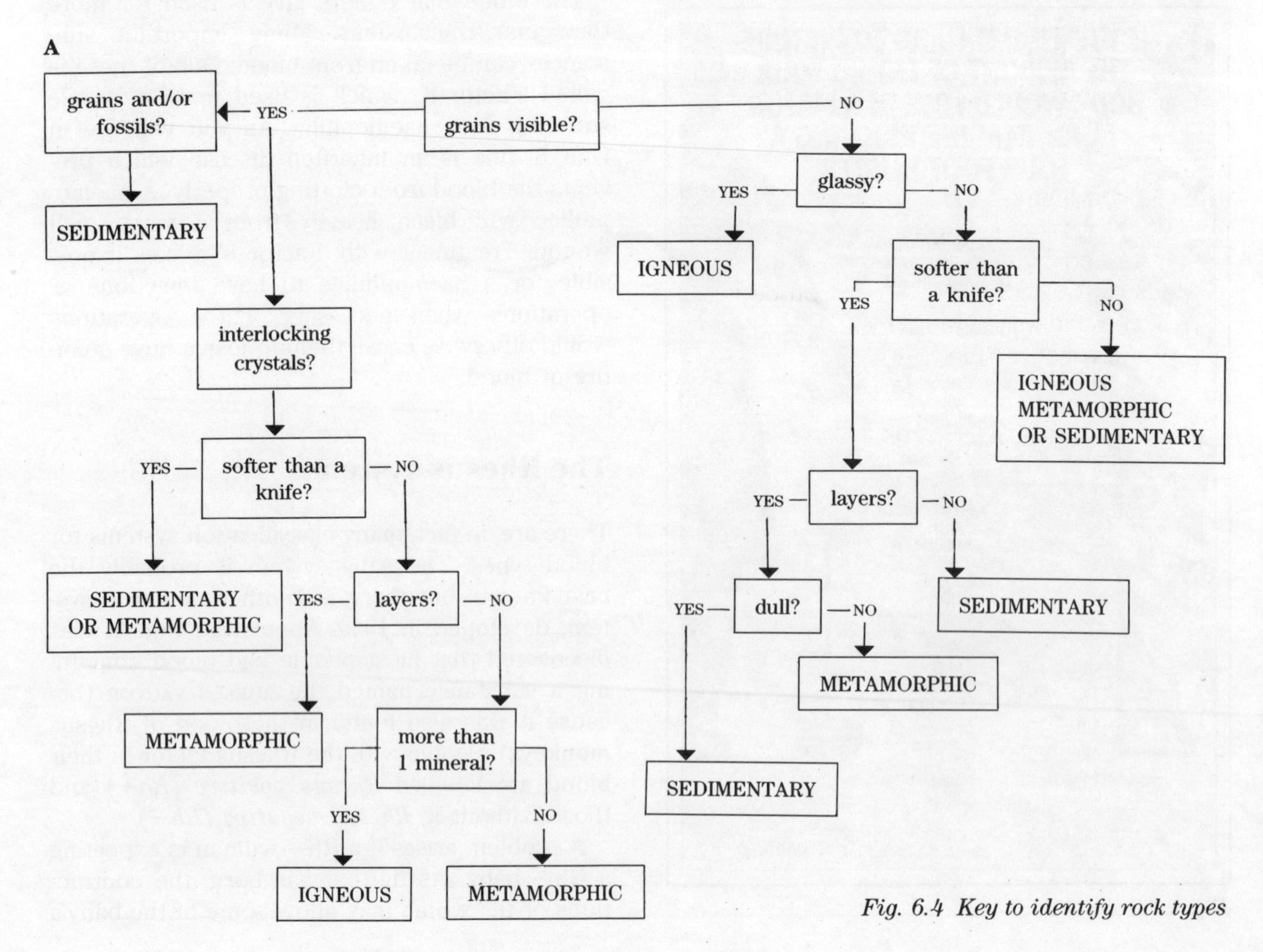

Fig. 6.4 Key to identify rock types

B

Question	Answer	Result
1 Grains visible?	yes	Go to 2
	no	Go to 6
2 Clear grains and/or fossils?	yes	SEDIMENTARY
	no	Go to 3
3 Harder than a knife blade?	yes	Go to 4
	no	SEDIMENTARY or METAMORPHIC
4 Layers visible?	yes	METAMORPHIC
	no	Go to 5
5 More than 1 mineral?	yes	IGNEOUS
	no	METAMORPHIC
6 Glassy rock?	yes	IGNEOUS
	no	Go to 7
7 Harder than a knife blade?	yes	IGNEOUS/METAMORPHIC/ SEDIMENTARY
	no	Go to 8
8 Layers visible?	yes	Go to 9
	no	SEDIMENTARY
9 Dull rock?	yes	SEDIMENTARY
	no	METAMORPHIC

shows some of the properties of each of them. Both of the keys in Fig. 6.4 ask the same set of questions of any rock sample you may want to identify. Study the two keys and you will see that each question 'asks' if the rock has a particular property. There are seven different questions, one for each property, but the layout of the key may mean that the same question appears more than once. You will probably find that you prefer one of the keys, because the pattern of properties seems easier to see. Divide your classmates into two groups:

(a) those who prefer the same key as you;
(b) those who prefer the other key.

When you have to make a key you must choose the properties that will separate the information clearly and quickly. As a rule, if you

Assignment – Identifying minerals

graphite

galena

zinc blende

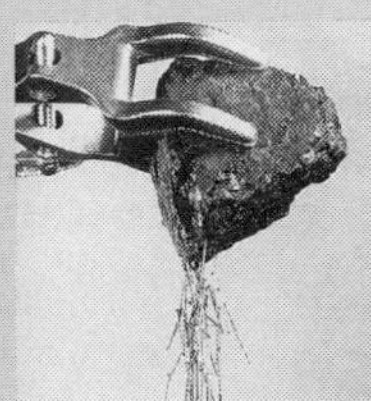
magnetite

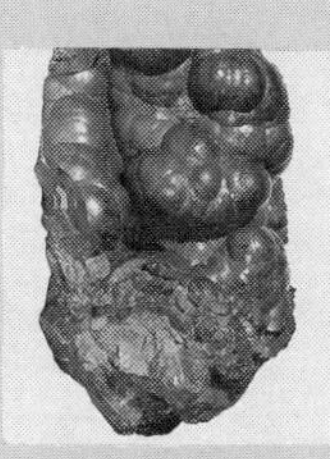
haematite

garnet

asbestos

As you learned in Unit 1, minerals are the building blocks of most rocks. They can be identified by using five of their properties. The table summarises these properties for seven minerals. Use the information in it to make a key to the minerals.

Lustre – minerals that shine like a metal have a *metallic lustre*.
Streak – the colour of the powdered mineral.
Cleavage – the mineral breaks into flakes or plates.

Mineral	Hardness	Colour	Metallic lustre?	Streak	Cleavage easy?
Graphite	softer than knife	lead grey	yes	dark	no
Galena	softer than knife	grey	yes	dark	yes
Zinc blende	softer than knife	yellow/brown	yes	light	yes
Magnetite	harder than knife	black	yes	dark	no
Haematite	softer than knife	red/brown	no	no	no
Garnet	harder than knife	red, brown, green crystals	no	no	no
Asbestos	softer than knife	green/white fibres	no	no	yes

have seven 'bits' of information to identify, your key will need to have seven questions. Six 'bits' will need six questions, and so on. Then you will need to try to arrange those questions in a way that means each question is asked only once. As you can see in the two versions of the rock key, this has not been possible. Many people have tried to make rock keys, but they have found that it is not easy to do in a single key. Figures 6.4 A and B show why, for there are several sorts of each rock type. We really need further keys to each of metamorphic, igneous and sedimentary rocks if we want to be able to name individual rocks within these groups. However, if we only need to discover whether a rock is sedimentary, then the keys work well.

Assignment – Grouping ions by their reactions

The tables give you some information on the way the most common ions react. Some of the information may not be new to you and your teacher may give you some extra information. If you were given a chemical whose name you did not know, you could identify the ions it contained by doing the tests shown in the table. However, it would take quite a time to do all of them and you might not have a big enough sample of the chemical to do every one.

Use the information given to you to make two keys: one to identify the negative ions, and the other to identify the metal ions. Use the smallest number of steps you can in your key. Some of the terms in the table are explained, but you may want to refresh your memory by looking at the work on ions that you did from Unit 5.

Your teacher may let you try out your key with some 'nameless' compounds, in which case you will be given instructions on how to carry out the tests. Remember to wear eye protection.

A. Reactions of some cations

Metal	Charge on ion	Flame test (on solid)	Sodium hydroxide solution added to solution of the ion	Ammonia solution added to solution of the ion
Al	3+	–	white solid, dissolves in excess to clear solution	white solid, insoluble in excess
Ba	2+	apple green	–	–
Ca	2+	brick red	white solid, soluble in excess	–
Cu	2+	blue-green	blue solid, insoluble in excess	blue solid, dissolves in excess to *dark* blue solution
Fe	2+	–	green jelly-like solid, insoluble in excess	green jelly-like solid, insoluble in excess
Fe	3+	–	brown jelly-like solid, insoluble in excess	brown jelly-like solid, insoluble in excess
K	1+	lilac	–	–
Na	1+	yellow	–	–
Zn	2+	–	white solid, dissolves in excess to clear solution	white solid, dissolves in excess to clear solution

B. Reactions of some anions

(Wear eye protection if carrying out these tests.)

Test (NB: All chemicals used are corrosive)	Anion	Symbol	Reaction
A few drops of dilute nitric acid and a few drops of silver nitrate solution	chloride bromide iodide	Cl^- Br^- I^-	white solid, silver chloride creamy solid, silver bromide yellow solid, silver iodide
Dilute hydrochloric acid	carbonate sulphite sulphide	CO_3^{2-} SO_3^{2-} S^{2-}	carbon dioxide given off sulphur dioxide given off (choking gas) hydrogen sulphide given off (poisonous gas)
A few drops dilute hydrochloric acid and equal volume barium chloride solution (poison)	sulphate	SO_4^{2-}	white solid, barium sulphate
Brown ring test*	nitrate	NO_3^-	brown ring where two layers meet

* Equal volume of iron(II) sulphate solution is added to test solution. A small volume of concentrated sulphuric acid is poured in, with *very great care*, so it makes a layer on the bottom.

Classification systems

WHAT YOU SHOULD KNOW

1 A classification system is a set of labels that allows us to put things in groups.

2 There are different types of classification systems and the one we pick depends on how we want to use it.

3 Classification systems can change if our needs change, or if our information changes.

4 A good system will let us
   find information quickly;
   add new information easily;
   highlight patterns to help us make predictions.

5 One example of a system that does not change is that of blood grouping. There are four blood groups, A, B, AB and O.

6 Blood can contain other substances, amongst them 'Factor 8' and, sometimes, the Rhesus factor.

7 'Factor 8' can be taken from donated blood, and given to haemophiliacs to help their blood.

8 A key is one version of a classification system. It is usually used where things need identifying. There is often more than one way of arranging a key.

9 To make a key, list the properties about which you have information. Then arrange a table, or a list of questions where each box in the table, or question, refers to just *one* property.

10 Rock types can be identified using a key.

11 It is sometimes not possible to make a complete identification in just one key. So far no-one has been able to do so for rocks.

QUESTIONS

1 Find out about some other classifying systems such as the ones for classifying books in libraries, for railway engines or for films. Write a few sentences to describe how the classification works, its good points and its disadvantages.

2 Make a list of the different types of dwelling in your area. Use the information to classify the types of living accommodation.

3 Early experiments with blood transfusions were either very successful or produced distressing symptoms.

   (a) If, in one set of experiments, all the patients had blood group A, which blood groups would give a successful transfusion?

(b) What happens if the blood mixture is unsuccessful?

(c) Explain why an unsuccessful mixture may cause 'distressing symptoms'. Your teacher may help you.

4 A medical report says 'Rh− women of child-bearing age should not be given Rh+ blood in transfusion'.

(a) Explain what Rh− and Rh+ mean.

(b) What happens if an Rh− person is given Rh+ blood?

(c) Why is it important *women* of child-bearing age should not be given Rh+ blood?

5 Use the information in the table to make a key to identify gases.

Gas	Colour change with litmus	Other tests
ammonia (NH_3)	red turns blue	White smoke with HCl fumes
bromine (Br_2)	blue turns red	Is brown. Starch/iodide paper turns blue
carbon dioxide (CO_2)	blue turns pink	Limewater goes milky
carbon monoxide (CO)	no change	Burns with a blue flame
chlorine (Cl_2)	blue turns red then white	Is green. Starch/iodide paper turns blue
hydrogen (H_2)	no change	Lights with a 'squeaky' pop
hydrogen chloride (HCl)	blue turns red	White smoke with NH_3
hydrogen sulphide (H_2S)	blue turns red	Lead nitrate paper goes black
nitrogen (N_2)	no change	No tests work
sulphur dioxide (SO_2)	blue turns red	Acidified dichromate paper turns green

Classification

CLASSIFYING LIVING ORGANISMS

For thousands of years people have invented classification systems for living organisms. Different people used different groupings. By the sixteenth century most systems started by dividing organisms into animals and plants. These two groups, or KINGDOMS, were very large and, as more new plants and animals were discovered, they got even larger. Most of the people studying these plants or animals invented their own classification, which made life difficult when they needed to compare notes or to talk about their work. At last, in 1735, a Swede called Carolus Linnaeus came up with a classification system which worked so well that everyone used it, and we use it still. As you have just seen in the previous topic, a good classification is one that helps you sort out information, add new information and spot patterns. That is why Linnaeus' system has lasted. With just a few words you can identify an organism and see how it is related to all the others.

LABELS IN THE LINNAEAN CLASSIFICATION

Table 6.3 Group labels in the classification of living organisms

KINGDOM	a group of phyla
PHYLUM	a group of classes
CLASS	a group of orders
ORDER	a group of families
FAMILY	a group of genera (the plural of genus)
GENUS	a group of species
SPECIES	individuals can breed

The groups are arranged so that they make a 'ladder'. Each level has a special group name, as you can see in Table 6.3. Organisms from the same level have more in common than those on a higher level. The smallest group in the classification system is called the SPECIES. Organisms in one species are so alike that they can mate to produce fertile young (see Unit 8).

Often a number of different species share quite a few properties. As you can see in Table 6.3, such a group is called a GENUS. Working up the ladder, which group will contain organisms with the least number of common features?

Assignment – Mnemonics

To remember the order of the groups in Table 6.3, a student used '*K*ing *P*runes *C*ome *O*nly *F*rom *G*ood *S*tones'. You can see that each initial letter is the same as the initial letter of the group labels in Table 6.3. Such a memory aid is called a *mnemonic*. Can you make up mnemonics for:

(a) the colours of the visible spectrum (Unit 3);
(b) the planets in the solar system (Unit 1);
(c) one other group that you need to remember?

This system is easy to use, but at first it caused a few problems until everyone agreed to use the same sort of properties for deciding where an organism belonged. For example, bats, birds and some insects could all belong to the same group because they can all fly. However, if you look at Fig. 6.7 you can see that they are not even in the same phylum! Over the years people who studied many living organisms discovered that the sensible way to decide how to group them was to compare how they were built.

In Unit 8 you will learn about the work of

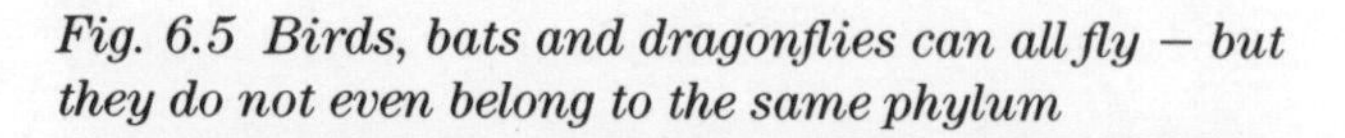

Fig. 6.5 Birds, bats and dragonflies can all fly – but they do not even belong to the same phylum

Charles Darwin and his theory of evolution. His efforts also helped to improve the classification system for living organisms. Most versions of the classification system now put organisms into one of *three* kingdoms. Organisms that do not easily fit into either the animal or plant kingdom are grouped together in the PROTISTA kingdom.

In Fig. 6.7 you can see examples of organisms in each of these three kingdoms. In each phylum the features that its organisms have in common are written at the top. As you can see, some phyla (the plural of phylum) have been divided into classes; in particular there are six classes of vertebrates (which includes the two classes of fish).

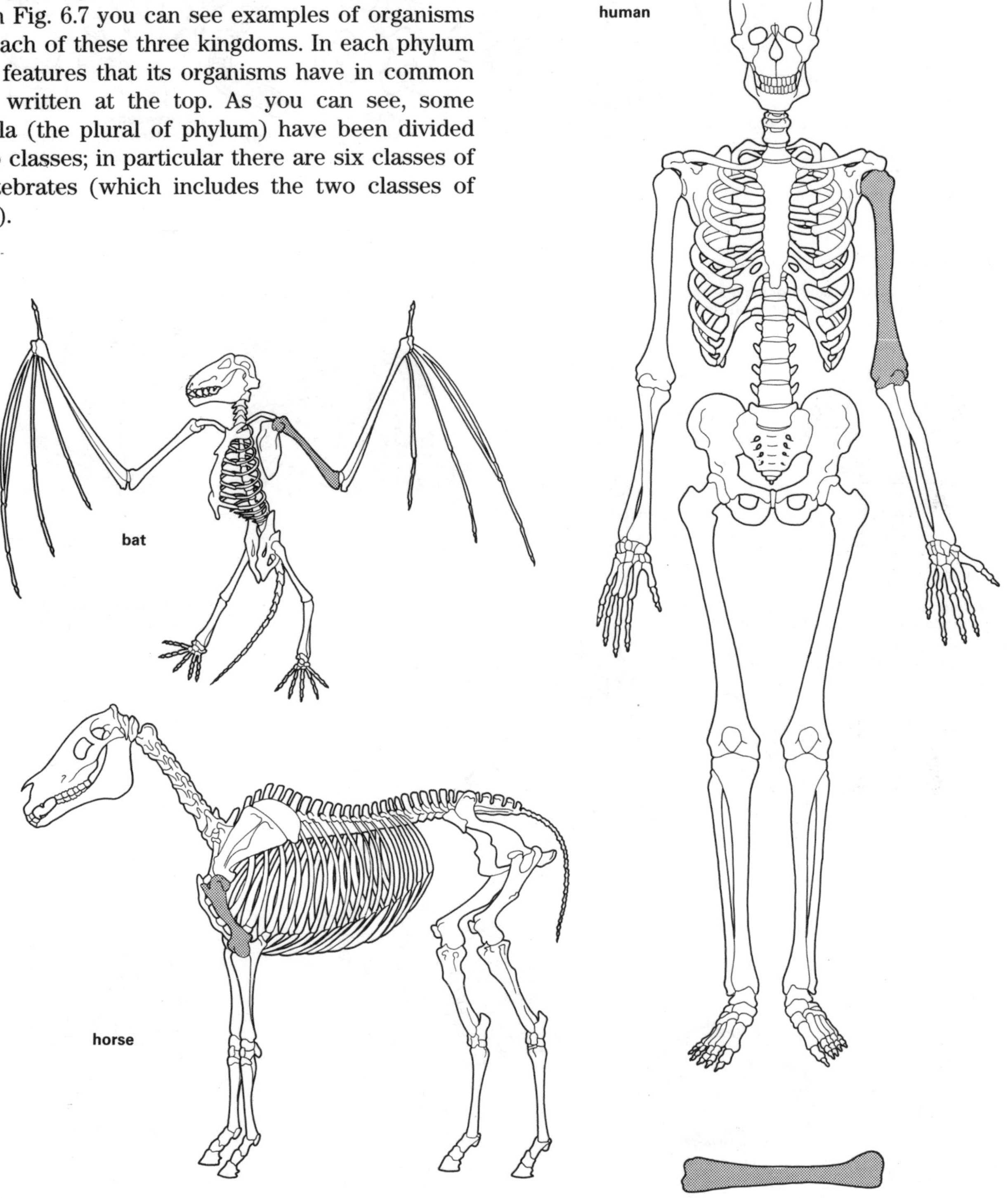

Fig. 6.6 Can you see how alike the bone structures of different animals can be?

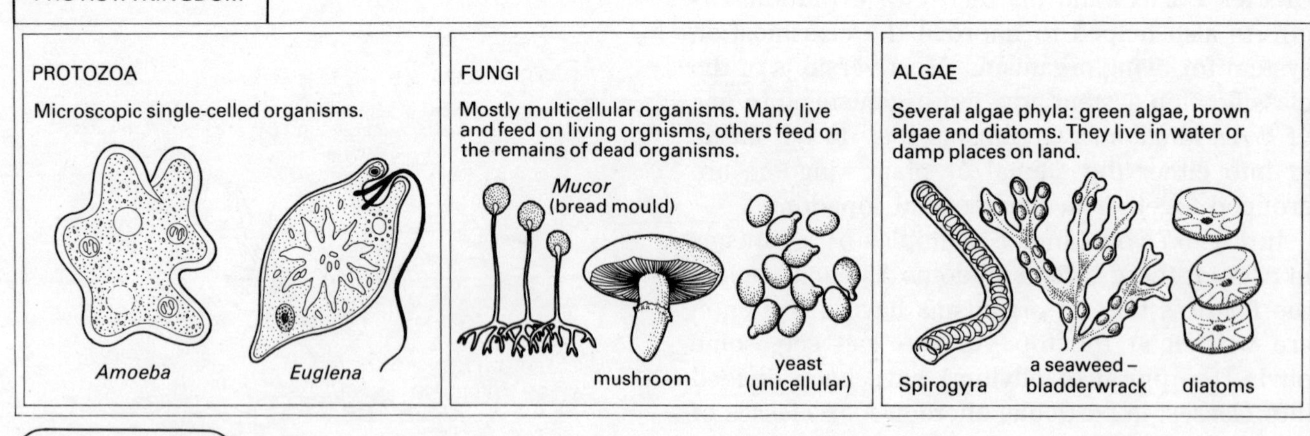

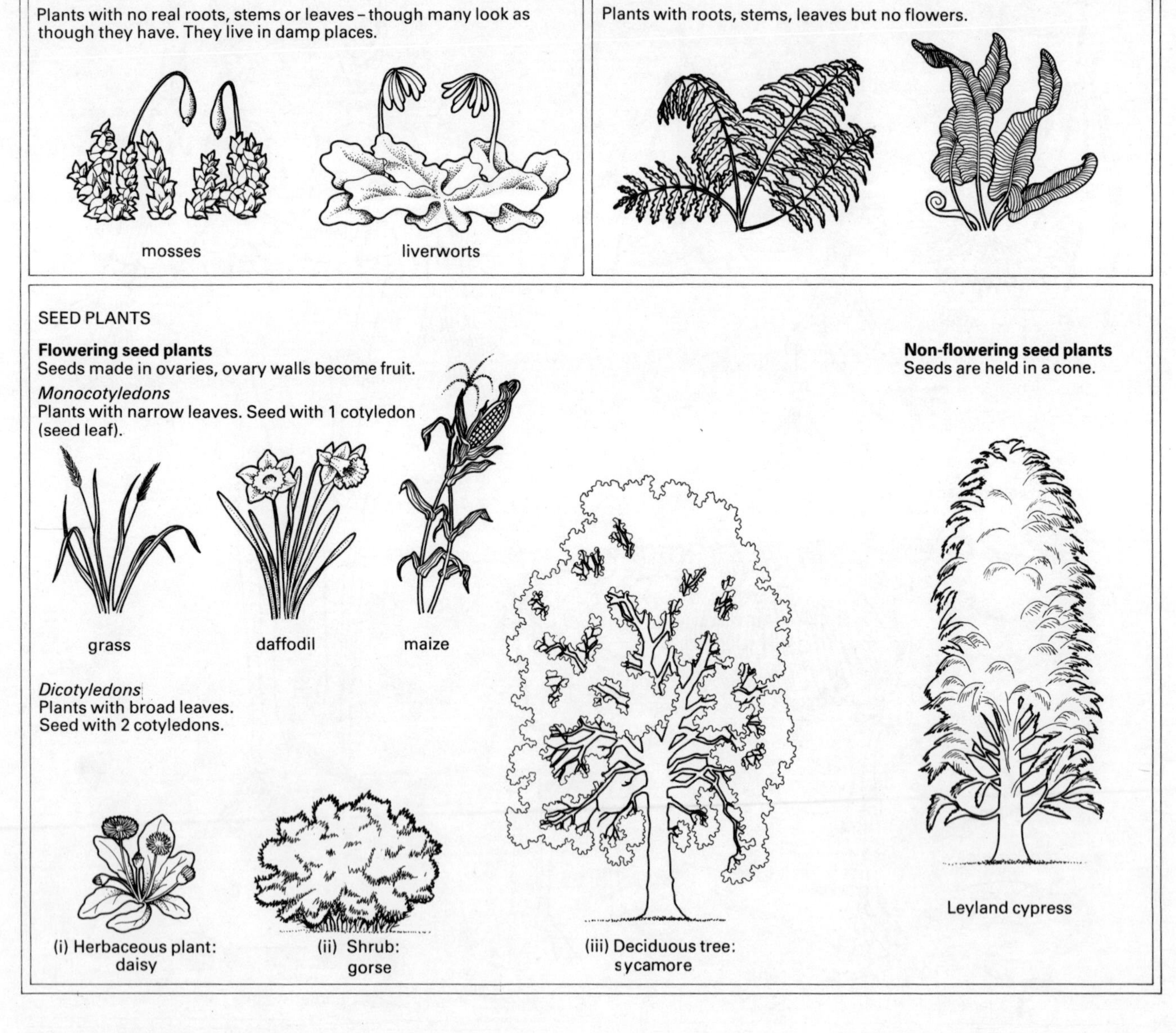

Fig. 6.7 Classifying living organisms

ANIMAL KINGDOM

INVERTEBRATES

COELENTERATES

Most live in the sea. Have mouth at one end, tentacles around.

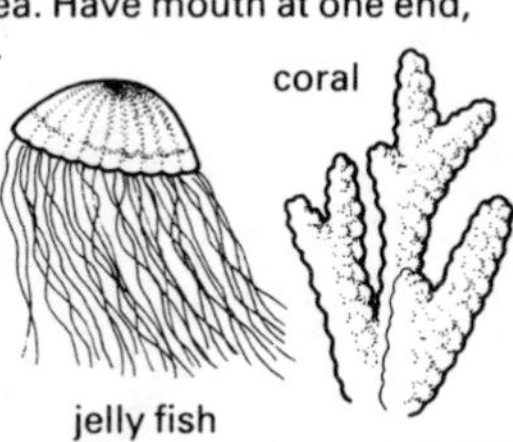

TRUE WORMS

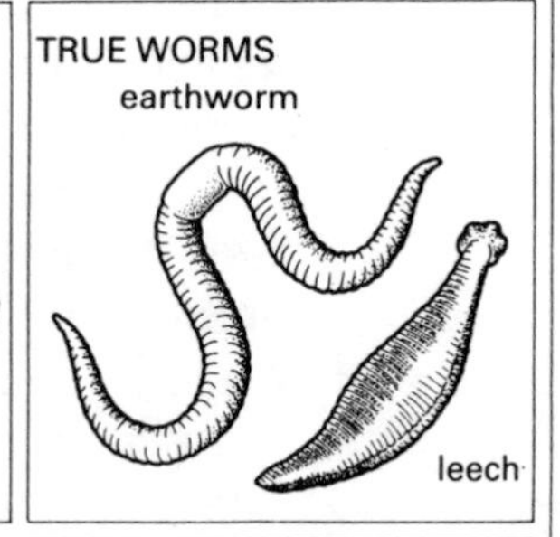

MOLLUSCS

Have soft bodies inside one or two shells

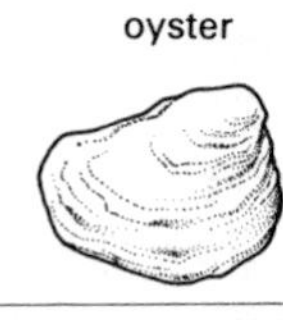

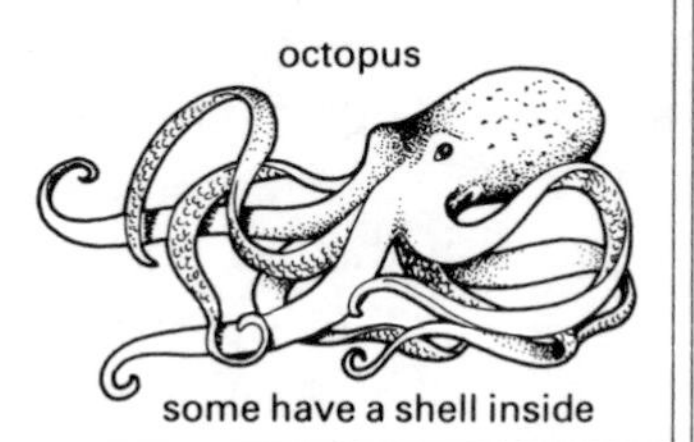

ECHINODERMS

Sea animals with spiny body, often in 5 segments arranged like wheel spokes.

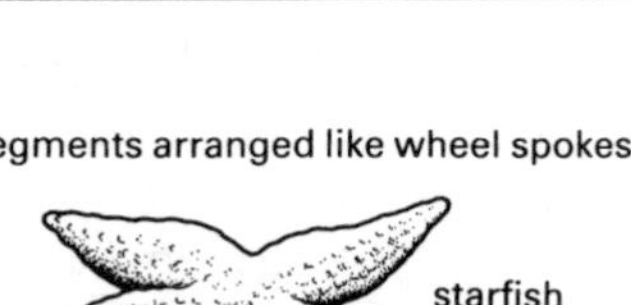

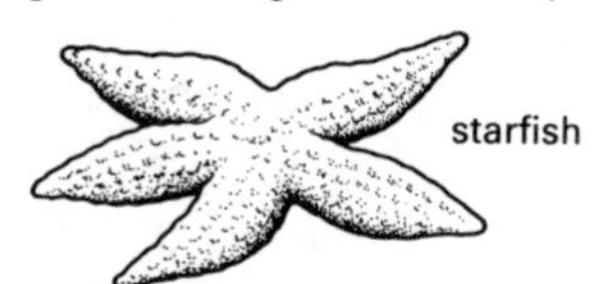

ARTHROPODS

Have body made in segments, protected by a hard jointed covering called an exoskeleton. Legs are jointed and come in pairs.

Crustaceans
Have 2 pairs of antennae

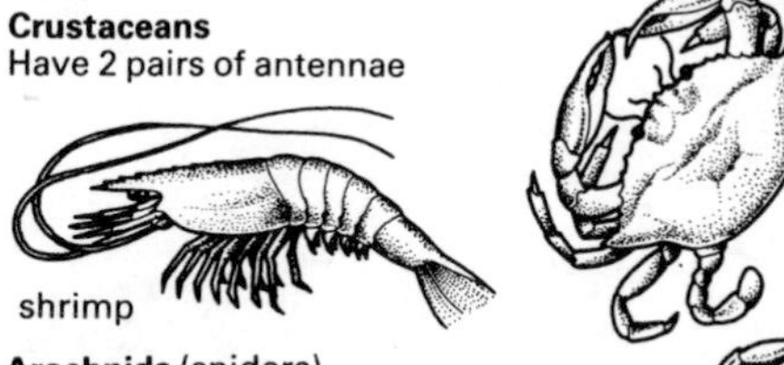

Arachnids (spiders)
Have no antennae; 4 pairs of legs.

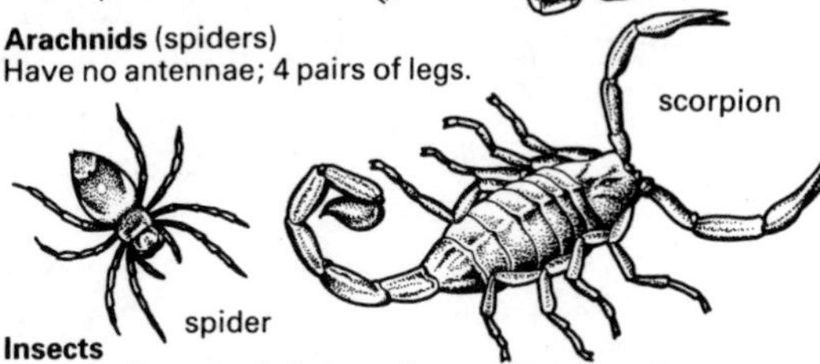

Insects
3 pairs of legs, body in 3 sections, often have wings.

Myriapods
Long body in many sections, each with a pair of legs.

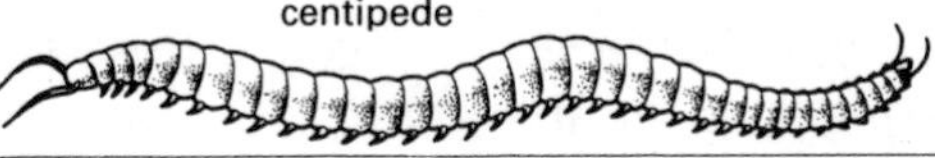

VERTEBRATES All have backbones.

COLD BLOODED VERTEBRATES: body temperature can change

FISH
Some have a skeleton of cartilage:
ray

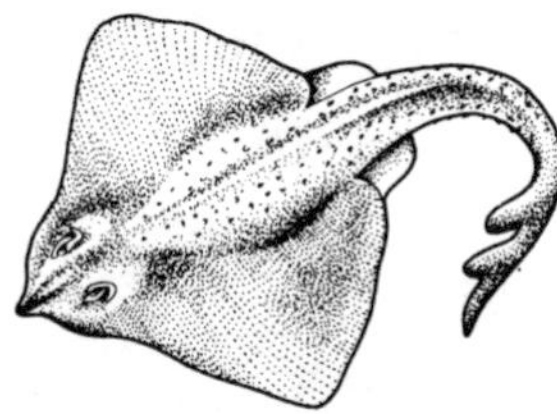

Some have a skeleton made of bone: trout

AMPHIBIANS
Live on land, lay eggs in water.

REPTILES
Dry scaly skin, lay eggs with leathery skin, on land.

WARM BLOODED VERTEBRATES: constant warm body temperature

BIRDS
Feathers, front limbs are wings.
Most can fly.

sparrow

MAMMALS
Hairy skin, young are born alive and feed on mother's milk.

dolphin

human

monkey

bat

kangaroo

horse

Assignment – Using the Linnaean classification system

Background information

Every organism is given a two-part name, which we use as its scientific name.

	1st part	2nd part
Name :	1st part	2nd part
Tells you :	Genus	species

Examples:

Mus	*domesticus*	house mouse
Gallus	*domesticus*	domestic chicken
Canis	*familiaris*	domestic dog
Canis	*lupus*	wolf
Canis	*latrans*	coyote

Notice that we often say the name the other way: *Mus domesticus* becomes house mouse. Notice also that the genus name always has a capital letter: it can be shortened, or used alone, but it must never be left out. For example, you can talk about organisms in the genus *Canis*, or even about *C. familiaris*, but to talk about *domesticus* does not make clear whether you mean a mouse or a chicken!

Here are some house plants, with their scientific names. Study the diagrams and then answer the questions which follow.

1. Use the information in the diagram, and your understanding of the classification system, to put the house plants in their correct groups.
2. The information in the diagram can be found in a plant catalogue, which always gives the scientific name as well as the common name. Explain why the makers of the catalogue feel it is sensible to give the two names rather than just the simpler common name.
3. Use a library, reference books in the science department, or even a plant and seed catalogue, and find out:
   (a) the scientific names for some of the plants you know (they could be house plants or ones growing in a garden);
   (b) four plants which have more than one common name.

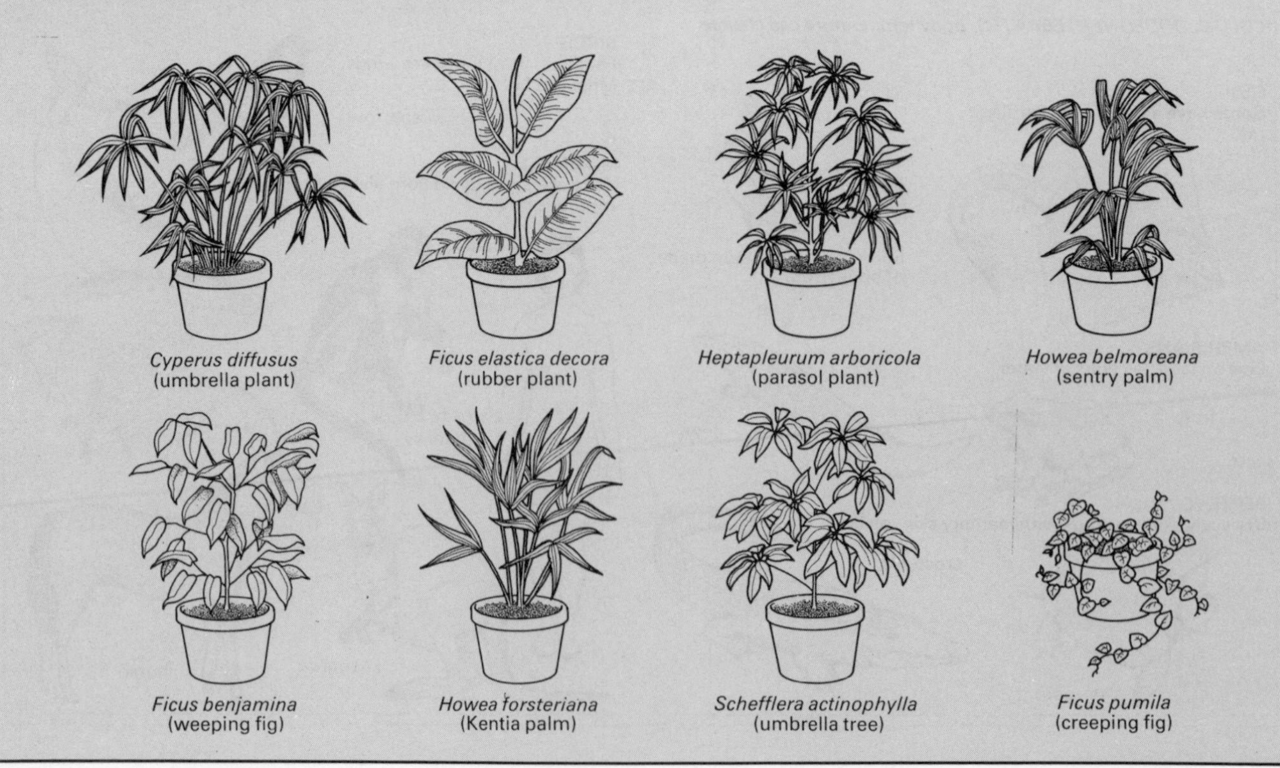

Cyperus diffusus (umbrella plant)
Ficus elastica decora (rubber plant)
Heptapleurum arboricola (parasol plant)
Howea belmoreana (sentry palm)
Ficus benjamina (weeping fig)
Howea forsteriana (Kentia palm)
Schefflera actinophylla (umbrella tree)
Ficus pumila (creeping fig)

WHAT YOU SHOULD KNOW

1 Living organisms vary in many ways and we can make information about them easier to handle by grouping them.
2 Organisms can be classified into groups on the basis of the differences between them.
3 Organisms in the same group have more in common than organisms in different groups.
4 Our classification system is based mainly on the structure of organisms.
5 The groups in the system are:

KINGDOM
PHYLUM
CLASS
ORDER
FAMILY
GENUS
SPECIES

Organisms in each level have less differences than those in the levels above them.
6 The system we usually use has three kingdoms: plants, animals and protista.
7 By international agreement each organism is given two names. The first of these indicates its genus, the second its species.

QUESTIONS

1 Which of these do not have flowers: grass, fern, daisy, oak, moss?
2 David Wellaway found a new plant on his travels. When he next appeared on TV he said it was a *dicotyledon*, and pointed out the features that helped him classify the plant.

Which features should he have found if the plant has been correctly classified?

Classification

CLASSIFYING THE ELEMENTS

ELEMENTS IN DISORDER

Unlike living organisms, there was no need to classify the elements until a couple of hundred years ago. It was about the middle of the eighteenth century that new elements began to be discovered nearly every year. Before long there was so much information that it needed sorting out to make it more useful. The first scheme sorted the elements into Metals and Non-metals, and Tables 6.4 and 6.5 show the features of each of these groups.

Table 6.4 Physical properties of metals and non-metals

Property	Metals	Nonmetals
Appearance	bright/shiny	dull
Conducts heat	yes	no
Conducts electricity	yes	no
Workable	bends without breaking, can be hammered into shape, drawn into a wire	brittle
Strength	yes	no
Melting temperature	high	low
Boiling temperature	high	low
Density	high	low
When hit	makes a ringing sound	does not 'ring'

Table 6.5 Some chemical properties of metals and non-metals

Property	Metals	Non-metals
Reaction with dilute acid	many give hydrogen	do not react
Oxides	are insoluble or make a basic solution	are acidic
Chlorides	unreactive	very reactive
Compounds with hydrogen (hydrides)	unstable and break down easily	stable

As you have learned, a classification may not work for every item. In this classification of the elements there are some that do not fit easily into either group. Here are two examples:

Boron – has most of the non-metallic properties but melts at a very high temperature.

Mercury – has most of the metallic properties but its melting temperature is so low that it is liquid at room temperature.

One of the main disadvantages of this classification system is that it tells us very little about how the elements react. It does not highlight the pattern of their *chemical* properties. As Table 6.5 shows, there are some chemical differences between elements in the two groups, but not enough to make a sensible classification of all the elements.

Assignment – Reactions of metals and non-metals

In a class exercise pupils burned samples of some elements in oxygen. Two pupils did not finish all the samples and their results are shown in the table.

	Element	How it burns	Oxide formed	Colour of pH paper in solution of the oxide
Non-metals	phosphorus	yellow flame	white smoke of phosphorus oxide	red
				--------
	carbon	glows red	--------------	--------
	sulphur	blue flame	--------------	--------
	hydrogen	blue flame	hydrogen oxide water (steam)	green
Metals	magnesium	bright flash	white powder – magnesium oxide	blue
	calcium	----------	--------------	--------
	sodium	----------	--------------	--------
	iron	----------	--------------	--------

1 Copy the table into your book and fill in the gaps. You should be able to do this if you look at Table A on page 6 and read the section above, especially Tables 6.4 and 6.5.
2 On the evidence of the pupils' results, where does hydrogen belong – with the metals or with the non-metals? Explain your answer.

FAMILIES OF ELEMENTS

By 1830 about 55 elements were known. This caused problems because their properties varied so much. The only solution was to find an efficient way to classify them. Careful searching revealed that some groups were alike in many ways and were given 'family' names. Several of these family names are still used today, though we have discovered more elements belonging to each family and know far more about all elements. One of these families of elements is the *Alkali Metals*. The first of these to be discovered were lithium, sodium and potassium. If you look closely at how these elements behave you can see why they belong together.

Family resemblances – the alkali metals

Lithium, sodium and potassium *all* have the following properties.

1 They are very soft and can be cut with a knife.
2 Freshly cut, they are shiny, but quickly go dull as they react with the air.
3 They float on water.
4 They react quickly with water giving off bubbles of hydrogen, leaving a basic solution of the metal hydroxide.
5 They burn in oxygen to make a white metal oxide.
6 They burn in chlorine to make a white metal chloride.

The only difference between the three alkali metals that was discovered was that potassium reacted faster than sodium and lithium reacted slowest.

Since that time, three new elements belonging to the same family have been discovered. They are rubidium, caesium and francium. Other families were found, most of which you will meet later in this topic. In each family the main difference between the elements is the speed with which they react. One family proved very different: the gases helium, argon, neon, krypton, xenon and (radioactive) radon. Their family resemblance was that they did not react at all. As a result they were called the Noble Gases or, sometimes, the Inert Gases: both names hint at their reluctance to react.

The family groups helped to classify many of the elements. There were still many which did not group together. Some elements could also belong to more than one family. A better classification system was needed, and the turning point came when it was realised that there were probably elements which had yet to be discovered. In 1860 a Russian scientist, called Dimitri Mendeleev, decided that several families contained elements that had not yet been discovered. The elements that did not fit belonged to families that contained many other undiscovered elements.

THE PERIODIC CLASSIFICATION

Mendeleev discovered that the properties of the elements changed, gradually, as their atomic number increased. But every now and again he came across an element that was very like an earlier one.

This regular, or *periodic*, likeness made him think of arranging the elements so that elements which were like each other came one under another in vertical columns. He published his Periodic Classification of the elements in 1869 and we still use his classification today.

As you now know, a good classification is one that helps you to sort information *and* highlights patterns so you are able to make predictions. That is why Mendeleev's system proved so successful because it did both of these. Where elements did not fit the pattern, he left gaps. He then used the pattern that the classification highlighted to make predictions about the properties of the undiscovered elements. When those elements were eventually found, his predictions proved to be very accurate. Table 6.6 shows how closely Mendeleev's predictions for the element he called eka silicon matched the element which was called germanium when it was discovered, 17 years later. (So far 105 elements have been discovered.)

Table 6.6 Mendeleev's predictions about germanium (Ge) (which he called 'eka silicon' (Es))

Property	Eka silicon – predicted properties	Germanium's actual properties
Appearance	light grey metal	dark grey metal
Atomic mass	73	72.6
Mass of 1 cm^3	5.5 g	5.3 g
Melting point	higher than tin ? 800°C	958 °C
Action in air	heated to a white oxide, EsO_2	heated to a white oxide, GeO_2
Oxide	slightly basic, less than tin	slightly basic
With acids and bases	hardly reacts	hardly reacts

GETTING TO KNOW THE PERIODIC TABLE

The periodic classification of the elements is usually shown in a table, the Periodic Table. You will find a copy of it on p. 283. Figure 6.8 shows the same table in outline. As you can see, there are eight vertical columns; these are called GROUPS and are given Roman numbers I to VII, and 0. In fact, the elements resemble the families you read about on page 17. Look at the full Periodic Table.

Group I contains the alkali metals.
Group VII contains the halogens.
Group 0 contains the noble gases.
You can probably guess why the noble gases were given the group number 0.

The horizontal rows are called PERIODS. To remember which is which, some people find this

G
R
O
U
P E R I O D helpful.

Looking along a period is useful because you can see how often a similar element appears. This is because an atom of the first element in a period has just one electron in its outermost shell, as you learned in Unit 5. Going along a period, each atom has one more electron in its outer shell than the atom of the element that is before it. So, at the end of a period the final element has atoms with full shells of electrons. If necessary, check back to your work on atomic structure from Unit 5. You can now see that having full electron shells not only explains why the gases in group 0 are each at the end of a period, but also why they are unreactive and have been given the name 'inert gases'.

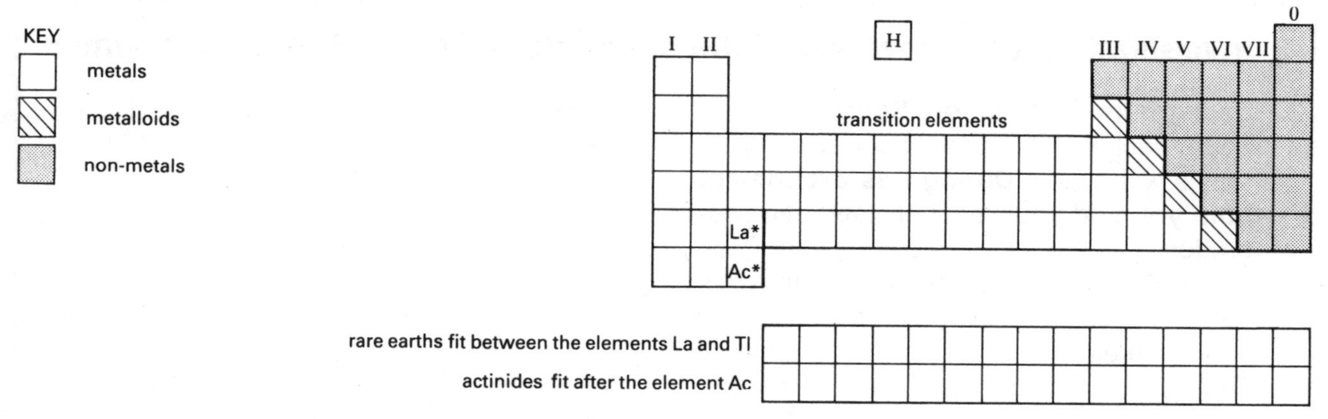

Fig. 6.8 Outline of the Periodic Table

The block of elements in the middle of the periodic table is called the TRANSITION ELEMENTS. Finally, there are two series of elements that are often shown apart, as in Fig. 6.8. These are called the RARE EARTHS (lanthanides) and the ACTINIDES. Many of these are radioactive, or too rare for study in school.

PATTERNS IN THE PERIODIC TABLE

Look at Fig. 6.8 and you will see the first pattern that is highlighted by this classification system. The metals and non-metals have their own areas, with a few borderline elements which are called METALLOIDS. As you have probably guessed, these are the elements that have some properties like metals and some like non-metals.

It would be impossible to investigate all the reactions of all the elements in the Periodic Table. You would not have enough time, and some experiments could not be carried out in school. However, if you do some of the investigations that follow, or similar ones suggested by your teacher, you should begin to see the main patterns in the table.

It is the *patterns* that are important, rather than how you found them.

Patterns across a period

Table 6.7 shows some of the properties of the elements in Period 3. If we look at these, how the elements react with water, and how their oxides behave, then a pattern becomes clear.

Table 6.7 Some properties of the elements in Period 3

	Sodium	Magnesium	Aluminium	Silicon	Phosphorus	Sulphur	Chlorine
Appearance	silvery metal stored in oil	silvery metal	silvery metal	black solid	yellow solid stored in water	yellow solid	greeny-yellow gas
Melts at	97.8 °C	650 °C	660 °C	1410 °C	44.2 °C	113 °C	−101 °C
Boils at	890 °C	1110 °C	2470 °C	2360 °C	280 °C	444 °C	−347 °C
Number of electrons in outer shell	1	2	3	4	5	6	7
Group number	I	II	III	IV	V	VI	VII
Formula of oxide	Na_2O	MgO	Al_2O_3	SiO_2	P_4O_6 (P_2O_3)	SO_2	–
Formula of chloride	$NaCl$	$MgCl_2$	$AlCl_3$	$SiCl_4$	PCl_3	S_2Cl_2	–

Investigation 6.1 How two of the elements in Period 3 react with water

Background information: Sodium

Your teacher will show you how sodium reacts with water. **During this experiment both you and your teacher need eye protection.** Once a small piece of clean, fresh sodium hits water, it starts to react. It reacts so energetically that you may see the sodium melt into a small ball. Glance at Table 6.7 and decide what the temperature of the sodium must be.

Gas bubbles escape as the sodium reacts, and they may make the sodium shoot across the water. If your teacher collects and tests the gas, it should be possible to identify it.

Collect

2 or 3 test tubes
samples of magnesium
container filled with water

What to do

1 Put a test tube into the water. Clean a strip of magnesium that is about as long as your thumb.

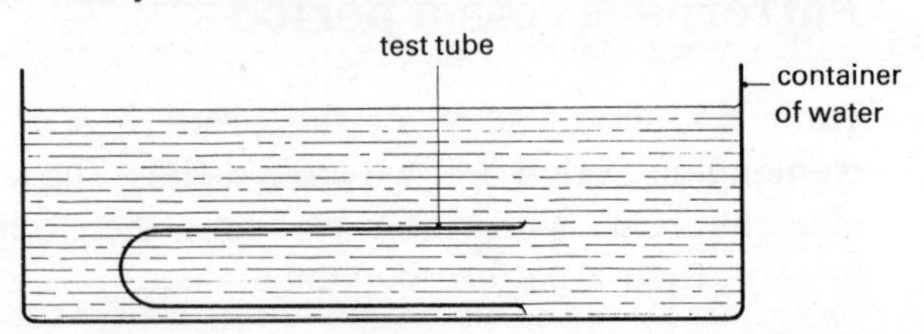

2 Twist the magnesium into a spiral and put it in the water.

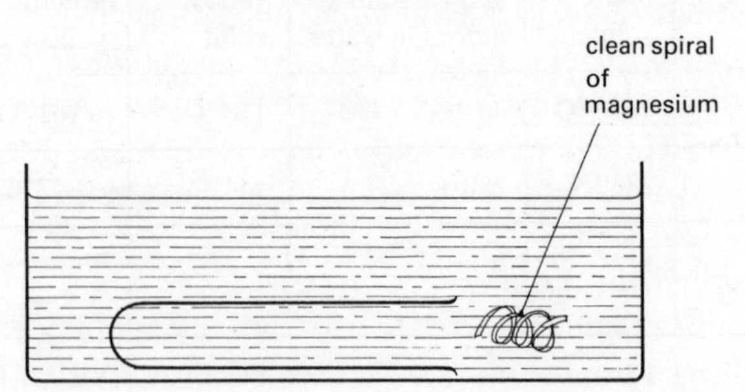

3 Place a test tube of water upside down over the magnesium spiral. Make sure no water escapes from the tube. Hold the test tube like that. How long does it take for bubbles of gas to collect?

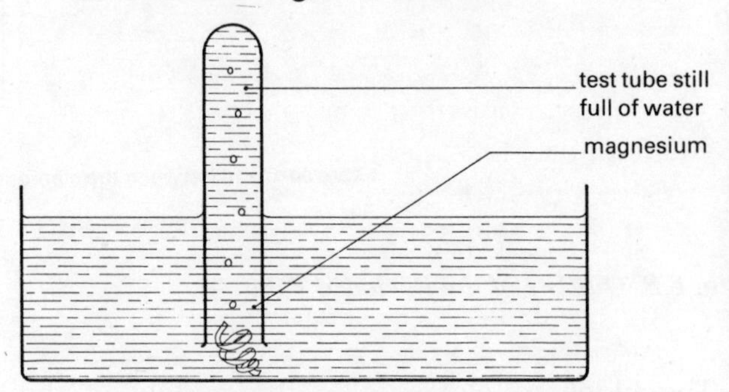

4 When there is enough gas to test, put your thumb over the end of the test tube, and take the tube out of the water. Test the gas in the same way as your teacher did after sodium reacted with water.

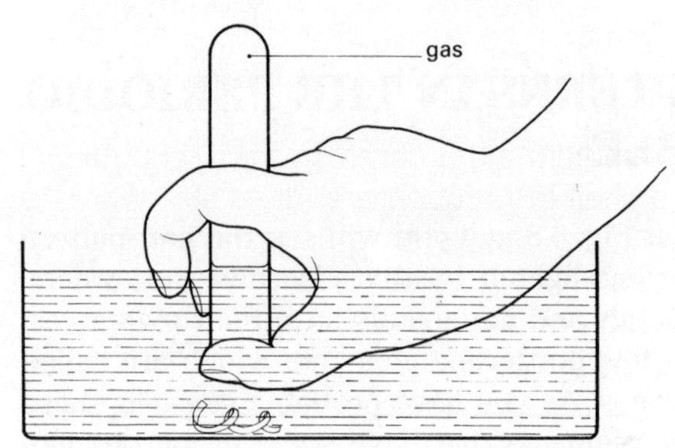

If you repeat this experiment with *hot* water, how would you expect the results to change? Your teacher may ask you to try it. You may be asked to use a temperature that is so high that the water is steam and the magnesium is red hot.

Questions

1 Why did the magnesium have to be cleaned?
2 Was the sodium cleaned for the same reason? Explain your answer.
3 Explain why such a small piece of sodium is used.
4 Which of the two metals is the most reactive?

Assignment – The reaction of sodium and magnesium with water

1 If you managed to make magnesium react with water, the gas you collected lit with a squeaky pop. What is the name of the gas that was given off?

2 If you add universal indicator to the water left after Investigation 6.1 it goes blue. What does this tell you about the substance that has been formed by the reaction?

3 The reaction that happens between magnesium and water can be summarised in an equation:

magnesium + water → magnesium hydroxide + hydrogen

In symbols this is:

$$Mg(s) + 2H_2O\ (l) \rightarrow Mg(OH)_2\ (s) + H_2(g)$$

(a) Write a word equation for the way sodium and water react together.

(b) Find out the number of electrons a sodium atom can lose from its outer shell to produce a sodium ion. What is the charge on a sodium ion? Use this information to turn your word equation into a balanced equation in symbols.

Aluminium and silicon

You can investigate the reactions of these two metals in the same way as you did with magnesium in Investigation 6.1. If you do, you should find that it is more difficult to get them to react with water. You will learn more about the reactivity of metals like sodium, aluminium and magnesium later on in this book.

Phosphorus

Look at Table 6.7 on page 19 and decide how phosphorus will react with water.

Sulphur

Remember the way that sulphur is taken out of the earth's crust? The Frasch extraction process was explained in Unit 5 and you will recall that it uses very hot water to force sulphur up to the surface. Can you think why?

Chlorine

Unlike the previous two elements, chlorine dissolves in water and makes an acidic solution.

Trends in the oxides of Period 3

Do Investigation 6.2 (page 22) now.

Aluminium oxide behaves like an acid when it is with basic substances (e.g. sodium hydroxide) and like a base when it is with acids. (You will learn more about acids and bases later in this unit.) Substances like aluminium oxide are called AMPHOTERIC substances, which means that they can behave as an acid and as a base. Look at the Periodic Table and at Fig. 6.8, and you will see that aluminium is one of the metalloids – those elements which show metallic and non-metallic properties. Does that help you to explain why aluminium oxide is amphoteric?

Now we have enough information to describe the pattern of reactivity in a period. For most properties, there is a *gradual change* as we go across the table. The best example of this is the way oxides of the elements at the start of a period (on the left of the table) are basic, those of the metalloids become amphoteric, and the oxides of elements near the end of a period are acidic. As usual, the noble gases do not react with oxygen at all.

Trends in the groups in the Periodic Table

Do Investigation 6.3 (page 23) now.

We can now summarise the pattern that we have found: in a metal group the elements get *more reactive* going *down* the group. This is because atoms lower down the group hold on to their electrons less tightly than atoms nearer the top and so react more easily.

We can now pick out another pattern: in a non-metal group the elements get *less reactive* going *down* the group.

For non-metals, their reactivity depends on how easy it is for the atoms to gain an electron. The nearer the bottom of a group the atom is, the harder it is to gain electrons.

For groups in the centre of the Periodic Table there is very little difference in the speed of reaction of the elements.

Investigation 6.2 Period 3 oxides and water

Collect

eye protection
a spatula
3–4 test tubes and bungs
test tube rack
test tube holder
universal indicator solution
dilute hydrochloric acid (HCl)
dilute sodium hydroxide solution (NaOH)
Bunsen burner.

What to do

A. You will be shown how some oxides react with water. Note down anything that happens, and the colour change for universal indicator when it is added to the water at the end.

For those oxides that you are given to investigate yourself, do the following.

1 Put half a spatula measure in a clean test tube.

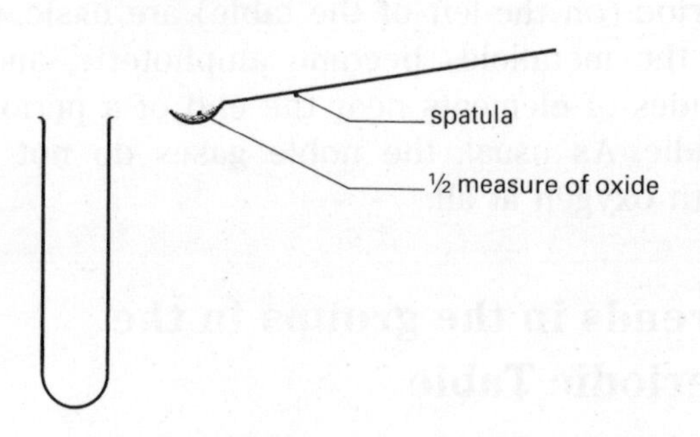

2 Add 2–3 cm of water, then three drops of universal indicator solution. Put a bung in the tube and shake it.

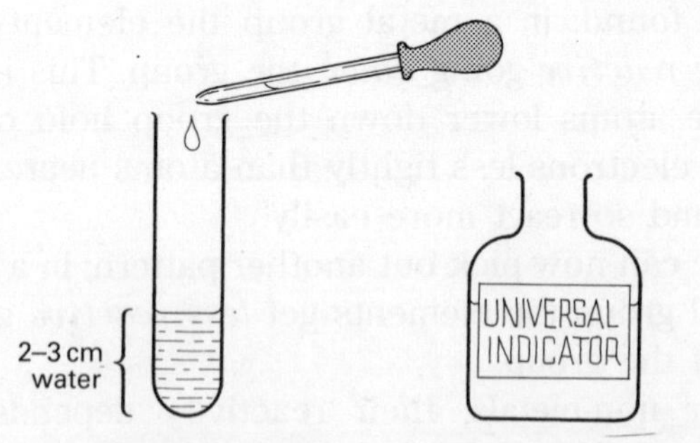

Note down the colour of the indicator.

B. Special treatment for aluminium oxide.

1 Put half a spatula measure of aluminium oxide in a clean test tube.

2 Add 2–3 cm of dilute hydrochloric acid.

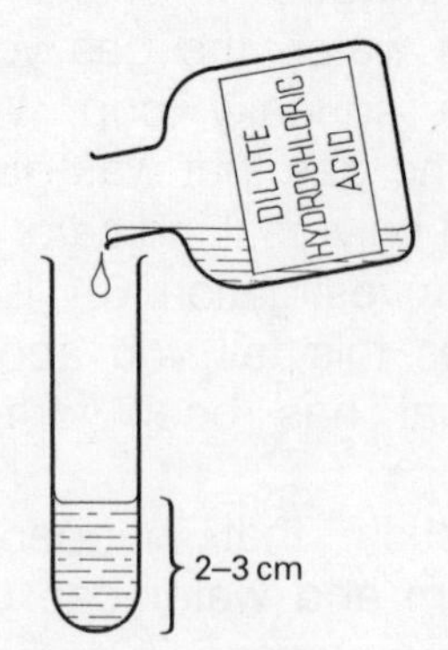

3 Warm the tube gently.

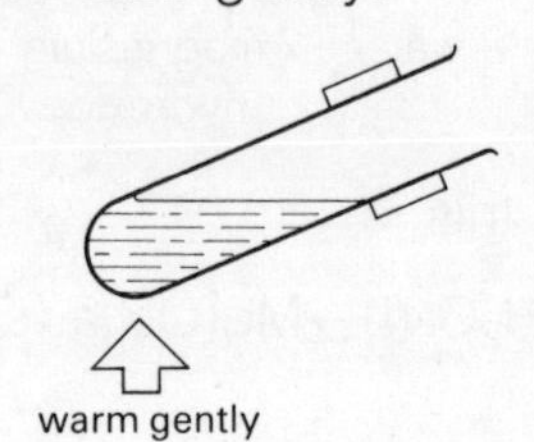

4 Test any gas that is given off.

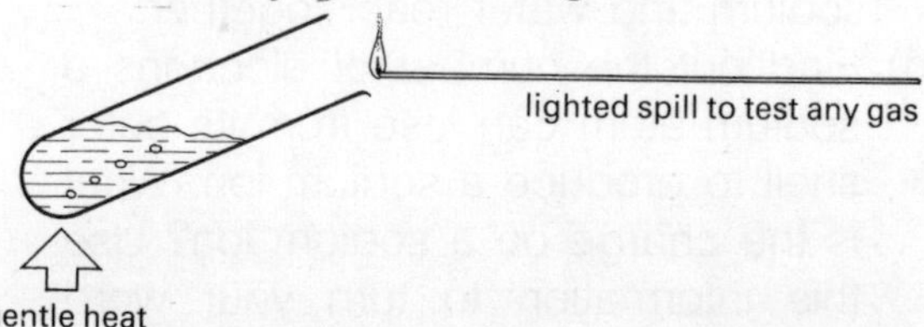

5 Repeat steps 1 to 4, but use dilute sodium hydroxide solution instead of the acid.

Questions

After doing a similar set of experiments, a pupil wrote:

'When I shook up a metal oxide with water, it nearly always went blue when I added universal indicator. The oxides of non-metals, like carbon, usually made the indicator go red. Aluminium oxide reacted with sodium hydroxide solution *and* with dilute hydrochloric acid.'

Use these results, and your own, to answer the following questions.

1 Which set of elements reacts with oxygen to make acidic oxides?

2 Which set of elements reacts with oxygen to make oxides that are basic?

Investigation 6.3 Reactions in a metal group

Look at the Periodic Table and pick out the elements in Group II. Group II is on the left of the table, so all the elements in the group are metals. By investigating how magnesium and calcium react with water, you can see the pattern of reactivity for all the elements in the group. Investigation 6.1 (p. 20) showed you how to investigate the reaction between magnesium and water. Look back at your results: if you did not do the experiment then, your teacher may ask you to do so now.

Repeat the same experiment, but using a sample of calcium. **Wear eye protection.**

Questions

1 All the elements in Group II are metals. If you were to add universal indicator to the water remaining after this experiment, what colour would you expect it to turn? Explain your reason for your answer.

2 Beryllium hardly reacts with water. From the pattern you have seen so far, and knowing its place in Group II, describe what you would expect to happen when a sample of strontium is placed in water.

3 Write a word equation for the reaction between calcium and water.

If you can, write the same equation in symbols and make sure it balances.

Assignment – Reactions in a non-metal group

Group VII is known as the halogens. Table 6.8 shows the results of some of the investigations carried out with three of the elements. Use the information in the table to answer the questions.

1 Which element is the most reactive?

2 Which element is the least reactive?

3 What appears to be the pattern of reactivity in Group VII?

4 One element has been left off the table and the table below shows how this element compares to the first three. What is the name of the mystery element?

Element	Appearance	Reaction with hydrogen	Reaction with hot iron	Number of electrons in outer shell	Charge on ion
?	pale yellow gas	explodes – even in the dark	bursts into flames	7	−1

Table 6.8 Properties of Group VII elements (the halogens)

Element	Appearance	Reaction with hydrogen	Reaction with hot iron	Number of electrons in outer shell	Charge on ion
Chlorine	greeny-yellow gas	Explodes in sunlight. Forms hydrogen chloride (HCl)	Reacts very easily. Forms iron chloride ($FeCl_3$)	7	−1
Bromine	red-brown liquid	Reacts easily when heated. Forms hydrogen bromide (HBr)	Reacts. Forms iron bromide ($FeBr_3$)	7	−1
Iodine	purple-black solid	Reacts very slowly, needs heat. Some hydrogen iodide (HI) formed	Reacts very little	7	−1

A SPECIAL SET OF ELEMENTS

The 30 transition elements are quite important: just look at some of the elements in the group and think of the number of ways we use them. These 30 metals all have similar properties:

They are hard and strong.
They sink in water.
They do not react easily with oxygen or water.
When they react with other elements, the compounds are coloured.
The atoms can form different ions: e.g. Fe^{2+}, Fe^{3+}.

AND ON ITS OWN: HYDROGEN

Hydrogen atoms have just one electron in their outer shell. That means that hydrogen can behave like a metal, by losing its electron, and *sometimes* it can react *with* a metal to make compounds called hydrides.

Usually, hydrogen is shown on its own at the top of the Periodic Table because of its very special properties.

Assignment – Explaining reactivity

1 Explain why elements are reactive if they can lose or gain electrons easily.
2 Explain why atoms near the bottom of a group
(a) lose electrons more easily if it is a metal group;
(b) find it harder to gain electrons if it is a non-metal group.

USING THE PERIODIC CLASSIFICATION

Knowing how the elements fit in the Periodic Table, and the patterns in it, can help us greatly. It makes it easy to find information and all we need to remember are the key patterns. This classification system was found to have another practical use. It could be used to make predictions, as you have seen with Mendeleev's 'eka silicon'.

Assignment – Mendeleev's predictions

Look back to Table 6.6 and Mendeleev's predictions. Find germanium in the Periodic Table, and use your understanding of the way the classification works to answer these questions.

1 Mendeleev gave eka silicon an atomic mass of 73 because it was the average of the atomic masses of silicon and tin $\left(\frac{28 + 118}{2}\right)$. Suggest a reason for this.
2 Why did he predict that it would be a metal?
3 Why did he say its oxide would be basic?
4 Suggest his reason for predicting that eka silicon would be less reactive than tin.

Assignment – How the lead got into petrol

In a car engine the mixture of petrol vapour and air should explode when the spark plug fires. This is when the piston has just reached the top of the cylinder. This explosion, just above the piston, pushes it down. The up and down movement provides the power to make the car move. In the early days of the motor car, many cars suffered from 'pinking' or 'knocking'. This meant that the petrol–air mixture was exploding while the piston was still on its way to the top of the cylinder.

Thomas Midgley, an American engineer, was given the task of solving the pinking problem by his employer in 1916. He tried adding different substances to petrol and found that iodine reduced 'knocking' quite well. However iodine and its compounds were expensive, so Midgley tried other halogens: bromine and chlorine were not as good as iodine. Then he moved left across the Periodic Table and found the pattern given in the table below. He predicted that lead would be the best for preventing 'pinking' and he was correct.

Group number	**IV**	**V**	**VI**	**VII**
increase in anti-'pinking' ability ↓	C	N	O	F
	Si	P	S	Cl
	Ge	As	Se	Br
	Sn	Sb	Te	I
	Pb	Bi	Po	At

← increase in anti-'pinking' ability

Today petrol has a lead compound added to make engines run smoothly. Unfortunately, the waste lead ends up in the exhaust fumes from petrol engines and is a major source of lead pollution.

1 Describe what happens when a car engine is 'pinking'.

2 What effect would you expect 'pinking' to have on:
   (a) the car's cylinders;
   (b) the petrol consumption of the car?
   Explain your answers.

3 Look at the Periodic Table and suggest an element with even better anti-pinking properties than lead.

4 There is evidence that even low levels of lead pollution can damage young children and babies. The lead levels in the blood of people living in towns and cities are increasing.
   (a) Suggest a reason for the increasing levels of lead in the blood of people living in towns.
   (b) Why have petrol engines become a *major* source of lead pollution?
   (c) Explain why lead pollution is thought to be more harmful to young children than to adults.

5 Find out about some of the other sources of lead pollution. (Your teacher may give you a hint.)

6 Write a letter to a newspaper expressing your views on lead pollution. Include some positive suggestions which could make people less worried.

WHAT YOU SHOULD KNOW

1 In the Periodic Table elements are arranged in order of their increasing atomic numbers.
2 Columns in the Periodic Table are called groups; rows are called periods.

G
R
O
U
PERIOD

3 Elements in the same group are called a chemical family; some groups have 'family' names.
4 Group number of an atom is the same as the number of electrons in its outer shell.
5 Most of the elements are metals or non-metals. A few borderline elements can have metallic and non-metallic properties.
6 In a metal group the elements get more reactive going down the group.
For a non-metal group it is the opposite.
7 The pattern of reactivity is explained by the ease of losing or gaining electrons. Metal atoms form positive ions by losing electrons. Non-metal atoms form negative ions. Atoms in Group IV tend to share electrons in covalent bonds.
8 Inert gases have full electron shells and so usually do not react.

QUESTIONS

1 This sketch of the Periodic Table shows the position of elements with atomic numbers from 3 to 18. The letters are showing some of those elements (they are *not* the symbols for those elements). Study the diagram and then answer the questions.

3 A	4	5	6	7	8 E	9	10 G
11 B	12 C	13	14 D	15	16	17 F	18

(a) Which of the letters represents an element that is
(i) a noble gas;
(ii) a halogen;
(iii) able to react most easily with chlorine?
(b) (i) Give the formula of the compound made between D and hydrogen (the hydride of D).
(ii) Give the formula of the oxide of C.
(c) Is the bonding in the oxide of C covalent or ionic?

2 Use your Periodic Table to answer these questions (explain each answer).
(a) Is astatine (At) likely to be solid, liquid or gas at room temperature?
(b) Is (Rb) rubidium a metal or a non-metal?
(c) Are vanadium compounds most likely to be colourless or coloured?
(d) (i) What is the formula of gallium oxide?
(ii) Would you expect this oxide to be basic or acidic?

3 Carbon and silicon belong in Group IV of the Periodic Table.
(a) Explain what this means.
(b) How does the behaviour of elements in the same group change:
(i) in a metal group;
(ii) in a non-metal group?
(c) Both carbon and silicon can show some of the properties of metals as well as those of non-metals. Suggest a reason for this.

CLASSIFYING COMPOUNDS

We have seen how the periodic classification proved to be a very good way of classifying the elements. You saw in the last topic how you can use the Periodic Table to tell you something about the way elements react and a little about their compounds. For example, the position of an element can help you to predict whether its oxide is acidic or basic. However, we now have much more information about many more compounds than in Mendeleev's time. So, we need a way of classifying them all. Like the elements, we can group the compounds by their properties: unlike the elements, we have not yet settled on a single way of grouping them. This topic deals with two ways of classifying the compounds. As always, the one you use will depend on what you want to do.

ACIDS, BASES, SALTS

There are many natural acids: the sharp taste of fruit, the pain of muscle cramp and pain from bee stings are all caused by acids. There are stronger, more powerful acids than these, but all acids have similar properties. Compounds called bases are the chemical opposites of acids. Though they react with acids, removing their acidity, some bases can cause as much harm as acids. The family of bases also has similar properties.

Table 6.9 shows some of the acids and bases you will probably meet in school. Look closely at the table and you may see a pattern for the acids, and another for the bases.

One important thing to notice is that all the solutions of these compounds look like water. You must be very careful as you use them not to muddle your liquids or to mistake them for water. **You should always wear some protection, especially for your eyes, when you have to use acids and/or bases.**

If you do spill any on you, rinse it off straight away, with plenty of water. *Then* report the accident to your teacher.

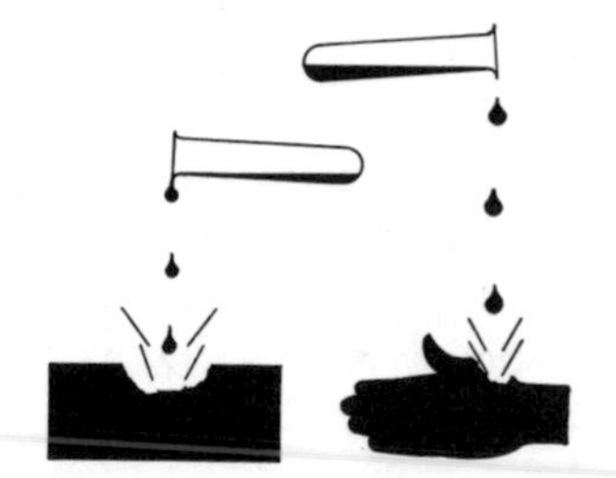

Fig. 6.9 This symbol warns that a substance is corrosive. It is found on containers of strong acids and bases

Table 6.9 Laboratory acids and bases

Name	Formula	Dilute solution looks like	Ions it produces	
Sulphuric acid	H_2SO_4	water	H^+	SO_4^{2-}
Hydrochloric acid	HCl	water	H^+	Cl^-
Nitric acid	HNO_3	water	H^+	NO_3^-
Ethanoic (acetic) acid	CH_3COOH	water, smells vinegary	H^+	CH_3COO^-
Sodium hydroxide (caustic soda)	$NaOH$	water	Na^+	OH^-
Potassium hydroxide (caustic potash)	KOH	water	K^+	OH^-
Ammonium hydroxide (ammonia solution)	NH_4OH	water, may smell of ammonia	NH_4^+	OH^-
Calcium hydroxide (limewater)	$Ca(OH)_2$	water, may be cloudy	Ca^{2+}	$OH^- \times 2$

Later you may have to dilute some concentrated acid (or other dangerous solution). Always add the acid to the water, NEVER the other way around.

Acids

All acid substances dissolve in water to make a solution that turns universal indicator red or orange. You have probably met this already, but check with the pH chart, Fig. 6.11, to remind yourself that this means acids have a pH between 7 and 1. The lower the pH number, the stronger is the acid. Look at Table 6.9, and you can see that all the acids contain hydrogen. When an acid dissolves in water it splits up into its ions. This is called IONISATION (see Unit 5).

Fig. 6.10 (right) An acid and a base that can be found in the kitchen

Fig. 6.11 Colour scale of universal indicator

pH	1	2	3	4	5	6	7	8	9	10	11	12	13	14

colour of universal indicator: red ← green → blue

We can describe what happens by an *ionic equation*, e.g.

$$\text{hydrogen chloride(g)} \xrightarrow[\text{in water}]{} \text{hydrogen ion + chloride ion}$$

$$HCl(g) \xrightarrow[\text{in water}]{} H^+(aq) + Cl^-(aq)$$

$$\text{ethanoic acid} \longrightarrow \text{hydrogen ion + ethanoate ion}$$

$$CH_3COOH(l) \longrightarrow H^+(aq) + CH_3COO^-(aq)$$

When an acid ionises, hydrogen ions are released. The pH scale is linked to this, as a low pH means that a large number of hydrogen ions are released. In the two examples above, hydrochloric acid releases its hydrogen ions more easily than ethanoic acid does. Ethanoic acid is described as a *weak* acid, and hydrochloric acid as a *strong* acid.

Using an indicator is an easy way of recognising whether a liquid is an acid or a base. However acids have several reactions in common.

Reactions of acids

1. With reactive metals like magnesium, acids lose their hydrogen as gas. You will find out more about this sort of reaction in Unit 9, but the pattern of reaction is like this example:

$$\text{sulphuric acid + magnesium} \rightarrow \text{magnesium sulphate + hydrogen}$$

$$H_2SO_4(aq) + Mg(s) \rightarrow MgSO_4(aq) + H_2(g)$$

If you wanted to prove that your liquid was an acid, then you could test the gas to see if it was hydrogen (see page 8).

2. When carbonates meet acids, bubbles of carbon dioxide are given off. You could identify the carbon dioxide by testing with limewater, which goes cloudy. The pattern of reaction between acids and carbonates follows this example:

$$\text{hydrochloric acid + calcium carbonate} \rightarrow \text{calcium chloride + water + carbon dioxide}$$

$$2HCl(aq) + CaCO_3(s) \rightarrow CaCl_2(aq) + H_2O(l) + CO_2(g)$$

3. Acids also react with bases as you will see on page 32.

Table 6.13 on page 37 summarises the main reactions of the acids.

Assignment – Acids: concentration or strength?

The strength of an acid is related to how easily it releases hydrogen ions:

strongest acid: HNO_3, HCl, H_2SO_4
H_2CO_3 (carbonic acid)
↓ CH_3COOH
weakest acid: H_2SO_3 (sulphurous acid)

Concentration tells you how much of a substance is dissolved in 1 dm^3 of solution. This is usually given as the number of moles of substance in 1 dm^3. Two acids of the same *concentration* can be of different *strengths*.

For example:

1.0 M HCl contains 1 mole of HCl per dm^3

1.0 M CH_3COOH contains 1 mole of CH_3COOH per dm^3

but hydrochloric acid is a stronger acid – nearly all of it is ionised to H^+ and Cl^- ions in the solution.

Ethanoic acid is much weaker – very little of it ionises. A litre of 1.0 M CH_3COOH contains many fewer H^+ ions than a litre of 1.0 M HCl.

Try to answer these questions.

1 What is the mass of 1 mole of HCl?
2 How many grams of HCl are in 1 dm^3 of 1.0 M HCl?
3 How many grams of HCl are in 500 cm^3 of 1.0 M HCl?
4 What is the molarity of a solution of HCl that contains 3.65 g HCl?
5 What is the mass of 1 mole of CH_3COOH?
6 How many grams of CH_3COOH are in 1 dm^3 of 1.0 M CH_3COOH?
7 How many grams of CH_3COOH are in 500 cm^3 of 1.0 M CH_3COOH?
8 What is the molarity of a solution of CH_3COOH that contains 6.0 g CH_3COOH?
9 Which solution will have the lowest reading on the pH scale, 0.1 M HCl or 0.1 M CH_3COOH?

Making acids

As you saw in the last topic, many oxides of non-metals dissolve in water to make an acidic solution:

e.g. sulphur dioxide + water → sulphurous acid

$$SO_2(g) + H_2O(l) \rightarrow H_2SO_3(aq)$$

carbon dioxide + water → carbonic acid

$$CO_2(g) + H_2O(l) \rightarrow H_2CO_3(aq)$$

Making acids on a large scale sometimes takes a roundabout route, and you will learn in Unit 7 of the route taken to make sulphuric acid. Nitric acid is made in a roundabout way from ammonia, as you will see in Unit 9.

Carbonic acid is special. It is a *very* weak acid and breaks up into carbon dioxide and water very easily. You will not find any bottles of this acid on the shelves though it has an important part to play in the weathering of rocks. Small amounts of carbon dioxide dissolve in rain, making an acid solution. This solution attacks carbonate rocks (limestone, for example) and buildings made of them. The reaction follows pattern 2 above. For more information about the problems caused by acid rain, read pages 274–5.

Bases

Only a few bases dissolve in water. When they do so, the solution is called an ALKALI. Such solutions are described as alkaline or basic because they make universal indicator turn blue. By now you probably know that a basic solution has a pH between 7 and 14.

At school you will probably use the soluble bases most often, because they can be used as a solution. Turn back to Table 6.9 and you will see that the soluble bases all produce hydroxide ions, OH^-. As these bases dissolve in water, they are ionised.

sodium hydroxide + water → sodium ion + hydroxide ion

$$NaOH(s) + H_2O(l) \rightarrow Na^+(aq) + OH^-(aq)$$

ammonia gas + water → ammonium ion + hydroxide ion

$$NH_3(g) + H_2O(l) \rightarrow NH_4^+(aq) + OH^-(aq)$$

Like acids, some of these bases release hydroxide ions more easily than others. In the first example, sodium hydroxide produces $OH^-(aq)$ ions more easily than ammonia. We describe that by saying sodium hydroxide is a *strong* base. Ammonium hydroxide is a *weak* base.

Fig. 6.12 A farmer spreads lime on his fields to neutralise the acid soil

Assignment – Bases: concentration or strength?

The stronger a base is, the more easily it releases OH^- ions:

strongest	NaOH, KOH
↓	ammonia solution
	$Ca(OH)_2$
weakest	$Mg(OH)_2$

Like acids (see the assignment on page 29), basic solutions of the same concentration can give different readings on the pH scale. Sodium hydroxide is a stronger base than ammonia solution because it releases OH^- more easily when it dissolves in water. Remind yourself about *concentration* by looking back to the assignment on acids, then answer the following questions.

1 What is the mass of 1 mole of KOH?
2 How many grams of KOH are there in 1 dm^3 of 1.0 M KOH?
3 What is the molarity of a solution containing 5.6 g KOH in 1 dm^3?
4 What is the mass of 1 mole of $Mg(OH)_2$?
5 How many grams of $Mg(OH)_2$ are there in 1 dm^3 of 1.0 M $Mg(OH)_2$?
6 What is the molarity of a solution containing 5.8 g $Mg(OH)_2$ in 1 dm^3?
7 Which of the solutions, in (3) and (6), will give the *highest* reading on the pH scale?

Assignment – Carbonic acid and hard water

Rainwater dissolves small amounts of carbon dioxide to make a weakly acidic solution of carbonic acid. This attacks rock as rain falls and makes a dilute solution of soluble compounds. In limestone areas the main chemical in the rock is calcium carbonate ($CaCO_3$).

1 Write a word equation for the reaction that happens when rain falls on limestone hills.

Most of the compounds wash away in rivers and streams and eventually reach the sea. If we take our water supply from one of those rivers, then our tap water will contain a large amount of the dissolved salts, which makes it very hard to make a lather with soap. We call this HARD WATER and it makes lots of scum with soap before a decent lather is obtained. Two sorts of compounds cause this hardness. The first group is:

magnesium hydrogencarbonate, $Mg(HCO_3)_2$
calcium hydrogencarbonate, $Ca(HCO_3)_2$
} which cause 'temporary hardness'

When water containing either of these is heated this sort of reaction happens:

$$Ca(HCO_3)_2(aq) \xrightarrow{\text{heat}} CaCO_3(s) + CO_2(g) + H_2O(l)$$

2 Write an equation for the reaction that happens when water containing magnesium hydrogencarbonate is heated.
3 Where are these reactions likely to happen in your home?

The insoluble calcium carbonate drops to the bottom and does not prevent soap lathering any longer. It is now called LIMESCALE.

The other group of water 'hardening' compounds is calcium and magnesium sulphates, $CaSO_4$ and $MgSO_4$. If they are in the water it is called 'permanently hard', because they do not break down when the water is heated. These compounds are removed by using a water softener, or by adding chemicals that react with them, like sodium carbonate (washing soda).

Limescale in a water pipe

4 Give two reasons why limescale is a nuisance.
5 Kettle descalers are usually powders that you add to a kettle of water and leave for a while. There is a great deal of fizzing and when it stops the kettle should be free from 'fur' or scale. What sort of substance must be used to make the descaler?
6 The photograph shows one of the features of a limestone area. These formations are often spectacular. Find out how stalagmites and stalactites are formed from hard water.

Reactions of the soluble bases

WARNING: solutions of bases are MORE dangerous than acid solutions.

1 If a solution of base is mixed with an ammonium compound and warmed, the ammonia gas that is given off can be smelt. The reaction pattern is like this:

ammonium chloride + sodium hydroxide → sodium chloride + ammonia + water

$$NH_4Cl(s) + NaOH(aq) \rightarrow NaCl(aq) + NH_3(g) + H_2O(l)$$

2 The soluble bases react with acids like this:

hydrochloric acid + sodium hydroxide → sodium chloride + water

$$HCl(aq) + NaOH(aq) \rightarrow NaCl(aq) + H_2O(l)$$

During this reaction the acid's properties are destroyed by the base, and vice versa. This type of reaction is called a NEUTRALISATION, and the final solution is neutral. If you put in too much base, or too much acid, the final solution is not neutral.

As you will see in Unit 7, you can use an indicator to show you when you have added exactly the right amount of base to neutralise the acid.

What about the other bases?

Bases include the oxides, hydroxides and carbonates of metals and ammonia. The only property that all bases share is their ability to neutralise acids. For example:

copper(II) oxide, a black insoluble base, neutralises sulphuric acid like this:

$$H_2SO_4(aq) + CuO(s) \rightarrow CuSO_4(aq) + H_2O(l)$$

The blue copper(II) sulphate formed makes it easy to see when all the copper(II) oxide has reacted.

Salts

You now know that when an acid and a base react together they neutralise each other. If you use exactly the right amount of acid and base you will make a neutral solution with a pH of 7.

Assignment – Using neutralisation

You now know that acids and bases are chemical opposites; they neutralise each other, destroying each other's properties.

Acids can cause problems in everyday life. Indigestion is caused when there is too much hydrochloric acid in your stomach. You can cure this by taking stomach powders that contain sodium hydrogencarbonate or magnesium hydroxide. Farmers may be unable to grow some crops if the soil is acidic. They add lime to the soil to neutralise it.

Try the following experiment

two acid drop sweets
some sodium hydrogencarbonate (bicarbonate of soda)
a small amount of stomach powder

As this involves tasting, be doubly sure that you are using the correct chemicals!

What to do

1 Put an acid drop in your mouth and suck it. What does it taste like?
2 Put a little sodium hydrogencarbonate on the *clean* palm of your hand.
3 Lick the palm of your hand so that you pick up a little of the sodium hydrogen carbonate on your tongue. Does it have any effect on the taste of the acid drop?
4 Repeat the experiment with the other acid drop, and using some stomach powder instead of the sodium hydrogencarbonate.

Questions

1 What effect did the stomach powder and sodium hydrogencarbonate have on the taste of the acid drop?
2 Is the *acid* drop correctly named? Explain your answer.
3 Explain the reaction which happened when you mixed the two substances in your mouth.
4 Look at some sweet packets and see if you can name the substance in the acid drop which reacted with the stomach powder.

Look back (page 32) at the equation for one neutralisation; you can see that the substances made are sodium chloride and water. Sodium chloride is one member of a family of compounds, called SALTS.

All neutralisation reaction follow the same pattern:

ACID + BASE → SALT + WATER

Assignment – Writing ionic equations for neutralisation reactions

Why does a neutralisation reaction always produce water? Let us look at the pattern of reaction in a little more detail. We can summarise the reactions of an acid with bases by writing equations. The following equations describe three different reactions of hydrochloric acid:

1 hydrochloric acid + sodium hydroxide → sodium chloride + water
$HCl(aq) + NaOH(aq) \rightarrow NaCl(aq) + H_2O(l)$

2 hydrochloric acid + potassium hydroxide → potassium chloride + water
$HCl(aq) + KOH(aq) \rightarrow KCl(aq) + H_2O(l)$

3 hydrochloric acid + magnesium hydroxide → magnesium chloride + water
$2HCl(aq) + Mg(OH)_2(aq) \rightarrow MgCl_2(aq) + 2H_2O(l)$

Remember that acids and bases *ionise* when they dissolve in water. If we look at reaction 1 we can write down the ions present:

$HCl(aq) \rightarrow H^+(aq) + Cl^-(aq)$ and
$NaOH(aq) \rightarrow Na^+(aq) + OH^-(aq)$
(you can check this with Table 6.9)

Now, when hydrochloric acid and sodium hydroxide solution are mixed, all four ions will be present. Some Na^+ ions will be attracted towards Cl^- ions, but will not join up. This is because sodium chloride is an ionic substance, as you saw in Unit 5. Ionic substances always exist as separate ions, which are free to move, in solution. However, when some H^+ ions are attracted to OH^- ions, these *will* join together to make an uncharged water molecule.

Now we can go back and check the particles present before and after the reaction:

Before: $\underline{Na^+}$ H^+, OH^-, $\underline{Cl^-}$
After: $\underline{Na^+}$, $\underline{Cl^-}$, H_2O

Equations are a way of summarising the *change* that takes place. In this reaction, there are some ions which do not change, so the only real change taking place is:

$$H^+(aq) + OH^-(aq) \rightarrow H_2O(l)$$

1 For reactions 2 and 3:
   (a) write down the ions present *before* the reaction;
   (b) write down the ions present *after* the reaction;
   (c) write an ionic equation which shows the real change taking place.
2 (a) Write a word equation for the reaction of nitric acid with sodium hydroxide.
   (b) Write the symbol equation for the same reaction.
   (c) Work out the ionic equation for the real change taking place.
3 Repeat 2 with as many acids and bases as you can.
4 Use your understanding of what happens, ionically, during a neutralisation reaction to explain why water is *always* produced.

Table 6.10 Some useful salts

Formula	Chemical name	Everyday name	Typical use
NH_4Cl	ammonium chloride		in dry cells
$(NH_4)_2SO_4$	ammonium sulphate		fertiliser
$BaSO_4$	barium sulphate		in white paint, as a pigment
$CaCO_3$	calcium carbonate	limestone, marble, chalk	cement, concrete; in buildings, in the blast furnace
$CaSO_4{\cdot}2H_2O$	calcium sulphate	gypsum	to make plaster of Paris
$CuSO_4{\cdot}5H_2O$	copper sulphate	in 'Bordeaux Mix'	to kill fungus
$FeSO_4{\cdot}7H_2O$	iron sulphate		making ink
$Mg(OH)_2$	magnesium hydroxide	magnesia/milk of magnesia	to treat indigestion
$MgSO_4{\cdot}7H_2O$	magnesium sulphate	Epsom salts }	in medicines as purgatives –
$Na_2SO_4{\cdot}10H_2O$	sodium sulphate	Glauber's salt }	to 'clean the system'
KCl	potassium chloride		fertiliser
KNO_3	potassium nitrate	saltpetre	in gunpowder
$K_2SO_4{\cdot}Al_2(SO_4)_3{\cdot}24H_2O$	potassium aluminium sulphate	potash alum	as a mordant – to help dyes hold on
$Na_2CO_3{\cdot}10H_2O$	sodium carbonate	washing soda	to soften water
$NaHCO_3$	sodium hydrogencarbonate	bicarbonate of soda	to cure indigestion
$C_{17}H_{35}COONa$	sodium stearate	soap	cleaning
$NaCl$	sodium chloride	common salt	cooking, an important raw material in the chemical industry
$ZnCO_3$	zinc carbonate	calamine	treating sunburn
$ZnCl_2$	zinc chloride		as a flux in solder
$ZnSO_4{\cdot}7H_2O$	zinc sulphate		as an emetic – to make people vomit

Look carefully at the formulae of the salts in Table 6.10. Can you spot what they all have?

Cover up the metal symbol: what you have left is an 'acid without its hydrogen'. That means that you can make salts in *any* reaction that swaps the hydrogen of an acid for a metal ion (or the ammonium ion). Figure 6.13 shows what happens during the reaction!

A salt is a substance that is formed when the hydrogen of an acid is replaced by a metal, or the ammonium ion (which behaves like a metal).

Table 6.11 shows how salts get their names.

Table 6.11 Salts are named according to the acid they came from

Acid	Salts made
acetic (ethanoic)	acetates (ethanoates)
hydrochloric	chlorides
hydrobromic	bromides
hydroiodic	iodides
nitric	nitrates
phosphoric	phosphates
sulphuric	sulphates

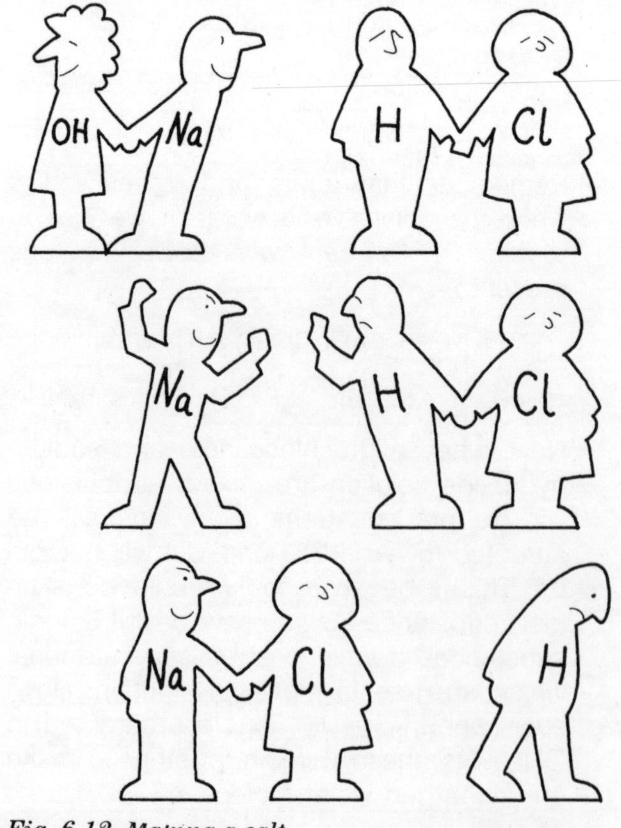

Fig. 6.13 Making a salt

Making Salts

As Fig. 6.13 shows, you can make any salt if you swap the hydrogen of an acid for a metallic ion. By reading back over this topic you should be able to pick out most of the ways that you can do this swapping. In each you make a solution of the salt that you want, in water. Some salts are insoluble in water (see Table 6.12) so they have to be made another way.

Making Soluble Salts

Each of these reactions will make a salt:

ACID + CARBONATE → SALT + WATER + CARBON DIOXIDE

ACID + REACTIVE METAL → SALT + HYDROGEN

ACID + BASE → SALT + WATER

The salt will be dissolved in water, and you can get a sample of the salt on its own by letting a little water evaporate and leaving the solution to crystallise.

Investigation 6.4 Making copper sulphate

Collect

- eye protection
- spatula
- filter funnel
- evaporating basin
- beaker
- stirring rod
- copper oxide
- filter paper
- Bunsen burner, tripod and gauze

What to do

1 Gently warm some dilute sulphuric acid in a beaker. Remove from heat before adding half a spatula of black copper oxide. Stir until it has dissolved.

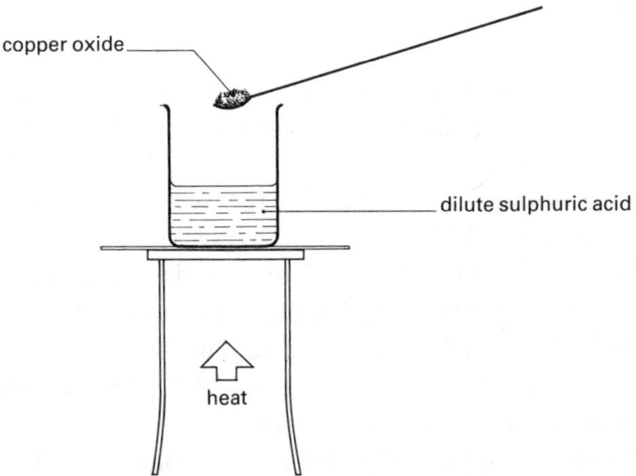

2 Keep adding copper oxide, and stirring, until no more will dissolve. Heating is not necessary at this stage.

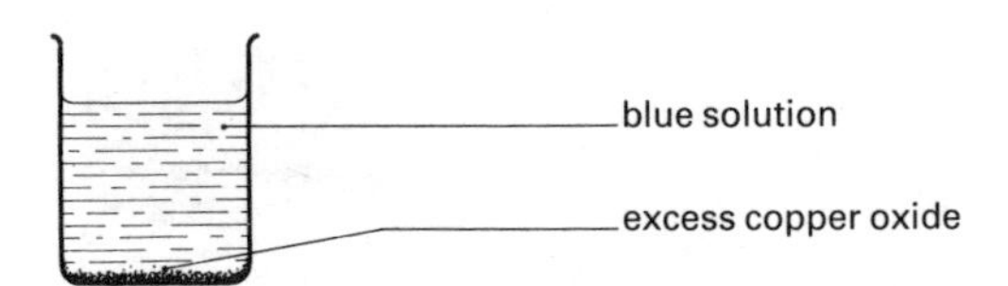

3 Pour the warm mixture through the filter paper in the funnel into the evaporating basin. Leave the basin to stand for a few days: the water evaporates to leave blue crystals of copper sulphate. (Warming gently first will speed it up.)

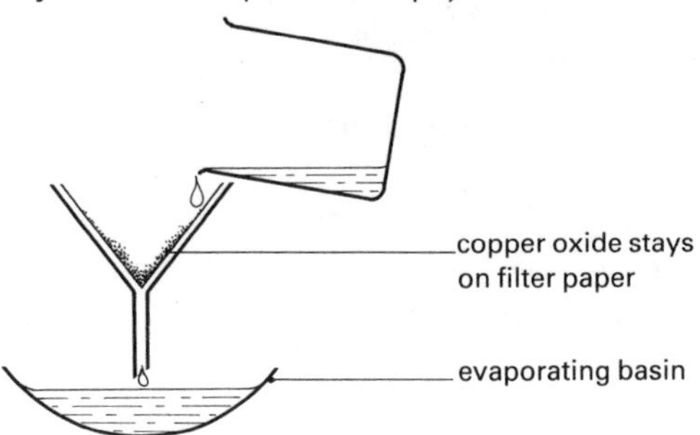

The reaction can be summarised in this equation:

copper oxide + sulphuric acid → copper sulphate + water

$CuO(s) + H_2SO_4(aq) \rightarrow CuSO_4(aq) + H_2O(l)$

Questions

1 The black copper oxide dissolved until the end of stage 2. Why did it stop dissolving and sit on the bottom of the beaker?

2 Why did stage 3 suggest using a container like an evaporating basin?

Assignment – Water of crystallisation

Many salts use water molecules to help build their crystals. Such crystals are called *hydrated* and the water used is called *water of crystallisation*. Blue copper sulphate crystals contain water. Early in your school life you may have heated some copper sulphate crystals. Remember how they went white? That was because water of crystallisation had been driven off. The white powder you had left is called *anhydrous copper sulphate*, because it has lost its water.

The whole happening can be summed up by this equation:

hydrated copper sulphate $\xrightarrow{\text{heat}}$ anhydrous copper sulphate + water

$$CuSO_4{\cdot}5H_2O(s) \longrightarrow CuSO_4(s) + 5H_2O(g)$$

You probably added water to this white powder and found that the blue colour came back.

Your teacher may let you do this if you cannot remember doing it before.

1 Look at Table 6.10 and write down the names and formulae of five hydrated salts.
2 Find out the name of a salt which has two hydrated forms. What are they?
3 Write down the names and formulae of five anhydrous salts.

Table 6.12 Pattern of solubility in salts

Soluble in water	Except
All salts containing Na^+, K^+, NH_4^+	–
All salts containing NO_3^-	–
All salts containing SO_4^{2-}	Ag_2SO_4, $BaSO_4$, $CaSO_4$, $PbSO_4$
All salts containing Cl^-	$AgCl$, $PbCl_2$
Insoluble in water	**Except**
All salts containing CO_3^{2-} All salts containing S^{2-} All salts containing SO_3^{2-}	any of these that also contain Na^+, K^+, NH_4^+

Making insoluble salts (precipitation)

When the salt you want to make does not dissolve in water you must use a different method. You need to find soluble compounds; one that contains the positive ion and the other the negative ion that make up the insoluble salt. For example:

silver chloride (AgCl) is an insoluble salt; sodium chloride (NaCl) and silver nitrate ($AgNO_3$) are soluble salts. } check these in Table 6.12

Mix the solution of one soluble salt with the solution of the other soluble salt.

silver nitrate + sodium chloride → silver chloride + sodium nitrate

$$AgNO_3(aq) + NaCl(aq) \rightarrow AgCl(s) + NaNO_3(aq)$$

Points to notice:

(a) The reaction makes solid silver chloride. It drops to the bottom, so is called a PRECIPITATE.
(b) The reaction produces a precipitate, so this way of making insoluble salts is often called PRECIPITATION.
(c) Sodium nitrate stays in the solution because it is soluble.

You would get out the silver chloride by filtering the mixture. The chloride stays in the filter paper, ready for you to wash and dry it.

If you wanted, you could get the sodium nitrate, after filtering, in the same way as you obtained copper sulphate in Investigation 6.4.

Figure 6.14 is a key to the best way of making each of the salts. If you understand how it works, and you are aware of the information in Table 6.12, then you know all you need to know about salt-making.

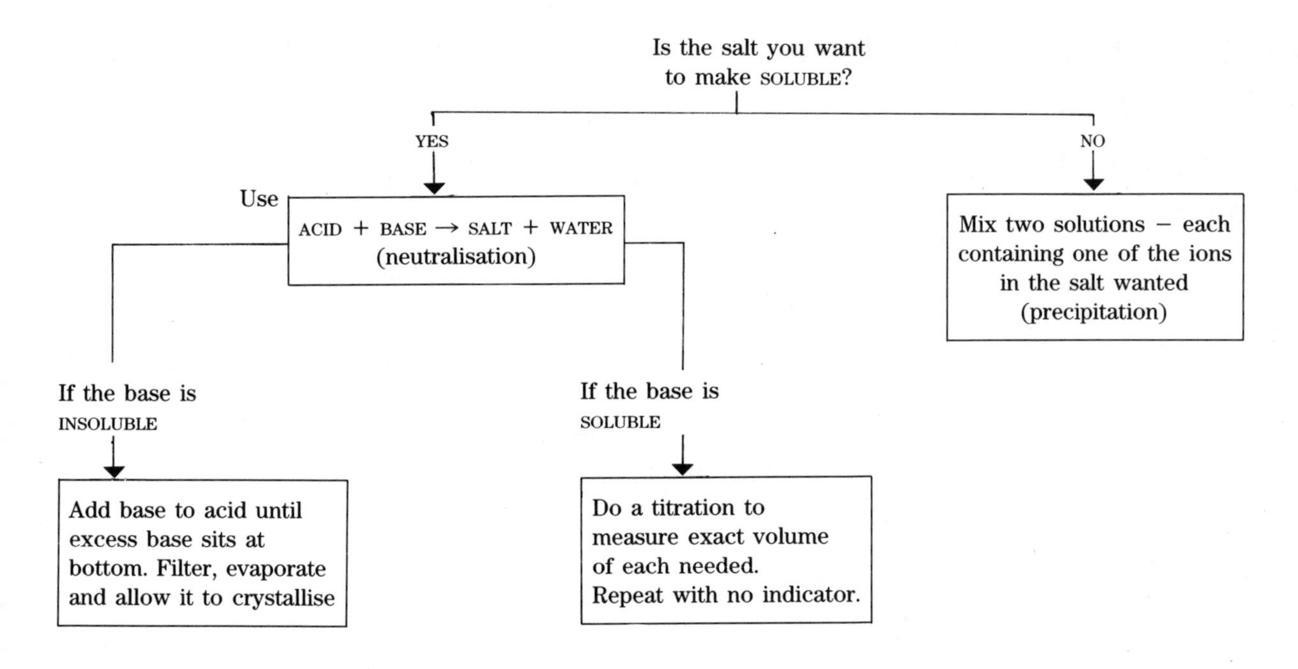

Fig. 6.14 Key to making salts

Table 6.13 Reactions of acids and bases – a summary

Reactions		Gas?		Salt made?		Any other compound	Comments
1 ACIDS							
acid + carbonate e.g. $H_2SO_4(aq) + CuCO_3(s)$	→	carbon dioxide $CO_2(g)$	+	yes $CuSO_4(aq)$	+	water $H_2O(l)$	calcium carbonate + sulphuric very slow
acid + metal e.g. $H_2SO_4(aq) + Zn(s)$	→	hydrogen $H_2(g)$	+	yes $ZnSO_4(aq)$		– –	only reactive metals
acid + base e.g. $2HNO_3(aq) + CuO(s)$	→	– –		yes $Cu(NO_3)_2(aq)$	+	water $H_2O(l)$	– –
2 BASES							
base + acid – see 1 (above)							
base + ammonium salt e.g. $CuO(s) + 2NH_4Cl(aq)$	→	ammonia $NH_3(g)$	+	yes $CuCl_2(s)$	+	water $H_2O(l)$	not a good way to prepare a salt
e.g. NaOH/KOH + amphoteric metal		hydrogen		–		complex compound	ONLY for amphoteric metal
3 OTHER SALT-MAKING REACTIONS							
element + element e.g. $2Na(s) + Cl_2(g)$	→	– –		yes $2NaCl(s)$		– –	only a few made this way
precipitation: mix solutions of two soluble compounds e.g. $AgNO_3(aq) + NaCl(aq)$	→	–		insoluble salt $AgCl(s)$	+	solution of a soluble salt $NaNO_3(aq)$	to make insoluble salts

COMPOUNDS CONTAINING CARBON ATOMS

Another way of classifying compounds

Early on (Unit 5, Book 1) you learned that carbon's place, in Group IV of the Periodic Table, means that carbon atoms make covalent bonds by sharing electrons with some other atoms. These can be atoms of different elements, or other carbon atoms. Carbon atoms can join together covalently to make chains, or rings, and some of the chains can be very long. Poly(e)thene is just one example of the gigantic molecules that can be made when thousands of carbon atoms join together.

Carbon atoms can join together in so many ways that there are about 50 times more carbon compounds than compounds of *all* the other elements put together! Most living organisms are

Investigation 6.5 Carbon in living organisms

You will be given samples of substances made of organic material, to carry out two reactions that should break them down to release any carbon they contain.

Collect

eye protection
crucible lid
spatula
some test tubes
Bunsen burner
tongs
copper oxide
limewater

What to do

1 Put a small sample on a crucible lid. Use tongs to hold the lid over a Bunsen flame. Heat gently, then more strongly. Look for the black carbon residue.

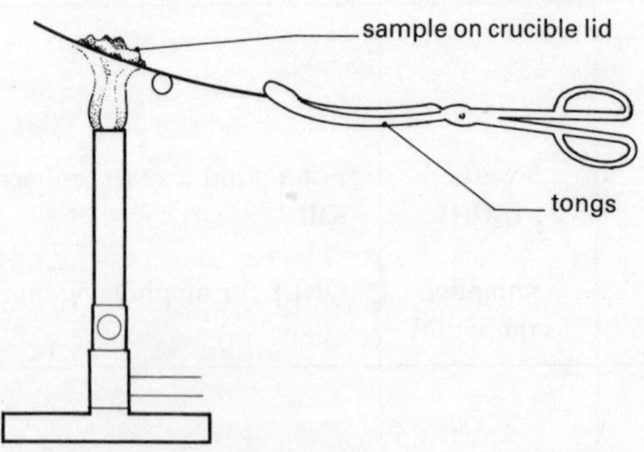

2 Mix one spatula-full of the substance with one spatula-full of copper oxide. Put the mixture into a test tube; put one more measure of copper oxide on top. Get ready to test for carbon dioxide. Your teacher will advise you how to do it.

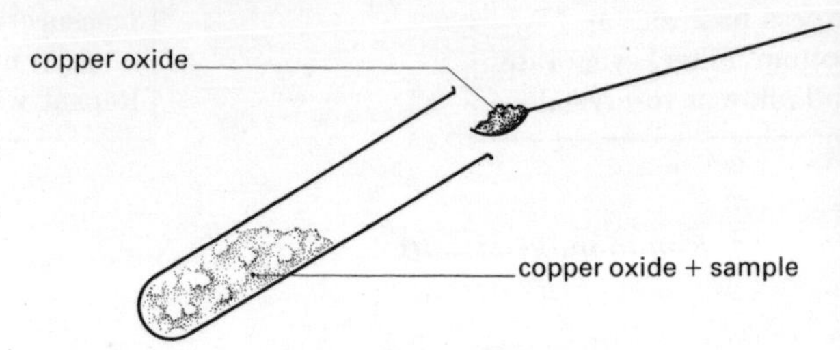

3 Heat the tube gently, then strongly. Continue heating until there is no more change. Test any gas that is given off. Repeat with all the other substances.

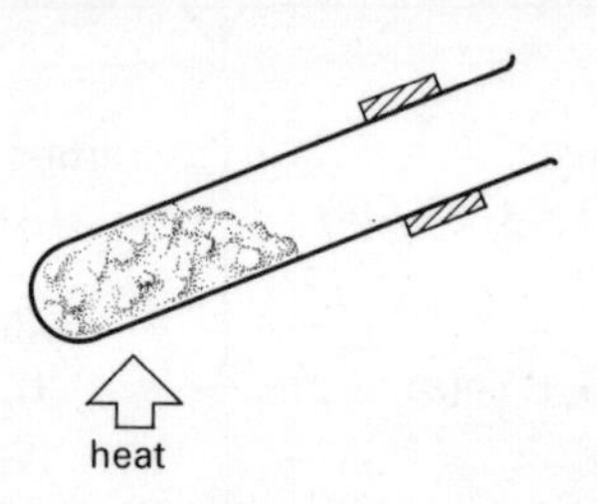

Questions

1 The limewater in the gas test (stage 3) will go cloudy. What gas has been produced?

2 Carbon reacts with copper oxide like this:

carbon + copper oxide → copper + carbon dioxide

$$C(s) + 2CuO(s) \rightarrow 2Cu(s) + CO_2(g)$$

Where does the carbon come from?

3 Did you notice some moisture on the top part of the test tube? What is this? How could you test it to find out? Do it, having first checked with your teacher.

4 What do the results of the two tests tell you about the sample?

made from carbon compounds (see carbon cycle, page 214), so these compounds are called ORGANIC. Compounds that do not contain carbon are called INORGANIC.

Fig. 6.15 Some of the many items made from polythene

We can now use another classification of compounds, by sorting them into groups labelled 'organic' and 'inorganic'. Like all classification systems, this one has a few 'oddities': carbon dioxide and the carbonates are usually put in the inorganic group.

Organic families

Once we have classified compounds as organic or inorganic, we have more than three million compounds in the organic group! Classifying such a large number of carbon compounds is easier than you would think, because groups of compounds have the same properties and the compounds differ only in the size of their molecules. The members of a family react in the same way because the molecules follow the same pattern.

Example: the ALCOHOL family

Table 6.14 The alcohol family

<table>
<tr><th>Structure</th><th>Formula and name</th><th>Functional group</th><th>Boiling point</th></tr>
<tr><td>H
|
H—C—OH
|
H</td><td>CH_3OH
methanol</td><td>−OH</td><td>64.1 °C</td></tr>
<tr><td>H H
| |
H—C—C—OH
| |
H H</td><td>CH_3CH_2OH
ethanol</td><td>−OH</td><td>78.5 °C</td></tr>
<tr><td>H H H
| | |
H—C—C—C—OH
| | |
H H H</td><td>$CH_3CH_2CH_2OH$
propanol</td><td>−OH</td><td>97.4 °C</td></tr>
<tr><td>H H H H
| | | |
H—C—C—C—C—OH
| | | |
H H H H</td><td>$CH_3CH_2CH_2CH_2OH$
butanol</td><td>−OH</td><td>118 °C</td></tr>
<tr><td>H H H H H
| | | | |
H—C—C—C—C—C—OH
| | | | |
H H H H H</td><td>$CH_3CH_2CH_2CH_2CH_2OH$
pentanol</td><td>−OH</td><td>138 °C</td></tr>
</table>

Points to notice

- All have names ending −ol (this shows that they contain −OH).
- The difference between one molecule and the next is one $-CH_2-$ group.
- There is a slight difference in boiling points.
- All the alcohols react the same way, because they all have the functional group, −OH.

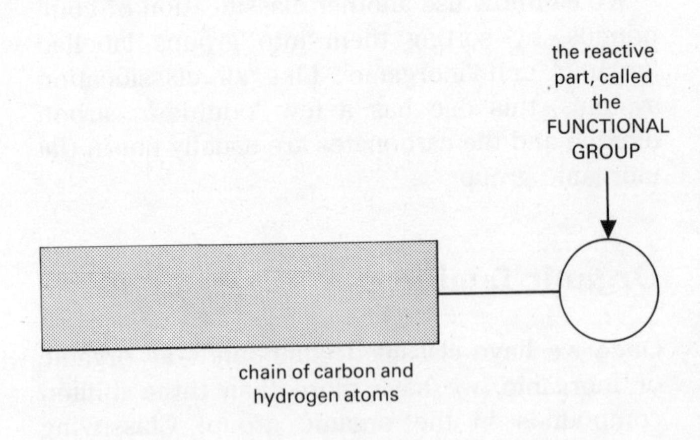

Fig. 6.16 Plan of an organic molecule

There are millions of carbon compounds that are made of carbon and hydrogen atoms only. Crude oil contains a vast number of them, and in Investigation 5.4, in Book 1, you saw how we can separate them. The large number of these compounds, which are called HYDROCARBONS, means that we can study them as a separate group.

The hydrocarbons

Compounds containing carbon and hydrogen only can be grouped into four families, each built a little differently. All the hydrocarbons have many properties in common, so if we study two of the families, you will see many of the patterns of the whole group. The families we shall examine are the ALKANES and the ALKENES, and you can see some of them in Table 6.15.

Points to notice

- Each alkane has a name ending –ane.
- Each alkane contains as many hydrogen atoms as the molecule can possibly contain, so the alkanes are often described as being SATURATED hydrocarbons.
- Each alkene has a name ending –ene.
- Each alkene is two hydrogen atoms short of its maximum number, so alkenes are often described as being UNSATURATED hydrocarbons.
- The first half of the hydrocarbon's name tells you the number of carbon atoms in the molecule.
- The second half tells you the family it belongs to.

Assignment – Another hydrocarbon family

The ALKYNES are another of the hydrocarbon families. Each alkyne molecule is four hydrogen atoms short of saturation; alkyne molecules have names ending –yne.

Alkynes

Formula	Name
C_2H_2	ethyne (acetylene)
C_3H_4	
C_4H_6	
	pentyne
	hexyne

Copy this table into your book. Use the pattern in Table 6.15 to help you complete the table.

Most of the alkanes come from crude oil, and 'cracking' (or breaking up) larger alkane molecules into smaller bits produces a mixture of small alkane *and* alkene molecules.

Reactions of the hydrocarbons

Alkanes are not very reactive, but they do burn and release a great deal of energy, as you will see in Unit 9.

Alkenes are much more reactive and are used to make many useful compounds. In one reaction, alkene molecules can join to make long chain molecules called polymers (see Unit 5).

If we study how the first one or two members of an organic family behave, then we can see the pattern for the whole group.

Table 6.15

Alkanes		Alkenes	
Name	**Formula**	**Name**	**Formula**
methane	CH_4	—	—
ethane	C_2H_6	ethene	C_2H_4
propane	C_3H_8	propene	C_3H_6
butane	C_4H_{10}	butene	C_4H_8
pentane	C_5H_{12}	pentene	C_5H_{10}
hexane	C_6H_{14}	hexene	C_6H_{12}
heptane	C_7H_{16}	heptene	C_7H_{14}
octane	C_8H_{18}	octene	C_8H_{16}

Like all hydrocarbons, alkenes will burn producing only carbon dioxide and water.

e.g. ethene + oxygen → carbon dioxide + water

$$C_2H_4(g) + 3O_2(g) \rightarrow 2CO_2(g) + 2H_2O(g)$$

Alkenes are not used as fuels because they are much more useful for making other compounds. As alkene molecules are unsaturated, they have space for other atoms to add on. Therefore most of the alkenes' reactions are described as ADDITION reactions. Here are some examples of such reactions.

(i) ethene + bromine → dibromoethane

$$C_2H_4(g) + Br_2(aq) \rightarrow C_2H_4Br_2$$

(ii) ethene + hydrogen → ethane

$$C_2H_4(g) + H_2(g) \rightarrow C_2H_6(g)$$

(iii) ethene + steam $\xrightarrow[\text{acid}]{\text{with sulphuric}}$ ethanol

$$C_2H_4(g) + H_2O(g) \longrightarrow C_2H_5OH(l)$$

(iv) many ethene molecules $\xrightarrow[\text{catalyst}]{\text{great pressure}}$ part of a polythene molecule

$$\ldots CH_2-CH_2-CH_2-CH_2 \ldots$$

Reaction (i) is useful as a test for an unsaturated compound. Shaking the compound with bromine dissolved in water will make the 'bromine water' lose its brown colour.

Organic compounds that are not hydrocarbons

There are many organic compounds which contain other atoms as well as carbon and hydrogen. By the end of your course, you will have met several groups of them, among which are proteins, carbohydrates, fats, amino acids, cellulose and polymers. For most of them, their structure and reactions are too complex to study at this stage. However, there is one family that we can examine here – the alcohols. Alcohols were some of the first compounds that humans made, and one well known alcohol is ETHANOL. Ethanol is the scientific name for the 'alcohol' in alcoholic drinks. Although ethanol can cause brain damage if taken in excess, many people enjoy drinks containing it.

Table 6.16 Ethanol content of some alcoholic drinks

	% ethanol	'proof' rating
beer & cider	2–4	5–9
wine	10–12	20–25
sherry & port	17	35
brandy & other spirits	up to 40	70 or more

Ethanol can be made from ethene (reaction (iii)) or by fermenting carbohydrates. The next investigation follows a fermentation process.

Fig. 6.17 Large copper stills are used to distil whisky

Investigation 6.6 Making ethanol from larger molecules by fermentation

This reaction takes several days, and needs to be left in a warm place. It is an example of fermentation, during which large molecules of sugar are broken down to ethanol molecules. How this happens is explained in Unit 9, but it can be summarised by:

$$\text{sugar (carbohydrate)} \xrightarrow{\text{yeast}} \text{ethanol} + \text{carbon dioxide}$$

Collect

conical flask with bung and delivery tube
spatula
beaker
15 g sugar
yeast
(ammonium phosphate)
limewater
stirring rod

What to do

1 Dissolve about 15 g sugar in 50 cm^3 water in a flask. Put some limewater in a beaker.
2 Add two spatula measures of yeast to the flask. (You can also add a pinch of ammonium phosphate, which helps the yeast to act.)

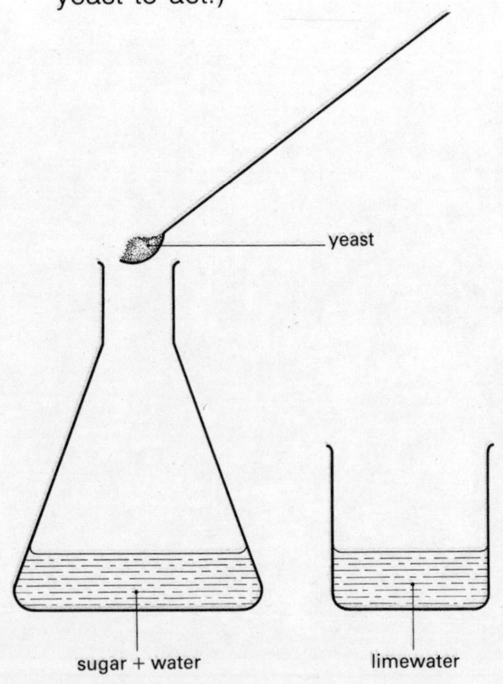

3 Set up the flask and beaker like this and leave them in a warm place.

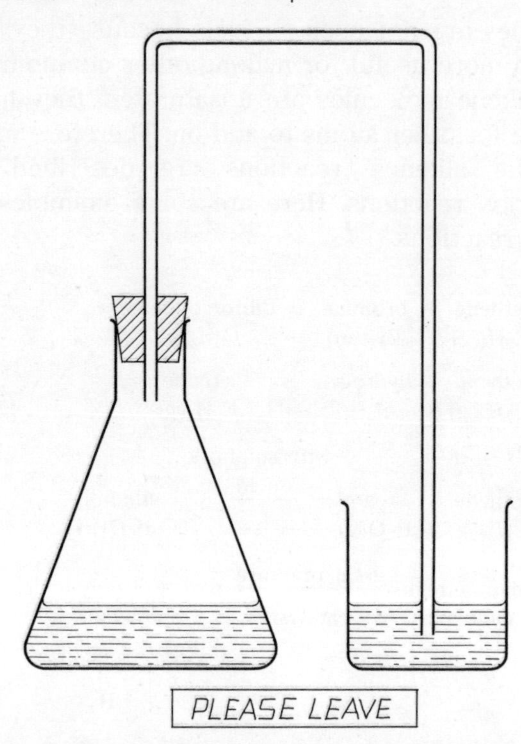

When the fermentation is nearly finished you will notice that the limewater is milky, and there is an 'alcoholic' smell in the flask.

Questions

1 The word fermentation comes from a Latin word that means 'to boil up'.
   (a) Explain why fermentation is the name given to the reaction that happens in this experiment.
   (b) Is fermentation a chemical reaction? Explain your answer.
2 Fermentation is the name given to reactions caused by micro-organisms. In this experiment the micro-organism is yeast. Such micro-organisms produce enzymes which speed up the chemical reactions.
   (a) What are enzymes? (Your work from Unit 2 may help.)
   (b) Explain why the experiment had to be left in a *warm* place.
   (c) Give the possible reasons why the limewater turned milky in this experiment.

Assignment – Concentrating the alcohol

The alcohol you made in Investigation 6.6 is very dilute. The best way of separating the ethanol from the water is to distil the mixture. Ethanol boils at a lower temperature than water so it will boil over first and can be collected.

WARNING – This experiment must be done only by a teacher and not by you. Eye protection must be worn.

1 Take the flask from Investigation 6.6, stage 3. Pour off the liquid leaving the solid bits behind.
2 Put the liquid in a distillation flask (it may be wise to put in the 'alcohol' samples from several fermentations).
3 Set up a fractional distillation column over the flask.
4 Put some of the liquid you have collected into an evaporating basin. Touch it with a lighted spill, ensuring that the flames are kept well away from the distillation apparatus. When the apparatus has cooled, try to do the same with a little of the liquid that is left in the flask.

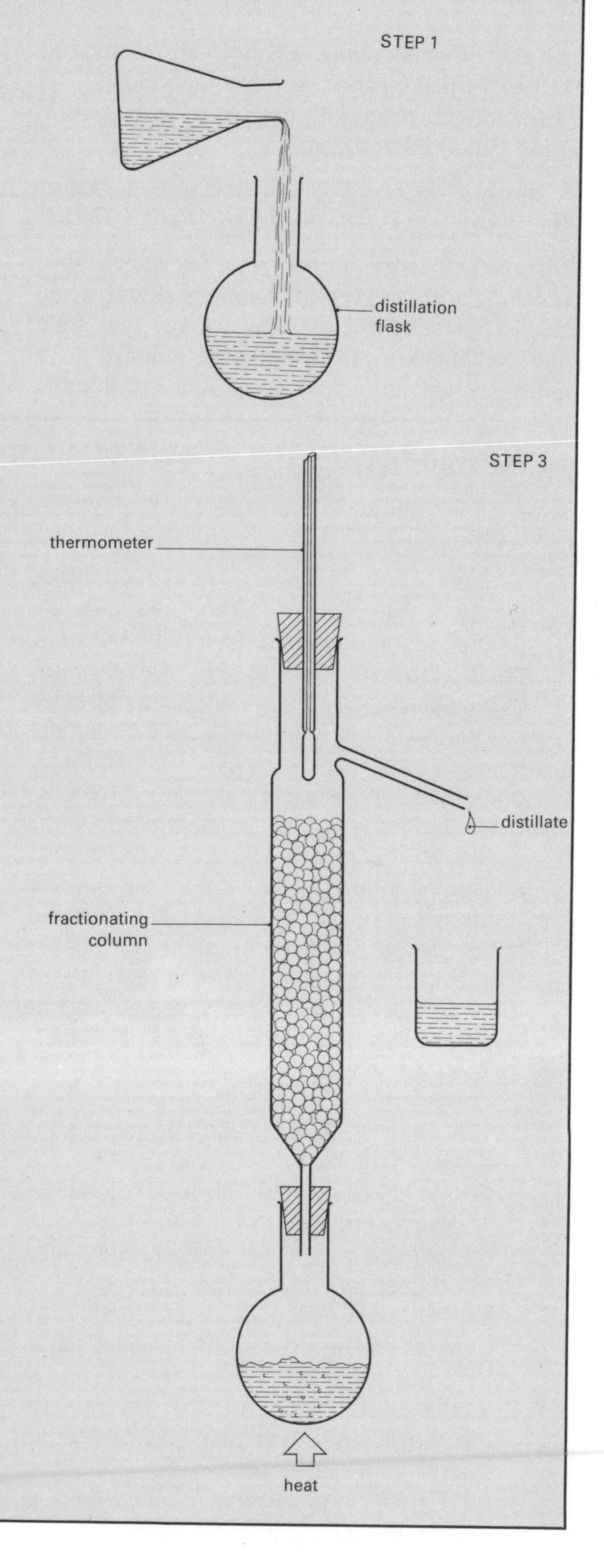

Questions

1 At what temperature did the first liquid come over? Compare the temperature with the boiling point of ethanol. Is your sample made of ethanol?
2 If your teacher managed to light the liquid that collected, describe the flame. Write an equation to describe how ethanol burns.
3 Commercial alcohol production involves distilling to make some alcoholic drinks. Which alcoholic drinks are distilled at least once?

   Why would some drinks need to be distilled more than once?

From ethanol to vinegar and pear drops!

Like the other alcohols, ethanol can be used as itself or to make other useful compounds.

Ethanol can react with oxygen to form a compound called ethanoic acid:

$$\begin{array}{ccccccc} \text{ethanol} & + & \text{oxygen} & \rightarrow & \text{ethanoic acid} & + & \text{water} \\ CH_3CH_2OH(l) & + & O_2(g) & \rightarrow & CH_3COOH(aq) & + & H_2O(l) \end{array}$$

Ethanoic acid is the proper name for vinegar (see Table 6.9) and this reaction happens slowly when wine is opened, which is why we say that stale wine is 'vinegary'. Ethanoic acid belongs to a 'family' called the CARBOXYLIC ACIDS (alkanoic acids) – they all have carbon, hydrogen and oxygen in them.

Ethanol and carboxylic acids will react together, and make the members of another family – the ESTERS. We can follow the pattern with ethanol and ethanoic acid:

$$\begin{array}{ccccccc} \text{ethanol} & + & \text{ethanoic acid} & \rightarrow & \text{ethylethanoate} & + & \text{water} \\ C_2H_5OH(l) & + & CH_3COOH(aq) & \rightarrow & CH_3COOC_2H_5(l) & + & H_2O(l) \end{array}$$

Compare this with a neutralisation reaction (page 32) and you can see that esters are made in a similar way to salts.

Most esters have a 'fruity' smell. Ethyl ethanoate smells of pear drops. For this reason esters are widely used as flavourings, or 'smells', in food stuffs, perfumes, etc.

Assignment – The effects of alcohol on the human body

The following article appeared in a newspaper:

'Alcohol has complex effects. Small amounts act as a stimulant and can be helpful. Many people enjoy a drink before a meal because it increases the flow of digestive juices and stimulates the appetite. Large quantities of alcohol make the nervous system less sensitive. Alcohol was once used to deaden pain during operations before modern anaesthetics were available. Alcohol slows our reactions and excessive intake over a long time causes damage to liver, kidneys and brain. A study in the United States of America has shown that even "moderate" drinking can affect mental agility and memory. It also shows that the brain can recover, but it may take as long as two years.'

1 The article says 'alcohol stimulates digestive juices'.
   (a) What does the phrase 'digestive juices' mean?
   (b) Explain why it is helpful to stimulate digestive juices before a meal.
2 Explain why alcohol was useful to patients during operations in years gone by.
3 In the poster you can see just one of the measures being taken to persuade people to avoid alcohol.
   (a) Explain how alcohol could affect a driver's ability to drive safely.
   (b) What other measures do you think should be taken to reduce accidents caused by 'drunk driving'?
4 In England and Wales it is against the law for anyone under 18 to buy alcohol. However, there is much concern about the increasing number of teenagers who drink alcohol, and often have a 'drink problem'.
   (a) What is meant by the phrase 'drink problem'?
   (b) Suggest two reasons for the concern over the number of teenagers who drink alcohol in England.
   (c) How could the number of teenagers with 'drink problems' be reduced?

WHAT YOU SHOULD KNOW

1 Acids release H^+ ions when they dissolve in water. Bases release OH^- ions when they dissolve in water.
2 Acidic solutions have a pH between 1 and 7. Basic solutions have a pH between 7 and 14.
3 A strong acid releases H^+ ions easily. A strong base releases OH^- easily.
4 Acids and bases react as follows:

$$\text{acid} + \text{base} \rightarrow \text{salt} + \text{water}$$

This is described as a neutralisation reaction.
5 There are several ways of making salts. All of them involve replacing the H^+ ion of an acid with any positive ion.
6 Carbon compounds are often called *organic*. There are millions of them because carbon atoms can join in many different ways.
7 There are several families of organic compounds. Members of a family all behave the same chemically.
8 Hydrocarbons are compounds of carbon and hydrogen only. Alkanes, alkenes and alkynes are three hydrocarbon families.
9 Other families include the carboxylic (or alkanoic) acids, of which ethanoic acid (vinegar) is well known; the alcohols and the esters.

QUESTIONS

1 Below are listed the stages in making copper sulphate crystals. Put them in the correct order.
  (a) Unused copper oxide powder filtered out.
  (b) Liquid heated to evaporate some water.
  (c) The mixture is warmed.
  (d) The remaining solution cools slowly, and crystals of copper sulphate form.
  (e) Black copper(II) oxide powder is added until no more dissolves.
  (f) The solution is now blue and clear.
2 Name all the substances formed in the reactions:
  (a) hydrochloric acid + sodium carbonate;
  (b) sulphuric acid + sodium hydroxide.
3 (a) Which element is present in all acids?
  (b) Explain the difference between a weak acid and a strong acid.
  (c) What do all alkalis contain?
  (d) Give the name of three alkalis.
  (e) What is the difference between an alkali and a base?
4 Use the information on pages 40–41 to make a key to identify the alkanes and alkenes.
5 (a) (i) What are hydrocarbons?
  (ii) Give the names and formulae of four hydrocarbons.
  (b) (i) What are carbohydrates?
  (ii) Name one.
  (iii) Find out the formula of at least one carbohydrate.
  (c) (i) What are alcohols?
  (ii) Give the names of three alcohols.
  (iii) Write down the formula of the alcohol which is found in drinks.
6 Copy this table into your books, and put pH values from the list below in the best place in the table.

pH values 12, 9, 7, 3, 1

1M HCl	
1M NaOH	
1M CH_3COOH	
tap water	
limewater*	

* See Table 6.9

7 Write one or two short sentences to say how you would make these salts:
  (a) calcium chloride;
  (b) lead sulphide;
  (c) copper nitrate.
  (d) Write a word equation for each of the reactions you used in (a), (b) and (c).
8 Copy out and complete these equations for salt-making reactions:
  (a) potassium + ... → potassium chloride
  (b) ammonium chloride + ... → lead chloride + ...
  (c) sulphuric acid + ... → sodium sulphate + ...

7. Change and Stability

This unit is about changes: how things change and why they change.

Figure 7.1 shows two situations which are not stable. In each case, a change could take place to make it more stable. Try to work out what the change is.

Are the situations stable once the change you have suggested is complete?

CHANGES IN COMPOUNDS

Chemical changes are an important part of our lives. In our own bodies, chemicals are altered all the time. The large molecules of the food you eat are broken up by the digestive system. They are rearranged into the chemicals needed for growth and energy. Chemical changes outside the body are even more obvious. Cooking an egg is a clear chemical change. So is letting off a firework.

WHAT IS A CHEMICAL CHANGE?

In a chemical change, atoms, ions or molecules are rearranged. The bonds between some particles are broken and new bonds are made. It is usually easy to tell that a chemical change has happened.

Fig. 7.1 Unstable situations

Energy is given out or taken in by the chemicals. The energy may be in many forms: for example, heat, light or electrical energy.

Fig. 7.2

The substances produced look different from the chemicals that made them.

Fig. 7.3

Chemical changes are difficult to reverse. It is often hard to get back to what you started with.

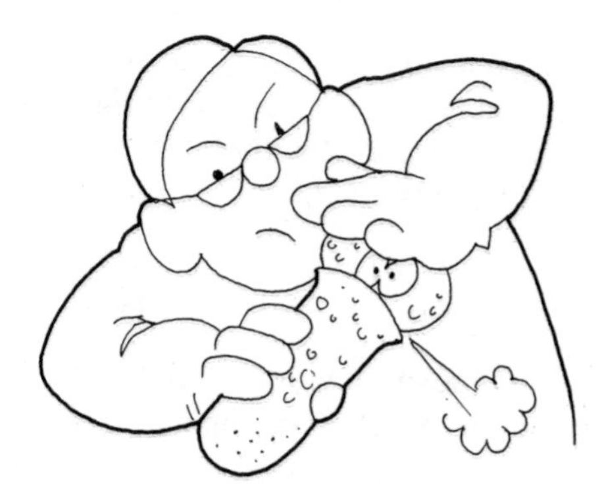

Fig. 7.4

HOW DO CHEMICALS COMBINE?

Before doing the experiments in this section, you must be sure that you understand chemical symbols. The assignment you did in Unit 5 on chemical symbols may help you. In this section we will be using symbols as a shorthand way of describing chemical reactions. So it is important that you understand them. Here is a reminder:

$$H_2SO_4$$

This formula normally represents one mole of sulphuric acid. For the moment, we will imagine it as just one molecule. The formula tells us that one molecule of sulphuric acid contains:

H_2 two atoms of hydrogen
S one atom of sulphur
O_4 four atoms of oxygen

In an equation, the symbols represent moles of particles, not individual atoms. The result is the same, though. This equation shows the reaction between magnesium and oxygen:

$$2Mg(s) + O_2(g) \rightarrow 2MgO(s)$$
magnesium + oxygen → magnesium oxide

This is what the equation means:

$2Mg(s)$ two moles of magnesium which is solid
$O_2(g)$ one mole of oxygen gas molecules (the $_2$ says that each molecule is made of two atoms)
→ react together to make
$2MgO(s)$ two moles of magnesium oxide which is solid. Each mole of magnesium oxide contains one mole of magnesium atoms and one mole of oxygen atoms.

Chemical equations also tell us how much of each substance will react, and how much will be produced. Using the Relative Atomic Masses, $A_r(Mg) = 24$, $A_r(O) = 16$:

$$2Mg(s) + O_2(g) \rightarrow 2MgO(s)$$
$$2 \times 24\,g + 16\,g \times 2 \rightarrow 2 \times (24\,g + 16\,g)$$

So 48 g of magnesium would react with 32 g of oxygen to make 80 g of magnesium oxide. The total mass of the reactants (48 g + 32 g) equals the mass of the products (80 g).

Assignment – Working out equations

For each equation below, do two things.
(a) Say what it means in words.
(b) Work out how much of each reactant in grams is needed to make 1 mole of product.

If you do not know any of the chemical names, have a guess.

Relative Atomic Masses (A_r): H = 1, N = 14, O = 16, Na = 23, S = 32, Cl = 35.5, Fe = 56, Cu = 63.5

1. $H_2O(l) + SO_3(g) \rightarrow H_2SO_4(l)$
   (SO_3 is sulphur(VI) oxide or sulphur trioxide)
2. $4Fe(s) + 3O_2(g) \rightarrow 2Fe_2O_3(s)$
   (Fe_2O_3 is iron(III) oxide)
3. $N_2(g) + 3H_2(g) \rightarrow 2NH_3(g)$
   (NH_3 is ammonia)
4. $2H_2(g) + Cl_2(g) \rightarrow 2HCl(g)$
   (HCl is hydrochloric acid)
5. $CuSO_4(s) + 5H_2O(l) \rightarrow CuSO_4 \cdot 5H_2O(s)$
   ($CuSO_4$ is copper(II) sulphate – white)
   ($CuSO_4 \cdot 5H_2O$ is copper(II) sulphate-5-water – blue)
6. $2Na(s) + Cl_2(g) \rightarrow 2NaCl(s)$
   (NaCl is sodium chloride, known as common salt)

EQUATIONS FROM EXPERIMENTS

Chemical equations are worked out by doing experiments. In this section you will meet several types of different chemical change. For each change, the correct equation can be worked out.

Investigation 7.1 A reaction which produces a gas

In this experiment you can find the equation for the reaction between magnesium and hydrochloric acid. The only measurements needed are the mass of magnesium used and the volume of gas produced.

Collect

150 cm^3 conical flask
bung and connecting tube
gas syringe
clamp and stand
access to top-pan balance
10 cm of magnesium ribbon
a medium test tube
25 cm^3 of 2M hydrochloric acid
emery paper
eye protection

What to do

1. Clean the magnesium ribbon until it shines on both sides.
2. Record the mass that you use as accurately as you can.
3. Put 25 cm^3 of acid in the conical flask. Clamp the gas syringe firmly, but not so tightly that it sticks.
4. Push the syringe plunger right in. Make sure that the bung will fit quickly and easily into the flask.
5. Now work fast. Drop the magnesium in the acid and *immediately* put the bung in the flask. Twist the syringe plunger to make sure that it does not stick.
6. When the reaction stops, measure how much gas has been produced. The plunger must not be stuck when you do this.
7. Disconnect the syringe carefully. Blow the gas into a test tube. Put your thumb over the end of the tube, and take it to a Bunsen burner. Try to light the gas.

CAUTION – Do not try to light the gas in the syringe. It might explode.

8. Wash away the acid carefully, then tidy up.

STEP 4

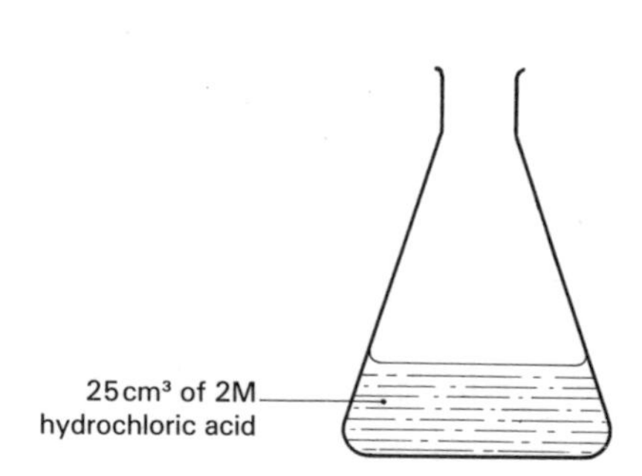

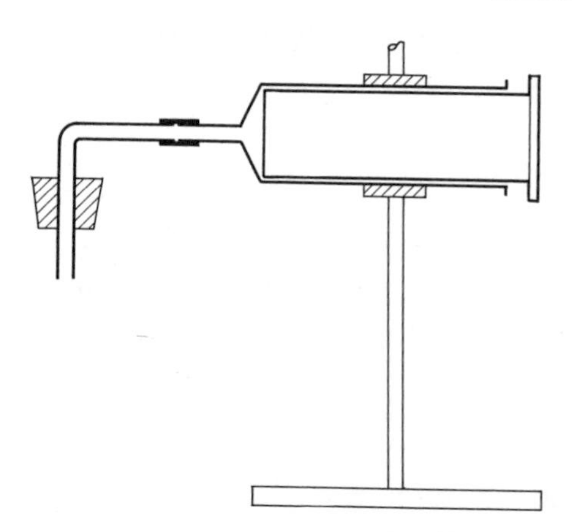

Questions

1 How do you know that there has been a chemical change?

2 What are the reactants (the chemicals that react together)?

3 What gas is produced?

4 Why does the reaction stop?

5 What is left in the conical flask at the end of the experiment?

6 One of the products is magnesium chloride. Where is this at the end of the experiment?

7 Copy the word equation:
magnesium + hydrochloric acid →
magnesium chloride + hydrogen

8 How many moles of magnesium did you use (A_r(Mg) = 24)? If you use a calculator, two significant figures, e.g. 0.0025 mol, is adequate. Why is this?

9 1 mole of any gas has a volume of about 24 000 cm^3 at room temperature and pressure. What volume of gas did you collect?

10 How many moles of gas is this? Hint: divide your volume by 24 000. You will get an answer very much less than 1. Again, two significant figures is accurate enough.

11 Copy and fill in this sentence:
___mol of magnesium produced ___mol of hydrogen.

12 Divide the moles of hydrogen by the moles of magnesium. This tells you how much hydrogen is produced by 1 mole of magnesium. Your answer should be near 1.0.

13 Copy the chemical equation:
$Mg(s) + HCl(aq) \rightarrow MgCl_2(aq) + H_2(g)$

We know from the results that 1 mole of magnesium produces one mole of hydrogen (if your results were accurate!). So the numbers in front of the magnesium and the hydrogen are both 1.

All that remains is to balance up the other chemicals. To do this, there must be the same number of moles of each element on the left and on the right of the equation.

There is 1 mole of magnesium on the left and 1 mole on the right, so this is correct.

There is 1 mole of hydrogen molecules (H_2) on the right, but only 1 mole of hydrogen atoms (H) in HCl on the left.

There are 2 moles of chlorine on the right (Cl_2), but only 1 mole on the left.

To balance things up, we can put numbers in front of the chemicals. This equation is balanced by putting a 2 in front of the HCl(aq):

$$Mg(s) + 2HCl(aq) \rightarrow MgCl_2(aq) + H_2(g)$$

The 2 in front of the HCl means that 2 moles of HCl react with 1 mole of magnesium. There are now 2 moles of H and 2 moles of Cl on the left hand side, and 2 moles of Cl and 2 moles of H on the right. This means that the equation is balanced.

14 Write down the balanced equation. Explain what it means in words.

15 Work out how many moles of HCl(aq) reacted with the mass of magnesium that you used.

Investigation 7.2 A reaction which produces a solid

In this experiment, potassium iodide solution is added to lead nitrate solution. A yellow solid, lead iodide, is formed. The equation for the reaction can be worked out by measuring the height of the solid which is made in a test tube.

Collect

6 identical test tubes
2 syringes, each 5 cm^3
250 cm^3 beaker of hot water
about 25 cm^3 of 1M lead nitrate
about 50 cm^3 of 1M potassium iodide

WARNING – All lead compounds are poisonous. Be sure to wash your hands after you have completed the experiment.

What to do

1 Label the lead nitrate beaker and label one syringe to use for lead nitrate only.
2 Label the potassium iodide beaker and the other syringe for the potassium iodide.
3 Mark the test tubes 1 to 6.
4 Use the lead nitrate syringe to put the amount of lead nitrate shown in the table into each tube.

Tube	1	2	3	4	5	6
Amount of lead nitrate/ cm^3	1.0	1.5	2.0	2.5	3.0	3.5
Amount of potassium iodide/cm^3	5.0	5.0	5.0	5.0	5.0	5.0
Height of precipitate/ mm						

5 Use the other syringe to add 5 cm^3 of potassium iodide to each tube.
6 Shake the tubes gently, then stand them in the beaker of hot water for at least 30 minutes to let the solid settle. Leaving for 24 hours or so is best.
7 When the solid has settled, very carefully measure the height of each precipitate.

Questions

1 Copy down the table. Fill in the precipitate heights.
2 How do you know that a chemical change has occurred?
3 What were the reactants in this experiment?
4 What are the products?
5 Make a line graph of your results, as shown in the diagram below.

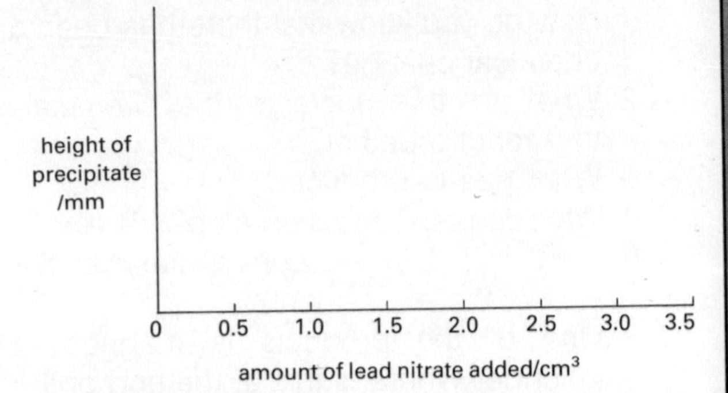

You can use (0,0) as a point on your graph. How do you know this?
6 Your graph should go up at first, then level off. Estimate how much lead nitrate was added when it started levelling off. This is the amount of lead nitrate that reacts exactly with 5 cm^3 of potassium iodide.
7 Explain why the graph levels off. If you are not sure, think back to Investigation 7.1. Why did the magnesium stop producing gas?
8 Your result should give you a simple ratio between the lead nitrate and the potassium iodide.

Both solutions were the same concentration, 1M. So if 5 cm^3 of lead nitrate reacted with 5 cm^3 of potassium iodide, 1 mole would react with 1 mole. If 2.5 cm^3 of lead nitrate reacted with 5 cm^3 of potassium iodide, 1 mole would react with 2 moles. Write down a sentence that gives your result as a simple ratio.

▸

9 Here is the chemical equation for the reaction:

$Pb(NO_3)_2(aq) + 2KI(aq) \rightarrow 2KNO_3(aq) + PbI_2(s)$

The equation has been balanced for you.

You can see that 1 mole of lead nitrate reacts with 2 moles of potassium iodide. This is the correct ratio.

The solid product is shown by the (s) in the equation. A solution in water, like lead nitrate, is shown by (aq).

Copy the equation, and put the correct chemical names beneath each formula.

10 Make a list of errors that may have affected your result. For each error, say how it is likely to have changed your result.

Investigation 7.3 A reaction which conducts electricity

Adding sulphuric acid to barium chloride produces barium sulphate and water. The two products are both poor conductors of electricity. When the correct amounts of the reactants are added, the solution will hardly conduct electricity. As this is a tricky experiment to set up, and soluble barium compounds are poisonous, it will probably be demonstrated to you. However, you can take readings and plot the graph. Try to work out why the experiment is so tricky.

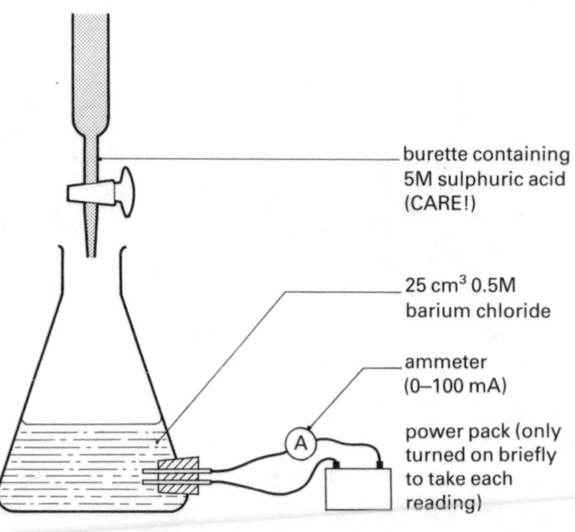

What to do

1 Draw up a results table and fill it in.

Amount of sulphuric acid added/cm^3						
Current flowing/mA						

2 Plot a graph of these results. Make the amount of acid added the *x*-axis, and the current flow the *y*-axis.

Questions

1 Write a word equation for the reaction.

2 How much sulphuric acid had been added when the current was lowest?

3 The sulphuric acid is 5M (5 moles in $1000 cm^3$ of solution). How many moles of acid had been added when the current was lowest? Hint: 5 mol in $1000 cm^3$ = 0.5 mol in $100 cm^3$ = 0.05 mol in $10 cm^3$.

4 At the start, there was $25 cm^3$ of 0.5M barium chloride in the beaker. How many moles of barium chloride is this? Hint: 0.5M means 0.5 mol in $1000 cm^3$.

5 Use your answers to 3 and 4 to complete the following:

___mol of sulphuric acid reacted exactly with

___mol of barium chloride.

6 How many moles of barium chloride would react with 1 mole of sulphuric acid?

7 The chemical formulae of the compounds in this experiment are:

sulphuric acid	H_2SO_4
barium chloride	$BaCl_2$
barium sulphate	$BaSO_4$
water	H_2O

Write the chemical equation for the reaction. Add the correct state symbols (s, l, aq). Make sure that the reactants are in the correct ratio. Check that the equation is balanced.

Investigation 7.4 A reaction between an acid and an alkali

When an acid is added to an alkali, the alkali is neutralised. The products are a salt and water. The point where the acid exactly neutralises the alkali is called the end-point. This can be found by adding an indicator which changes colour at the end-point.

This type of experiment is called a TITRATION. In this titration, hydrochloric acid in a burette is going to be added to sodium hydroxide in a conical flask. The indicator is phenolphthalein.

Collect

burette and stand
1M hydrochloric acid solution
25 cm^3 pipette and filler
1M sodium hydroxide solution
2 test tubes
phenolphthalein indicator
150 cm^3 conical flask
white tile

What to do

1 Put 1 cm depth of hydrochloric acid in one test tube, and 1 cm of sodium hydroxide in the other. Add 2 drops of phenolphthalein to each. Note the result.

2 Copy down this table for results.

Titration number	1	2	3
Final burette reading/cm^3			
Initial burette reading/cm^3			
Volume of acid added/cm^3			

You need to do three titrations to get an accurate result. The first one gives you a rough answer; the other two should be accurate to 0.1 cm^3.

3 Using a pipette and filler, place exactly 25 cm^3 of sodium hydroxide in the flask. Remember to measure at the *bottom* of the meniscus.

4 Add a few drops of phenolphthalein to the flask, and stand it on a white tile.

5 Put the burette in the stand and fill it with acid. Run a little acid out into a test tube, then take the initial reading. Remember to read the *bottom* of the meniscus.

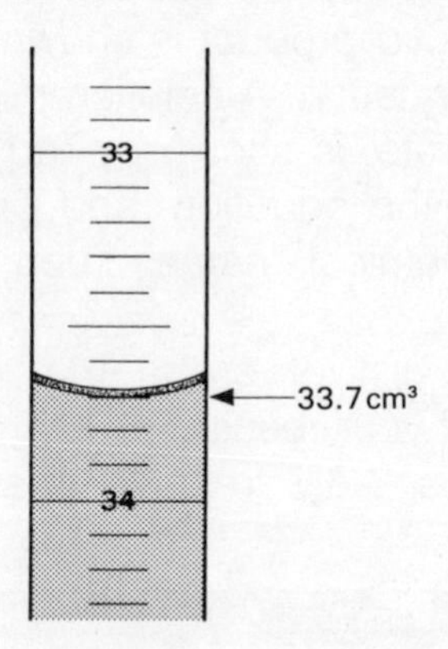

6 Stand the burette over the conical flask. Hold the burette in both hands, one either side of the tap. Run a little acid into the flask, then shake the flask. Repeat this until the colour starts to disappear. Then add the acid more carefully. With care you can add one drop at a time.

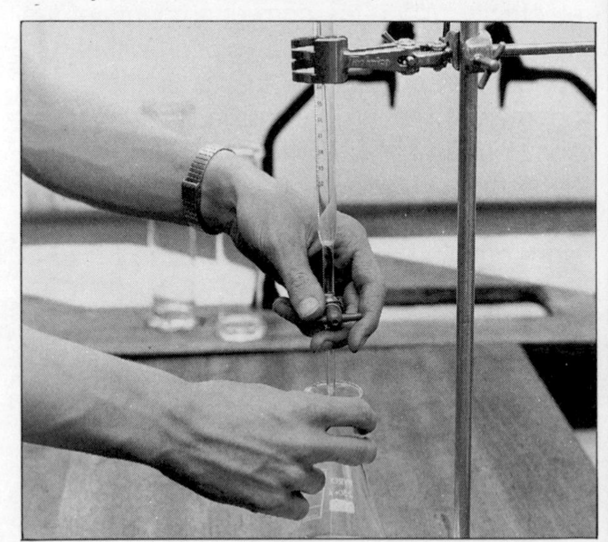

7 When the colour disappears and does not return, you have reached the end-point. Carefully read the burette and write the result in your table. Work out how much acid you added. Empty the flask and rinse it with water. Top the burette up.

8 Repeat the whole procedure twice more. This time you can let the acid run in until it is near your first answer. Then take great care over the last few drops. If you do it carefully, results 2 and 3 should be the same.
9 Clear up thoroughly. Be careful when washing the pipette and burette; they break easily.

Questions

1 What colour does phenolphthalein go with (a) acids and (b) alkalis?
2 Calculate the volume of acid that neutralises 25 cm^3 of sodium hydroxide. The best way to do this is to average your two accurate readings.
3 How many moles of acid react with one mole of alkali? As they are both the same concentration, the ratio of the two volumes will tell you this.
4 Write down a word equation for the reaction.
5 Write down the chemical symbols and the state symbols beneath the words.
6 Make sure that the reactants are in the same ratio as your results.
7 Make sure that the equation balances.
8 When you have added 10 cm^3 of acid, what is in the conical flask?
9 What is in the conical flask at the end-point?
10 If you added 10 cm^3 of acid after the end-point, what would be in the flask?

Assignment – Chemical pollution

THE OBSERVER, SUNDAY 27 FEBRUARY 1983

The town America wrote off for cash

from JOYCE EGGINTON in New York

IN THE WORDS of one of its leading residents, Times Beach, Missouri, now looks 'like the outskirts of Hiroshima after the atom bomb went off.'

Devastated by floods last December and subsequently threatened by chemical contamination which the waters unleashed, this working class community on the banks of the Meramec River has just earned a peculiar place in history. Last week it became the first town in the United States to be purchased in its entirety by the federal Government because it is considered too dangerous a place for anyone to live.

The Government plan calls for spending $33 million which will compensate the 2500 residents for the loss of their homes at a fair market price. If the flood had been the only disaster Times Beach could have been rebuilt for far less. But a wholesale evacuation of the area, perhaps for ever, has been deemed essential because the soil is so heavily contaminated by Dioxin, in some places 600 times the 'safe' level.

The contamination came from salvage oils sprayed on the roads in the early 1970s before they were paved. These oils were the refuse of a chemical plant in Missouri which manufactured the herbicide Agent Orange, of which Dioxin – widely regarded as the most toxic man-made chemical in existence – is a frequent contaminant.

The story behind the Times Beach disaster is only just being unravelled, and seems likely to have nationwide implications. There are believed to be at least 100 other Dioxin-contaminated sites in Missouri alone, all of them the consequence of careless dumping of toxic chemical waste during the 1960s and 1970s.

Environmentalists are asking whether the federal Government will be prepared to buy out the residents of all these places – or even to go to the expense and political embarrassment of identifying them. And if they were to do so, how much of America might be ruled uninhabitable.

Dioxin, like a host of similar man-made chemicals is not soluble in water and clings stubbornly to soil and can stay there for a century or as long as it takes to be washed out to sea.

Clearing up: Times Beach, lashed by floods and polluted by Dioxin.

Good Health

All things bright and beautiful,
All creatures great and small
Thrive on toxic chemicals
And seldom die at all.

Give thanks to Agent Orange
That bears a lovely name;
Once treated with dioxin,
You'll never feel the same.

All well-tried defoliants,
All poisons that they spray
Make our bodies healthier
And guard our DNA.

From Vietnam to Seveso
Their wonders are still sung:
They made the women fertile
And fortified the young.

ROGER WODDIS
Radio Times

1 Missouri is generally hot and dry. Why do you think that the roads were sprayed with oil before they were surfaced?
2 How did the dioxin come to pollute Times Beach?
3 Why has the dioxin not disappeared in the ten years since spraying?
4 Dioxin is found in Agent Orange, a herbicide. Find out what a herbicide is, and what it is used for.
5 In some places the dioxin level is '600 times the "safe" level'. Why is 'safe' in quotation marks? How do you think the 'safe' level is measured?
6 Suggest some ways that dangerous chemicals like dioxin could be dealt with. Give advantages and disadvantages for each method you suggest.

AN IMPORTANT CHEMICAL: SULPHURIC ACID

Sulphuric acid is a dangerous but important chemical. It is made from sulphur, which you learned about in Unit 5. Your teacher may show you how this can be done in the laboratory. The next assignment explains how sulphuric acid is manufactured in large quantities.

Assignment – How sulphuric acid is made

Sulphuric acid is made by the Contact process. There are three stages.

First, sulphur is roasted in a furnace with dry air to make sulphur(IV) oxide gas.

$$S(s) + O_2(g) \rightarrow SO_2(g)$$

Second, the sulphur(IV) oxide is converted to sulphur(VI) oxide by reacting it with more oxygen in a convertor.

$$2SO_2(g) + O_2(g) \rightleftharpoons 2SO_3(g)$$

The $\rightleftharpoons$ arrows mean that the reaction is reversible. Only some of the sulphur(IV) oxide becomes sulphur(VI) oxide.

Finally, sulphur(VI) oxide is reacted with water in an absorber to make sulphuric acid.

$$SO_3(g) + H_2O(l) \rightarrow H_2SO_4(l)$$

In a sulphuric acid plant, sulphur(VI) oxide is absorbed in sulphuric acid rather than water. The sulphur(VI) oxide then makes the sulphuric acid more concentrated. This method stops pollution from acid spray. The concentrated acid or oleum is diluted later if necessary.

The most difficult job is converting sulphur(IV) oxide into sulphur(VI) oxide. The reaction is slow, and the amount of sulphur (VI) oxide produced is small. The speed of the reaction can be increased by using a catalyst called vanadium(V) oxide. Increasing the temperature also helps. The catalyst and the higher temperature both make it easier for the molecules of oxygen and sulphur(IV) oxide to react. Unfortunately, heat also reduces the yield; less sulphur(VI) oxide is made at higher temperatures. A compromise temperature of about 750 °C is used.

Three other things are done to increase the amount of sulphur(VI) oxide made:

(a) The pressure is kept just above atmospheric pressure.

(b) More oxygen is used than is needed.

(c) The sulphur(VI) oxide made is removed from the gases by absorbing it.

There is one further serious problem. Any sulphur oxides or sulphuric acid mist that gets into the atmosphere would cause serious pollution. UK government regulations now insist on at least 99.5% efficiency in converting sulphur(IV) oxide to sulphur (VI) oxide. Modern plants have several absorbers and acid coolers to achieve this efficiency. In this way the environment is protected. (Look at the assignment on acid rain in Unit 10, page 274.)

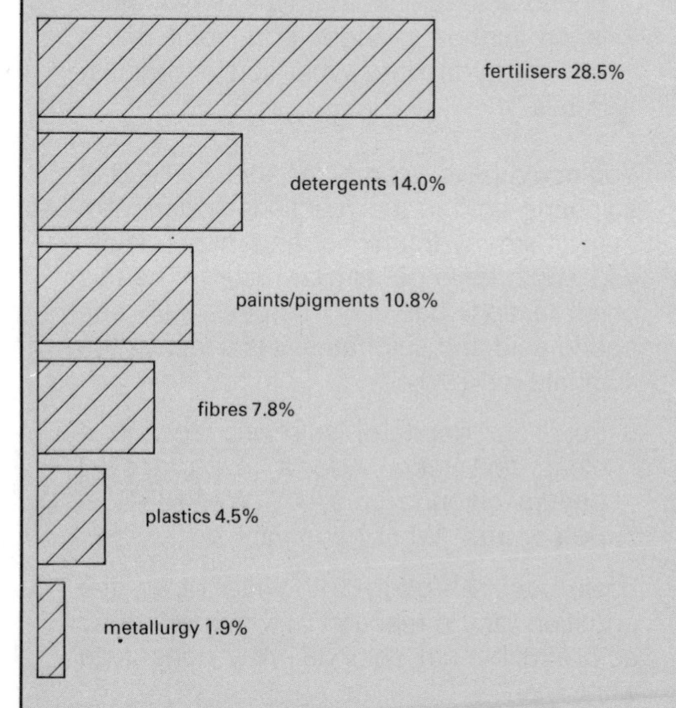

Questions

1 Draw a flow diagram to show the stages in the manufacture of sulphuric acid. Write in beneath each stage the chemicals that are present at that stage. The first step has been done for you.

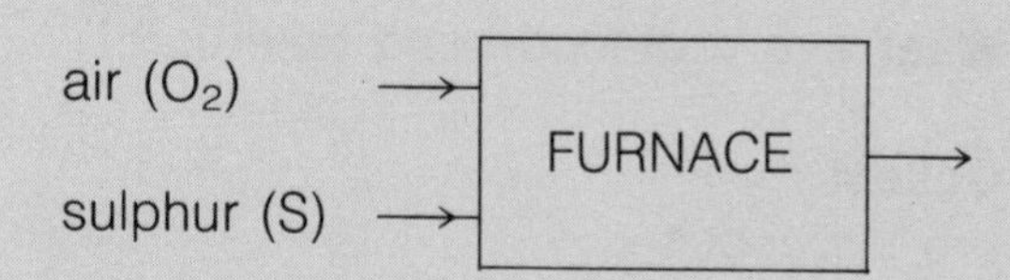

2 What happens in the convertor?

3 How is the rate of the reaction increased?

4 How is the amount of sulphur(VI) oxide made increased?

5 What happens in the absorbers?

6 Why are several absorbers used?

7 Why is the air dried before it is used?

8 What mass of sulphur would you need to make 1 mole of sulphur(IV) oxide if none is wasted?
($A_r(S) = 32$)

9 How many moles of sulphur(IV) oxide are needed to make 1 mole of sulphuric acid if none is wasted?

10 What mass of sulphur is needed to make 1 mole of sulphuric acid?

11 How much sulphur would you need to make 98 kg of sulphuric acid if none is wasted?
($A_r(H) = 1$, $A_r(O) = 16$)

12 Make a list of the safety precautions that the men and women working on a sulphuric acid plant would need to observe.

Some reactions of sulphuric acid

Concentrated sulphuric acid is extremely dangerous. Your teacher will show you why. In the demonstration, concentrated acid will be added to some of the following chemicals. Make notes of what you see.

1. Blue copper(II) sulphate, formula $CuSO_4 \cdot 5H_2O$.
2. Sugar, formula $C_{12}H_{22}O_{11}$.
3. Paper (which is made of cellulose molecules $(C_6H_{10}O_5)_n$).
4. Water

Questions

1 Write down what you saw in each experiment.

2 In the first three experiments, the acid removes a simple molecule from the other substance. What is the simple molecule? Hint: the fourth experiment may help you.

3 If you remove as many of the simple molecules as possible from each of the chemicals, what is left in each case?

4 Predict what will happen if concentrated sulphuric acid is added to methanoic acid (HCO_2H). What gas will be produced? Why would the experiment be dangerous?

Investigation 7.5 Dilute sulphuric acid

Dilute sulphuric acid is easier to handle than concentrated acid, but it is still dangerous.

Wear eye protection and take care!

Collect

5 test tubes
test tube rack
1M sulphuric acid
calcium carbonate powder
zinc granules
magnesium ribbon
0.5M barium chloride solution (**poisonous**)
0.5M copper(II) sulphate solution
limewater
5 cm^3 syringe
universal indicator solution
1M sodium hydroxide solution
eye protection

What to do

In some of the experiments you will need to identify a gas. Table 7.1 will help you. After each experiment, write down your observations and a conclusion.

1 Put 1 spatula of calcium carbonate in a test tube. Add 1 cm depth of dilute sulphuric acid. Test the gas – a syringe and limewater may help.

2 Put 2 zinc granules in a test tube. Add 1 cm depth of dilute sulphuric acid. Watch for a minute. Then add a few drops of copper(II) sulphate solution. Test the gas – covering the test tube with your thumb for a few seconds, then applying a lighted spill a few seconds later may help.

3 Put a short piece of magnesium ribbon in a clean test tube. Add 1 cm depth of dilute acid. Identify the gas.

4 Put 1 cm depth of sodium hydroxide in a test tube. Add a few drops of universal indicator solution. Add some dilute sulphuric acid, a little at a time. When there is no further change, add some more sodium hydroxide. What sort of reaction is this?

You could have used other acids instead of sulphuric acid in the reactions so far. With many, like hydrochloric and nitric acids, you would have got similar results. There is one reaction that only happens with sulphuric acid and sulphates. It is a test for the sulphate ion (SO_4^{2-}).

5 Put 1 cm depth of sulphuric acid in a clean test tube. Add a few drops of barium chloride solution. (**WARNING – poisonous**) What happens?

Look back at page 51. Write down the equation for the reaction between sulphuric acid and barium chloride. Why does it go white?

Table 7.1 Identifying gases

Gas	Symbol	Smell*	Acid/alkali	Other tests
Ammonia	NH_3	smelling salts	alkali	–
Carbon dioxide	CO_2	none	just acid	turns limewater cloudy
Hydrogen	H_2	none	neutral	'pops' when lit
Hydrogen chloride	HCl	choking	acid	fumes with ammonia
Chlorine	Cl_2	swimming baths	bleaches	–
Hydrogen sulphide	H_2S	bad eggs	acid	†
Nitrogen	N_2	none	neutral	unreactive
Oxygen	O_2	none	neutral	relights glowing spill
Sulphur(IV) oxide	SO_2	choking	acid	†

*** Do not smell any gas before you have been shown how to do it safely.**

† There are special tests for these gases.

MAKING THINGS REACT FASTER

Some chemical changes are very fast. When you strike a match, the phosphorus head of the match burns out straight away. Even the wood does not last long.

Other changes are slow. A piece of steel gradually rusts. This is because iron is turning into iron oxide. By treating or protecting the steel, the rusting can be slowed down even more, or stopped altogether.

Rusting is a nuisance, so slowing it down is useful. Other chemical reactions can be very helpful. Some of them, though, are far too slow to be useful. In these cases we have to find a way to speed them up. There are two main methods of doing this.

Fig. 7.5 A slow chemical change

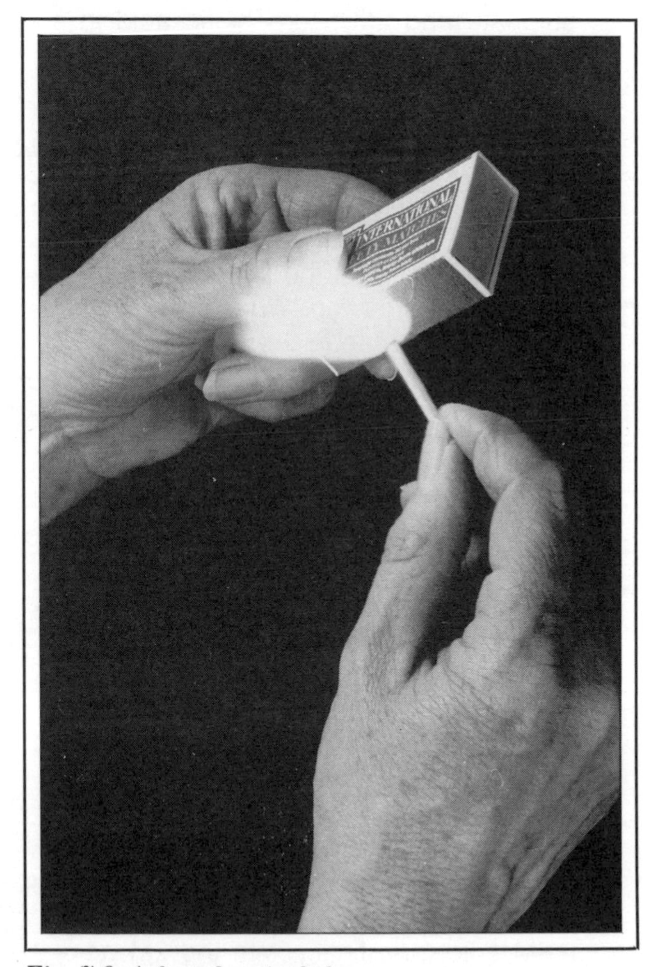

Fig. 7.6 A fast chemical change

(a) Make it more likely that the particles which we want to react will meet each other.
(b) Reduce the amount of energy needed for the particles to react once they have met.

The next experiments will show you how this can be done.

Investigation 7.6 Altering the concentration of the particles

Collect

thermometer 0–110 °C
150 cm^3 conical flask
2M sodium thiosulphate solution
0.1M, 0.5M, 2M hydrochloric acid solutions
watch or stopclock
2 measuring cylinders, 25 cm^3
250 cm^3 beaker
test tube

What to do

1 Mix equal volumes of sodium thiosulphate and 2M hydrochloric acid in a test tube. Watch carefully.

The reaction is:

thiosulphate ions + hydrogen ions → water + sulphur + sulphur dioxide

$$S_2O_3^{2-}(aq) + 2H^+(aq) \rightarrow H_2O(l) + SO_2(g) + S(s)$$

Questions

1 Smell the mouth of the tube (**carefully!**). Look at the equation. What is the smell?
2 What can you see in the tube?
3 What is the solid in the tube?

What to do

2 Measure exactly 25 cm^3 of sodium thiosulphate into the conical flask. Put a black cross on a piece of white paper, and stand the flask on it. Using the other measuring cylinder, measure out 25 cm^3 of 0.1M hydrochloric acid. When you are ready, tip the acid into the flask and start the clock. Give the flask a quick swirl to mix the reactants then stand it on the piece of paper with the cross. When the cross cannot be seen from above, note down the time.
3 Wash the flask out thoroughly with water. Repeat the experiment, first with the 0.5M acid, and finally with the 2M acid. Take care to use the same measuring cylinder for the acid each time.

Questions

4 Use your results to estimate what your result would have been with 1M acid. Hint: Drawing a graph could help.
5 What is the pattern in your results?
6 In this experiment, you have made it more likely that the particles of acid and thiosulphate will meet. Try to explain why this is true.

Investigation 7.7 Using a higher temperature

This experiment is almost the same as the previous one. The difference is that this time you are going to change the reaction temperature.

Collect

the same equipment as for Investigation 7.6, but you will only need 2M acid.
Bunsen burner, tripod and heatproof mat

What to do

1 Measure 25 cm^3 of sodium thiosulphate into the conical flask. Stand it on a cross again. Measure 10 cm^3 of 2M hydrochloric acid into a measuring cylinder. Pour the acid into the flask, swirl it round, and start timing. Take the temperature of the mixture, and note it down. When the cross disappears, record the time.
2 Wash the flask with water, and refill it with 25 cm^3 of sodium thiosulphate. This time, warm the solution up to about 40 °C. When it is hot enough, take it away from the flame and stand it on the cross. Then add 10 cm^3 of hydrochloric acid as before. Swirl it round, start the clock and take the temperature. Note down the time when the cross disappears.
3 Finally, do the whole experiment again, but this time warm the thiosulphate to 60 °C before adding the acid.

Questions

1 Estimate the result you would have got with the thiosulphate at 30 °C. Again, drawing a graph may help.
2 What is the pattern in your results?
3 What will happen to the particles in the flask when they are hotter?
4 Why does heat make it more likely that the particles will meet?

Investigation 7.8 Altering the size of the particles

This time we are going to look at a different reaction. When hydrochloric acid is added to calcium carbonate, carbon dioxide gas is produced. If the gas escapes, the rest of the mixture will have less mass. The speed of the reaction can be measured by seeing how fast the mixture loses mass.

The experiment may be a demonstration.

Collect

top-pan balance, 0.01 g
150 cm^3 conical flask
calcium carbonate (limestone) chips of three sizes
2M hydrochloric acid
watch or stopclock
50 cm^3 measuring cylinder
eye protection

What to do

1 Put about 25 g of the large chips in the conical flask. Measure out 50 cm^3 of acid. Stand the flask on the balance. Use a table like the one below for your results.
2 Add the acid to the chips and start the clock. If the balance has a 'tare', and will display negative readings, you can zero the balance now. If not, you must note down the reading. Then you have to take all the following readings away from it to get the loss in mass. Take readings of the mass every 15 s for a period of 5 minutes.
3 After 5 minutes, wash the flask out with water. Don't throw the chips down the sink!
Repeat the experiment with the medium chips.
4 Finally, do the experiment again with the smallest chips.

Questions

1 If your balance did not have a 'tare', you need to take all your readings away from the start reading for each size of chips.
2 On one graph, plot the three sets of results. Make the *x*-axis the time, and the *y*-axis the loss in mass. Try to use a different colour line for each set of results. Put a key on your graph.
3 Which chips reacted fastest? Suggest a reason for this.
4 Explain why the flask weighs less as the reaction goes on.
5 Why do the chips stop losing mass after some time?
6 Write an equation for the reaction in words (part 1 of Investigation 7.6 may give you some ideas).
7 Underneath the word equation, write in as many of the chemical symbols as you know. As a start, calcium carbonate is $CaCO_3$.
8 Add the state symbols (s, l, aq, g) after each of the chemical symbols.

Time		start	15 s	30 s	45 s	1 min	1 min 15 s	etc.
Mass/ g	large chips							
	medium chips							
	small chips							

Using a catalyst

A CATALYST is a substance that alters the speed of a reaction without getting used up. Catalysts usually work by reducing the energy that the particles need to react together.

They are particularly important in industrial processes. A reaction which is too slow may not pay for the costs of the process. A catalyst can make all the difference. Most of the chemical reactions in our bodies have catalysts called ENZYMES. Without the enzymes, we could not exist.

You may be able to see a demonstration of a catalyst. Hydrogen reacts violently with oxygen in the air to make water. When you test for hydrogen in a test tube with a lighted spill, this is the reaction that makes the 'pop'. It is also the reaction that made the zeppelin *Hindenburg* go up in flames in 1937. In your test tube, and in the *Hindenburg*, a flame or spark was needed to start the reaction.

However, a catalyst can also help hydrogen to combine with oxygen. If dried platinised ceramic wool is held in tongs above a gas jar of hydrogen, the energy needed to start the reaction is reduced. The hydrogen and oxygen will react without any spark or flame being necessary. The two gases explode on their own.

There are many other examples of catalysts in use. You can read about two of them in the manufacture of sulphuric acid (page 54), and in the manufacture of ammonia (page 90).

Changes in compounds

WHAT YOU SHOULD KNOW

1 A chemical change is a rearrangement of molecules, atoms or ions.
2 Chemical changes often produce energy changes.
3 Chemical changes are often hard to reverse.
4 A chemical equation tells you:
   (a) which chemicals react together, and what products are formed;
   (b) the state (solid, liquid, gas or solution) of the chemicals;
   (c) the amounts of the chemicals that react together;
   (d) the amount of products that will be made if all the reactants become products.
5 Chemical equations are worked out from experiments.
6 Sulphuric acid is an important industrial chemical.
7 Concentrated sulphuric acid removes the elements of water from anything it comes in contact with.
8 Dilute sulphuric acid shows the normal reactions of an acid.
9 Chemical reactions can usually be speeded up by:
   (a) increasing the concentration of the reactants;
   (b) making the reactants hotter;
   (c) increasing the surface area of the reactants;
   (d) using a catalyst.

QUESTIONS

1 Which of the following are chemical reactions and which are not? Explain your answer in each case. In some cases there is no 'right' answer!
   (a) making toast
   (b) burning a match
   (c) bending an iron bar
   (d) stirring sugar into coffee
   (e) sawing a piece of wood in half
   (f) making ice from water
   (g) boiling an egg
2 Explain, in full sentences, what the equations below tell you. In each case, work out what mass in grams of each reactant is needed to make 1 mole of the first product.

   (a) $2H_2(g) + O_2(g) \rightarrow 2H_2O(g)$
   (b) $CuO(s) + H_2SO_4(aq) \rightarrow CuSO_4(aq) + H_2O(l)$
   (c) $Na_2CO_3(s) + 2HCl(aq) \rightarrow 2NaCl(aq) + H_2O(l) + CO_2(g)$

   (Relative Atomic Masses are on page 283.)
3 In an experiment, seven test tubes were each filled with 20 cm^3 of 1M sodium carbonate solution. A different amount of 1M barium chloride solution was added to each tube, as shown in the table. The tubes were left to stand, and the height of the precipitate in each tube was measured.

Tube no.	1	2	3	4	5	6	7
Volume of 1M barium chloride/cm^3	5	10	15	20	25	30	35
Volume of 1M sodium carbonate/cm^3	20	20	20	20	20	20	20
Height of precipitate/mm	4	7	11	16	17	16	18

(a) Describe briefly how you would do the experiment.
(b) Plot a graph of the volume of barium chloride (x-axis) against the height of the precipitate (y-axis).
(c) What volume of barium chloride reacted exactly with 20 cm^3 of sodium carbonate?
(d) What is the ratio of the volumes of barium chloride and sodium carbonate which reacted exactly?
(e) Write a word equation for the reaction.
(f) Write a chemical equation for the reaction (sodium carbonate is Na_2CO_3, barium chloride is $BaCl_2$).
(g) What is present *in the solution* in test tube 7?
(h) Why are the last four results similar?
(i) Why are the last four results *not* identical?

4 A student did an experiment to find the equation for the reaction between calcium carbonate ($CaCO_3$) and hydrochloric acid (HCl). She added 50 cm^3 of acid to 0.5 g of powdered calcium carbonate. The reaction produced 120 cm^3 of gas.

(a) What was the gas?
(b) How could you identify the gas?
(c) Why was the calcium carbonate powdered?
(d) How many moles is 0.5 g of calcium carbonate? (Relative Atomic Masses on page 283).
(e) How many moles of gas were produced? (1 mole of gas at room temperature and pressure takes up about 24 000 cm^3.)
(f) Copy the equation below, and fill in the gaps:

___mol calcium carbonate + hydrochloric acid →

___mol ____ + calcium chloride + water

(g) Try to write the proper chemical equation.
(h) Do a labelled sketch to show how the experiment might have been done.

Change and Stability

CHANGES IN MOTION

Most of the laws that explain how things move are just commonsense. You have been living with moving objects since you were born, so you will already know many of the rules. They are simple to understand, but harder to explain. If you get lost in the maths of motion later in this topic, remember that the rules themselves are not hard.

THINGS STAY AS THEY ARE UNLESS . . .

You have already met this rule in Unit 1. Things stay as they are, moving or stationary, unless a force affects them. This is not quite as obvious as you might think. A moving car or a bike does not keep on going forever. But that is because a force acts on it to slow it down. We call this force FRICTION. If you can imagine sitting on a moving bike in outer space, where there is almost no friction, then you will get the idea. The

Fig. 7.7 Changes in motion can be dramatic . . .

Fig. 7.8 . . . or everyday.

bike will keep on going in a straight line. It will only change if a force affects the bike; for example, a hit by a meteorite, or getting too near a planet. Sir Isaac Newton was the first person to explain this clearly about 300 years ago. It is called Newton's First Law.

Even on the earth, where there is friction, this rule is important. If you travel in a car that stops suddenly, you will realise that you tend to keep moving forwards. In a car crash, passengers without a seat belt keep going at the same speed. This may result in very serious injuries.

The inertia-reel seat belt that prevents these injuries also uses Newton's First Law. When the car stops suddenly, a ball-bearing inside the reel mechanism keeps moving forwards. It knocks a small ratchet into the main reel of the seat belt so that it cannot turn. When a force pulls the belt, a spring then locks the reel securely.

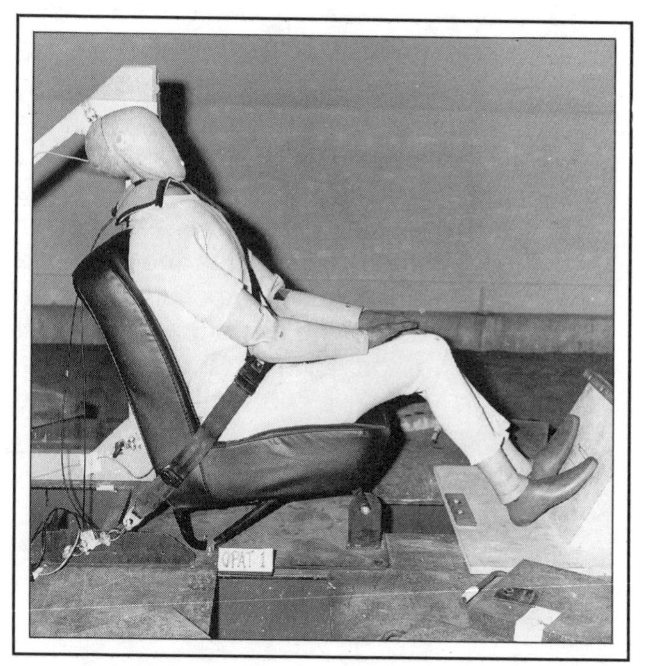

Fig. 7.9 How seat belts are tested

Assignment – Wearing seat belts

Seat belts were made compulsory for front seat passengers in the UK in 1982. This followed a long debate about whether or not seat belts are a good thing. Here is some of the information that was used in the debate

1 Table A does not give you a fair comparison between the results of wearing or not wearing a seat belt. Although injuries to belt wearers were less, there were fewer belt wearers in the survey. Make a new table of the same information which gives a fair comparison between belt wearers and non-wearers. Percentages could help!

2 Suggest some reasons for the differences shown in Table B.

B Usage of seat belts, 1973

	Number of drivers in every 100 wearing belts
on motorways	50
On 'A' class roads	40
In towns	20

3 Why did the government make seat belts compulsory for front seat passengers? You should be able to list a number of reasons.

4 Many people say that rear seat passengers should have to wear seat belts. Write a letter to a national newspaper (say which one) opposing compulsory seat belts for rear seat passengers.

A Injuries in accidents, 1973

	Total number of people	Deaths	Serious injuries	Slight injuries
People wearing belts	21 758	266	4273	17 219
People not wearing belts	112 860	2146	28 232	82 482

Assignment – Understanding velocity

VELOCITY is simply how fast something goes, but in a particular direction. It is always measured as a distance covered divided by a time taken. So the units are distance per unit of time: for example miles per hour (mph) or metres per second ($m\,s^{-1}$). The direction is important. Imagine running at 10 mph up a straight road, whilst a friend runs at 10 mph down the road. Although you have the same *speed*, you have exactly opposite *velocities*.

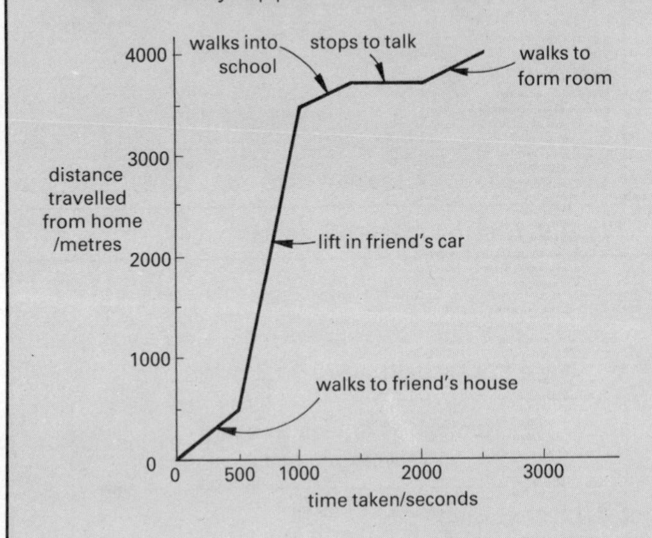

A graph of distance against time can help to explain changes in velocity. This graph represents a girl's journey from home to her form room at school.

To calculate velocity, just use velocity = distance/time.

1 At what velocity does the girl walk to her friend's house?
2 What is the average velocity of her friend's car? (Work out first how far she goes in the car and how long she spends in it.)
3 How long does she stop to talk for?
4 How far is her journey to school in total?
5 What time must she leave home to get to her form room by 8.45 a.m.?
6 What is the average velocity for her complete journey?
7 If Newton's First Law is right, where *must* there have been a force on the girl, ignoring gravity? There are at least five places. Try to explain your answer in each case.

HOW FORCE AFFECTS ACCELERATION

To understand this, you need to be clear what speed, velocity and acceleration are. You learned about speed and acceleration in Unit 1, and the assignment on this page will help you to understand the idea of velocity.

What is acceleration?

ACCELERATION tells you how the velocity changes. If the velocity is the same all the time, the acceleration is zero. A spacecraft in deep space with no rocket fuel may be travelling very fast (why?). But it will have a constant velocity and no acceleration. Look back at the beginning of this topic if you do not understand why.

Imagine that the spacecraft comes too close to a planet. Gravity will pull it towards the planet. Its velocity will increase because of this new force. This is called acceleration.

If the spacecraft continues towards the planet, and there is no atmosphere to stop it, it will keep accelerating. Eventually it would smash into the planet's surface. As it crashed, its velocity would change from very high to nothing. This is negative acceleration, or *deceleration*.

Measuring force and acceleration

If you remember Newton's First Law, you will see that velocity changes only when a force is involved. Acceleration is another way of saying 'velocity change'. So a force is needed to produce acceleration or deceleration. In Investigation 7.9, you will find out exactly how force alters acceleration.

Assignment – Understanding acceleration

Distance–time graphs help to explain and measure velocity. In the same way, velocity–time graphs help with acceleration. The graph shows a car journey from Manches-

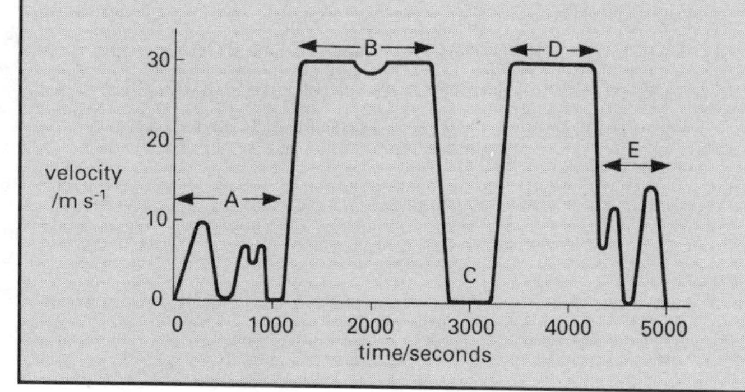

1 Sketch or trace the graph into your book.
2 How long did the journey take?
3 Leeds is 70 km from Manchester. What was the average velocity for the journey?
4 Mark on your copy of the graph places where the car was accelerating. Use green to mark it if possible.
5 Mark on your graph places where the car was decelerating. Use red.
6 What do you think the car was doing in each of the areas marked with the letters A to E?
7 Mark on your graph in blue all the places where the acceleration was zero.

Investigation 7.9 Force and acceleration

Collect

ticker-timer
power pack and leads
trolley and runway
3 identical pieces of elastic to pull the trolley
display paper for results
ticker-tape

What to do

The ticker-timer is a simple timing device that you used in Unit 1. It makes 50 marks on the tape every second.

1 Connect the timer to a power supply. Thread some ticker-tape through the timer, and turn on. Pull the tape through the timer and check that you are getting dots on the tape. If not, you may need to adjust the marking screw, or to replace the carbon disc. It depends on the type of timer you are using. In any case, do not leave the timer on when you are not using it.

2 You now have to arrange the runway so that friction will not ruin your results. Set it up so that it slopes away from the timer. Adjust the slope until the trolley just accelerates down it. Put the end of a 2 metre piece of tape through the timer and fix it to the trolley. Turn the timer on and give the trolley a gentle push. Take care to stop the trolley at the end! When you have the slope right, the tape will look like this:

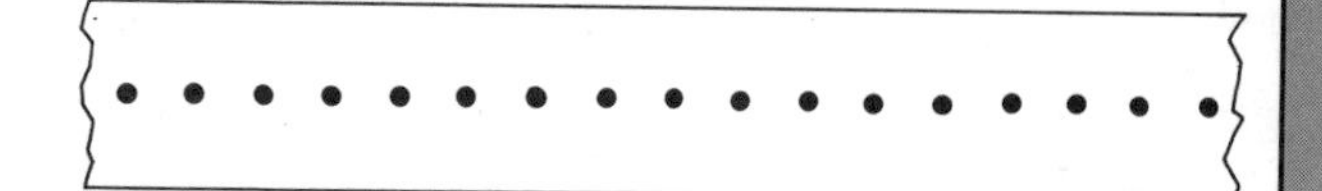

The timer dots should be the same distance apart at both ends of the tape. If it is not right, you will have to try again with a different slope.

A slope that gives a constant velocity is called 'friction-compensated'. The friction in the trolley and the tape is exactly balanced by the pull of gravity. This is vital, because only one force must affect the trolley during the experiment. That force is going to be you pulling it with some elastic.

3 A constant force is made by pulling the elastic so that it is always the same length. This takes some practice. The diagram shows you how.

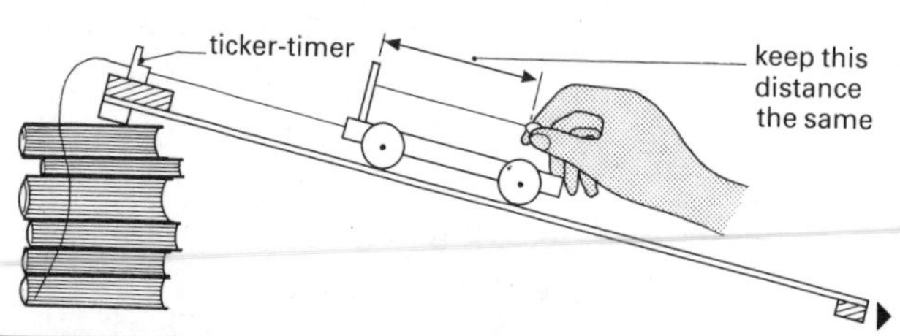

When you can do it, connect 2 metres of tape up through the timer to the trolley. Pull the trolley down the runway. Take care at the end! The tape should look like this:

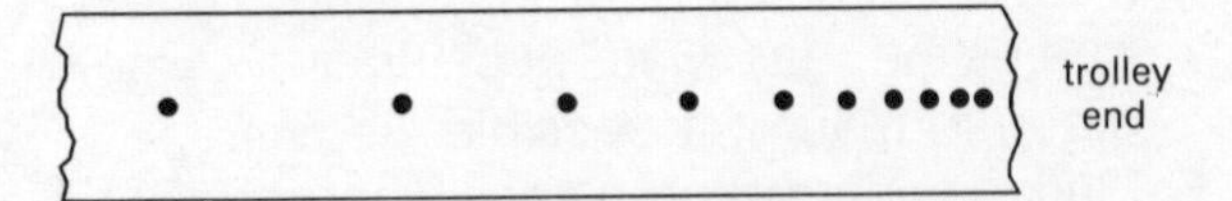

4 Cut the tape up, one piece at a time, at every fifth dot. Where the trolley is going fast, you may get 'echo' dots. They should be obvious as they do not fit the dot pattern. Ignore them, don't count them. Stick the pieces down in order as shown.

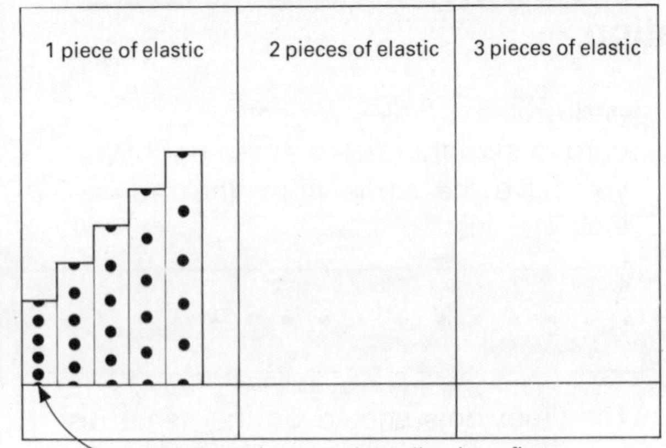

5 Repeat the experiment, but use two pieces of elastic side-by-side instead of one. When you are pulling, keep them the same length as the first one. Make a second graph, on the same paper as the first one.

6 Finally, repeat the procedure again, but with three pieces of elastic.

Questions

1 Each piece of '5-dot' paper represents 0.1 seconds. Work out how long it took the trolley to cover 2 metres for each set of results.

2 The velocity of the trolley is measured by distance divided by time. Work out the average velocity of the trolley for each run.
(Hint: total distance is 2 metres, total time you have already calculated.)

3 Look at your three graphs. What is similar about the shape of each one?

4 How do the shapes of the three graphs differ?

5 Write down a pattern in words that connects the force you used with the acceleration.

6 With a little maths, you can make the pattern more precise. Measure the gradient of each graph. This diagram should help you.

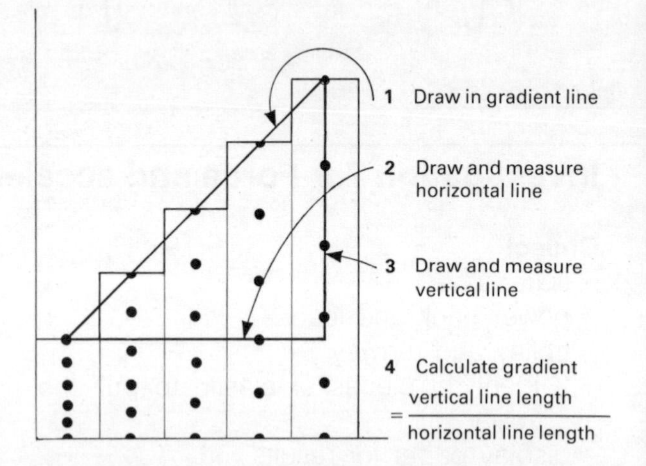

Copy the table below, and fill it in

Force units (no. of pieces of elastic)	**0**	**1**	**2**	**3**
Acceleration (graph gradient)				

Plot a graph of these results. Make 'force units' the x-axis, and 'acceleration' the y-axis. If you get a straight line, you have shown that more force produces more acceleration. In precise terms, the acceleration is proportional to the force.

7 For reasonable mathematicians (but do try it!): the graph you have just done should be a straight line. The point (0,0) should be on the line (why?). The line has an equation of the form $y = mx$ where y is force and x is acceleration. Work out the gradient (m) of your graph. Then write down an equation which connects force and acceleration.

Mass and acceleration

In the last investigation, you saw how force altered the acceleration of a trolley. In Investigation 7.10 you can alter the mass of the trolley and see how the acceleration is changed.

Investigation 7.10
Mass and acceleration

Collect

The same apparatus as for Investigation 7.9. You will need only one piece of elastic this time, but you will have to use three trolleys in the last part.

What to do

1 Connect the ticker-timer up as in the last experiment.
2 Adjust the slope of the runway until it is 'friction-compensated'.
3 The first run should be identical to the first run in Investigation 7.9. Use one piece of elastic only.
4 For the second run, stack a second trolley on top of the first one. Use one piece of elastic again. Keep it stretched as it was in the first experiment.
5 Finally, repeat the experiment with three trolleys stacked together. Take care!
6 Make three tape graphs of your results as before. Try to do them all on the same piece of paper.

Questions

1 Look at the shape of the three graphs. What is similar about each one?
2 How do the slopes of the three graphs differ?
3 Write down a pattern that connects the number of trolleys you used with the acceleration.
4 Predict what the tape graphs for these combinations of mass and force might look like:
(a) 1 trolley and 2 pieces of elastic;
(b) 2 trolleys and 2 pieces of elastic;
(c) 3 trolleys and 3 pieces of elastic.
If you have time, you could try combinations (b) and (c). Try to explain your results.

The last two investigations should have shown you two things:

more force ⟶ more acceleration
more mass ⟶ less acceleration

These two results give NEWTON'S SECOND LAW:

force = mass × acceleration

or $F = ma$

In other words, a force of 1 newton gives a 1 kg mass an acceleration of 1 metre per second per second.

Assignment – Force, mass and acceleration

A special Ferrari is fitted with a speedometer that records its speed every second. The car accelerates as fast as possible from a standing start. Here are the results:

Time from start/s	0	1	2	3	4	5	6	7	8
Velocity/m s^{-1}	0	5	10	15	20	24	28	31	34

1 Plot a velocity–time graph of these results. Make time the *x*-axis.
2 Is the acceleration constant? Explain your answer.
3 What is the change in velocity between 3 seconds and 4 seconds? This is the same as the acceleration between 3 and 4 seconds. The units are m s^{-2}.
4 What is the acceleration between 7 and 8 seconds?
5 Give at least two reasons why your answer to question 4 is less than your answer to question 3.
6 Imagine that the car contained enough heavy luggage to double its mass, and that the car's springs survived! Sketch on your graph, in a different colour, the shape of the new acceleration graph. Explain your answer.

Assignment – Braking force and distance

So far you have only used Newton's Second Law with trolleys and cars which accelerate. But it works just as well with things that decelerate. In this case:

braking force = mass × deceleration

So harder braking produces more deceleration, but a heavier object is harder to slow down. That is just commonsense. Try these questions.

1 Calculate the braking force needed to decelerate a 1000 kg car at $4\,\mathrm{m\,s^{-2}}$ (Don't forget the units of force after the answer.)
2 The car is fixed with a bar to a second identical car. What braking force is now needed to give the same deceleration?
3 Look at the table of shortest stopping distances from the Highway Code.

30 mph is about $13\,\mathrm{m\,s^{-1}}$
70 mph is about $30\,\mathrm{m\,s^{-1}}$
30 feet is about 9 metres

4 Explain what the phrases 'thinking distance' and 'braking distance' mean.

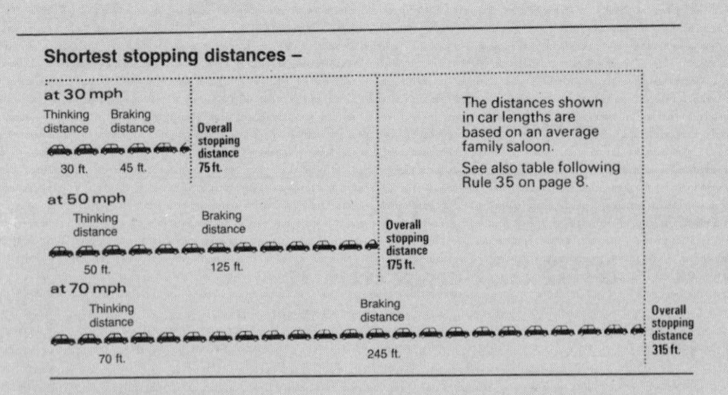

5 How does the thinking distance depend on the speed of the car?
6 The Highway Code says that these figures are measured under ideal conditions. What factors, apart from speed, might increase:
(a) the thinking distance,
(b) the braking distance?
7 How does the braking distance depend on the speed of the car?
8 If the speed of the car doubles, what effect does this have on the braking distance?
9 Suggest how the thinking and braking distances might be different for a motorcycle.

WORKING OUT HOW THINGS MOVE

There are some simple patterns that make it easy to work out how an object will move. They are called EQUATIONS OF MOTION. The equations are useful. If you know any *three* of these five quantities about an object, then the other two can be worked out.

Quantity	*Symbol*
initial velocity	u (often zero)
final velocity	v
distance travelled	s
acceleration	a
time taken	t

The equations that link these quantities are:

(1) $v = u + at$
(2) $v^2 = u^2 + 2as$
(3) $s = \frac{1}{2}(u + v)t$
(4) $s = ut + \frac{1}{2}at^2$

Here is an example to show you how to use the equations:

A sausage is dropped from the Post Office tower in London. The tower is 180 metres high, and we will ignore air resistance. The acceleration due to gravity is $10\,\mathrm{m\,s^{-2}}$. How long does the sausage take to hit the ground?

Working:
We know the initial velocity $u = 0$
acceleration $a = 10\,\mathrm{m\,s^{-2}}$
distance $s = 180\,\mathrm{m}$

We have to find the time t.
Equation (4) contains the four quantities u, a, s and t:

$$s = ut + \tfrac{1}{2}at^2$$

so

$$180 = 0t + \tfrac{1}{2}10t^2$$
$$180 = \tfrac{1}{2}10t^2$$
$$180 = 5t^2$$
$$180/5 = t^2$$
$$36 = t^2$$
$$6 = t$$

So the time taken is 6 seconds.

It is also possible to work out the force which accelerates the sausage. We will assume its mass is 0.1 kg.

Using $F = ma$:

$$F = 0.1 \times 10$$
$$= 1 \text{ newton}$$

A new definition of a newton, perhaps?

Fig. 7.10 Action and reaction

ACTION AND REACTION

You will already know about this rule because it is part of everyday life. You probably don't think about it though! The rule says that every force or action is balanced by an equal and opposite reaction. Figure 7.10 shows some examples.

In each example, try to work out what is the first force (the 'action'). Then think what the reaction to this force is.

Think carefully about the rocket. The force that pushes the rocket up is balanced by a force pushing the earth down. Is what you see the rocket taking off, or is the earth moving backwards? Fortunately the earth is a massive object and it needs a lot of force to accelerate it (remember $F = ma$). So the rocket, which is much lighter, accelerates fast, whilst the earth gets only a tiny acceleration.

The rule that action and reaction are equal and opposite is called NEWTON'S THIRD LAW.

Assignment – Rockets and jets

A jet engine produces thrust by burning fuel and air and then forcing it out of the engine.

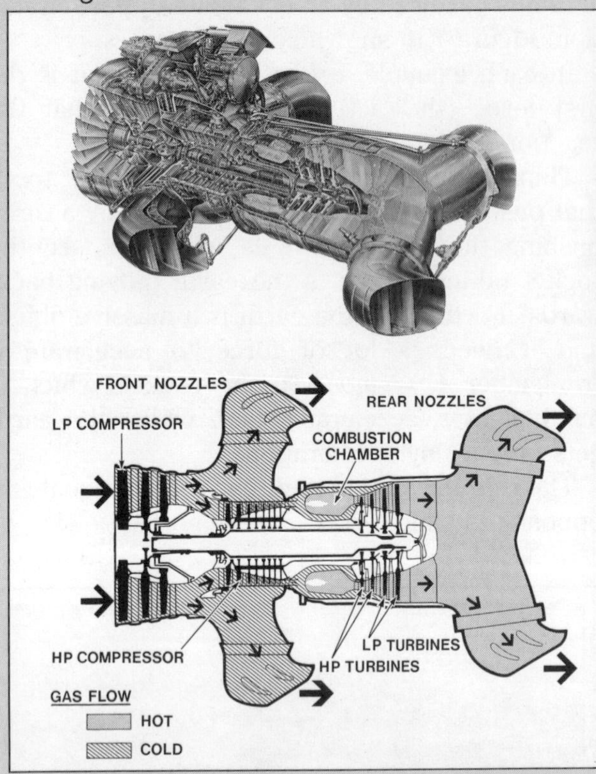

A.15 Pegasus jet engine used in Harrier jet

A rocket engine works on the same principle. However it carries its own air, or oxygen, supply.

Here is some information about two jets, the BAC 1–11 and the Hawker Harrier.

	BAC 1–11	Hawker Harrier short take-off	Hawker Harrier vertical take-off
Forward velocity at take-off/m s^{-1}	96	40	0
Take-off distance/m	1920	100	0
Engine thrust/N	108 000	100 000	100 000
Aircraft mass at take-off/kg	45 000	10 000	9000

1 Look at the figures for the BAC 1–11. Use the take-off velocity and distance to work out its acceleration down the runway. Hint: you know v and s, and the initial velocity u is 0. Choose the right equation of motion on page 68.

2 Use Newton's Second Law to work out the force needed to produce this acceleration in a BAC 1–11.

3 How does this force compare with the engine thrust in the table?

4 Work out the acceleration down the runway for a Harrier using a short take-off.

5 Work out the force needed to produce this acceleration in a Harrier.

6 Your result should be less than the total engine thrust for a Harrier. This is because the Harrier has a 'vectored thrust' engine. The exhaust from its jet is not forced out backwards. It is channelled out through four exhaust ducts which can be turned between horizontal and vertical. One of these ducts has been marked on the photograph. Explain, in your own words, why your answer to question 5 is less than the engine's thrust.

7 If the BAC 1–11 had 'vectored thrust' engines, could it take-off vertically? Explain your answer.

8 Why is the maximum safe mass at take-off less when a Harrier takes off vertically? Hint: think about the wings.

9 Outline some of the problems that you think the designer of the Harrier had to overcome.

MOMENTUM

As the number of cars on the road has increased, so has the number of road accidents. A great deal of time and money is spent each year trying to prevent serious crashes. The fencing which separates the two streams of traffic on a motorway is a good example. You may have noticed solid metal fences, cable fences and even hedges in the centre of the motorway. What do you think are the advantages of each type?

The problem that road safety experts have to deal with is this. Damage in accidents depends not only on the *velocity* of the vehicle, but also on its *mass*. The combination of mass and velocity is called MOMENTUM.

A heavy object has more momentum than a light one if they are going at the same velocity.

Fig. 7.11

A fast object has more momentum than a slow one if they have the same mass.

Fig. 7.12

One way to imagine momentum is to think how easy it is to start or stop an object moving. Which vehicles in the pictures would be easiest to stop moving?

Momentum is worked out by multiplying mass by velocity. In a collision, the total momentum before and after is always the same. This is not obvious, because friction usually transfers some momentum to the earth as heat. However if friction is eliminated, momentum conservation does work. An air track is a piece of equipment that has only a very small amount of friction. You may be able to see an investigation into momentum using an air track.

Assignment – Using momentum: the speed of an air gun pellet

The conservation of momentum can be used to work out how fast an air gun pellet travels.

A pellet is fired at a target mounted on a model railway truck. The truck stands on a railway line. When the pellet hits the target, its momentum is transferred to the truck which starts moving. The truck's velocity is measured by timing it.

1 Do a sketch to show how the experiment might be done.
2 Write down all the things you would need to measure, and explain how you would measure them.
3 Here is a set of results:

mass of pellet	0.5 g
mass of truck + pellet	400 g
time for truck to move 10 cm after being hit	2.0 s

Calculate:
(a) the velocity of the truck and pellet after the collision;
(b) the momentum of the truck and pellet after the collision (units?!);
(c) the momentum of the pellet before the collision;
(d) the velocity of the pellet before the collision.

Don't forget to make your working out clear.

4 What do you have to assume has happened to calculate the momentum of the pellet before the collision?
5 Do you think the actual velocity of the pellet is likely to be more or less than your calculation? Explain your answer.
6 Try to design an experiment that would let you work out the velocity of a rifle pellet directly. You may use ordinary laboratory equipment, including an electronic timer. Some aluminium foil might be useful. What safety precautions would be necessary?

Assignment – Air track collisions

Some experiments were done to find out if the momentum before and after a collision stays the same. In each experiment, one slider was pushed down the track and timed. A second slider was left in the middle of the track so that it did not move. The first slider hit and stuck to the second one. The two sliders then moved together, and were timed.

In each experiment, sliders of different sizes were used. Here are the results.

Collision	1	2	3	4
BEFORE COLLISION mass/g	500	1000	250	500
time to cover 10 cm/s	0.80	0.80	0.50	0.50
velocity/cm s^{-1}	10/0.8 = 12.5			
momentum/g cm s^{-1}	500 × 12.5 = 6250			
AFTER COLLISION mass/g	1000	1500	750	1500
time to cover 10 cm/s	1.61	1.25	1.50	1.51
velocity/cm s^{-1}				
momentum/g cm s^{-1}				

1 Copy the table into your book. Fill in the missing velocity and momentum figures. The calculations are the same as the first one.

2 Why is the mass after the collision always greater than the mass before the collision?

3 The Law of Conservation of Momentum says that the total momentum before a collision equals the total momentum after. Do the results agree with the law, within the accuracy of the experiment?

4 For the results which do not agree with the law, is there more momentum before or after the collision? Where has the 'missing' momentum gone?

5 All four collisions above were inelastic. This means that the two sliders stuck together when they hit. If the sliders bounced off each other, the collision would be 'elastic'. Try to predict what you would have seen in each of the four collisions if they had been elastic. It may help you to imagine the two sliders as marbles. When they have the same mass, they behave like one marble hitting another of the same size. If the first slider is lighter, it will behave like a small marble hitting a larger one.

Momentum is also related to Newton's Second Law. On page 67 it was written as:

force = mass × acceleration

As acceleration is the rate of change of velocity, this is the same as:

force = mass × rate of change of velocity

Mass × velocity is momentum, so this can be re-written:

force = rate of change of momentum

One use of this equation is to work out how a rocket will accelerate. The total mass of a Saturn V rocket (Fig. 7.10) is about 3 000 000 kg. The thrust of the first stage is about 36 000 000 N.

For the rocket:

force = mass × rate of change of velocity

So rate of change of velocity
= force/mass
= 36 000 000/3 000 000
= $12\,\mathrm{m\,s^{-2}}$

So after 1 second, the rocket's velocity will be about $12\,\mathrm{m\,s^{-1}}$, or nearly 30 mph.

Things become more complicated once the flight has started. The rocket uses up fuel and becomes lighter. This makes it easier for it to accelerate. It also means that both mass and velocity are changing. The motion of the rocket can be calculated quite easily, but it is best done with a computer.

Fig. 7.13 Why does the hammer move in a circle?

Assignment – Putting satellites into space

1 Why do rockets that put objects into orbit usually have three stages? Think carefully!
2 What are the advantages and disadvantages of having a re-usable rocket like the space shuttle? What extra problems does the engineer of a space shuttle have to overcome?

MOVING IN CIRCLES

You have probably not had the chance to throw the hammer. It is not easy, and can be dangerous. A hammer thrower needs to understand not only how things move, but how they move in circles.

The starting point is Newton's First Law. Can you remember it? Check up on page 63. When it is being swung by the thrower, the hammer ought to go in a straight line. It does not, because the thrower uses a force to keep it turning in a circle. Imagine doing this yourself. There would be a strong pull on your arms. Your muscles would have to work hard to hold on to the hammer. This is called CENTRIPETAL FORCE.

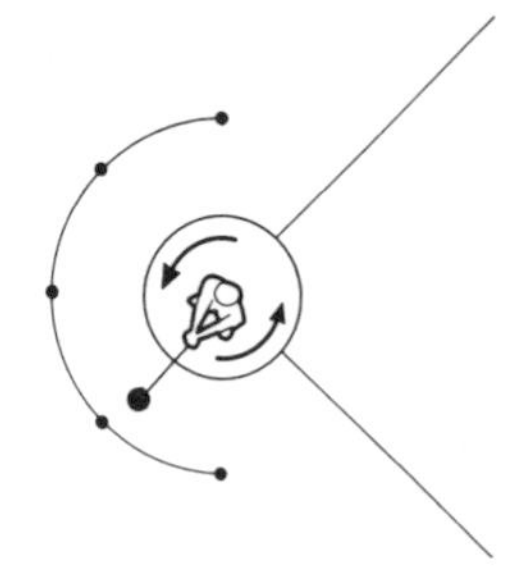

Fig. 7.14

What happens when the thrower lets go? Seen from above, it looks like this.

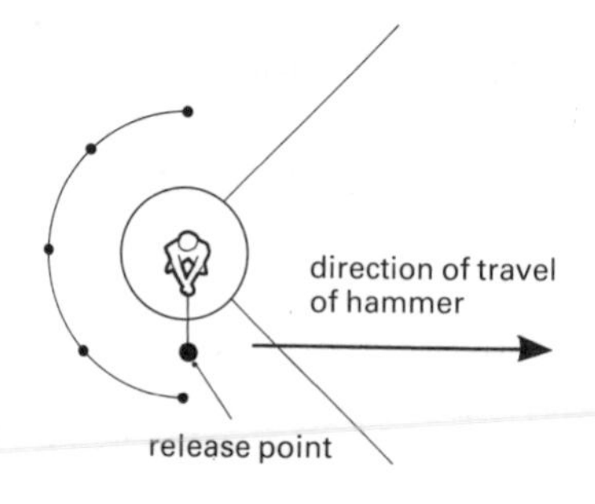

Fig. 7.15

The hammer is let go at the release point. It then travels in a straight line. No forces apart from gravity act on it, so it must go straight.

Top hammer throwers can transfer more momentum to their hammers by turning faster. This increases the force which they have to use to keep the hammer going in a circle. The force needed depends on both the mass and the velocity of the hammer. It also depends on the length of the chain attached to the hammer. In athletic competitions the mass of the hammer and the length of the chain are fixed by the rules. The differences between throwers are simply in technique.

MOLECULES ON THE MOVE

Any moving object has a certain amount of energy. You met this idea in Unit 3. The energy is called KINETIC ENERGY. It can be worked out using the formula:

$$\text{kinetic energy} = \tfrac{1}{2}(\text{mass of object}) \times (\text{velocity of object})^2$$

The mass of the train in Fig. 7.16 is about 400 000 kg, and its velocity is about $50\,\text{m}\,\text{s}^{-1}$

Its kinetic energy is about:

$$\tfrac{1}{2} \times 400\,000 \times (50)^2 = 500\,000\,000\ \text{J} = 500\ \text{MJ (megajoules)}$$

Investigation 7.11 Centripetal force

The measuring instrument in this experiment is going to be you. The experiment could be dangerous if you are not careful. It is best done out of doors, and away from people or objects.

Collect

about 2 metres of string
2 identical objects to tie on the string: plimsolls are ideal
stopwatch if possible

What to do

1 Tie a plimsoll to the end of the string. Hold the string 1 metre from the plimsoll. Swing the plimsoll around your head (**care!**) so that it stays the same height off the ground. Get it circling at 1 rotation every second. Once you have got it right, 'feel' the force you are using to stop it flying off: the centripetal force.
2 Double the speed of rotation. Aim for 2 turns per second. Compare the force on your hand with the force in part 1.
3 Increase the length of string between your hand and the plimsoll to 2 metres. Swing it at 1 turn per second. Compare the force with the force in part 1. You may need to repeat part 1 to get a good comparison.
4 Finally, put 2 plimsolls on the end of the string. Swing them at 1 turn per second. Assess how the force compares with the force in part 1.

Questions

1 Put your results into a table.

Experiment	1	2	3	4
No. of masses	1	1	1	2
Length of string/m	1	1	2	1
Speed/turns per second	1	2	1	1
Force on hand/'feelies'	1			

The forces you write down may be better done as 'same', 'more' or 'less' rather than using numbers.
2 Write down in words how each of these things affects the centripetal force:
(a) the mass (number of plimsolls);
(b) radius of rotation (length of string);
(c) speed of rotation.
3 Estimate how fast you would have to rotate a 2 m string with 2 plimsolls on to get the same force as you did with 1 plimsoll on a 1 m string. Try it if you can.
4 Design an experiment which uses a proper instrument to measure the centripetal force. What safety precautious would you adopt? You may be able to see an experiment of this type.

Fig. 7.16 How much kinetic energy does this high speed train have?

500 MJ is just a little more than the energy that a class of 15-year-olds transfers in a day.

Molecules also have kinetic energy. You learnt in Unit 5 that everything is made of particles, and that these particles move. This is called the KINETIC THEORY. If the particles move, they must have kinetic energy. The word 'kinetic' actually means moving.

How much energy do molecules have? Think about a large carboard box full of air molecules. It contains about a mole of molecules.

The mass of the molecules in the box will be about 30 g, or 0.03 kg.

The average velocity of the air particles will be about $500\,\mathrm{m\,s^{-1}}$ at room temperature.

The total kinetic energy of the particles in the box is:

$$\tfrac{1}{2} \times 0.03 \times (500)^2\ \mathrm{J} = 3750\ \mathrm{J}$$

This is about the same energy as you can get by completely burning a couple of matches.

As there are about 6×10^{23} molecules in the box of air, each molecule has only a very small amount of energy. This amount will be different for each molecule, as they move in different directions at different speeds. Because of this it is important to talk about the *average* kinetic energy, rather than the energy of each molecule.

HOW GASES BEHAVE

The constant movement of the particles explains the way that gases behave when pressure, volume or temperature changes.

First, it is important to understand what pressure, volume and temperature mean. Imagine a tin which contains ordinary air.

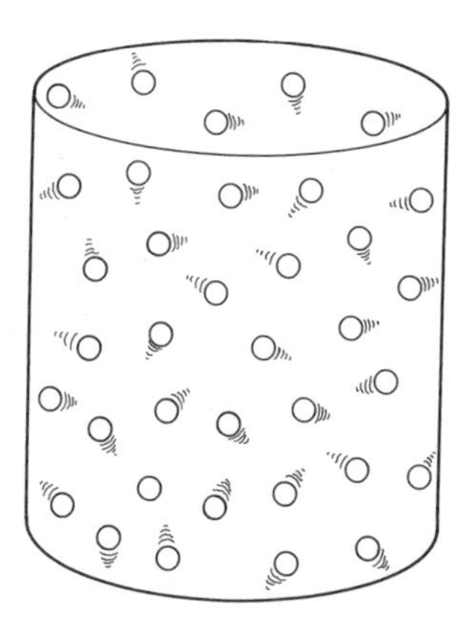

Fig. 7.17

The PRESSURE in the tin is the force of the molecules hitting the tin, divided by the area of the walls of the tin. It is a measure of the number of collisions, and the speed of the particles. If there are more collisions, or if the particles travel faster, the pressure will be greater.

The VOLUME is the space that the particles take up. In this case, it is the space inside the tin.

The TEMPERATURE is a measure of the kinetic energy of the particles. The hotter the air is, the faster the particles move, so the more kinetic energy they have. If the average kinetic energy is higher, then the temperature is higher.

If the gas is cooled down, the particles lose kinetic energy. Eventually they reach a point at which they can lose no more kinetic energy. Or we think they reach this point; no-one has ever managed to remove all the kinetic energy from a particle. Experiments show that the temperature at which this should happen is −273.16 °C, or ABSOLUTE ZERO.

This is the starting point of the Kelvin (or absolute) temperature scale. A Kelvin degree is the same as a degree Celsius, but 0 °C is +273 K (there is no ° sign with the K).

The pressure, temperature and volume of a gas are related. Changing one always changes at least one of the others. There are three laws that explain what will happen.

Boyle's Law

Imagine that the volume of the tin of air was halved by squashing it. There is still the same number of molecules in the tin, however. The same number of molecules in less space means more collisions. So the pressure increases.

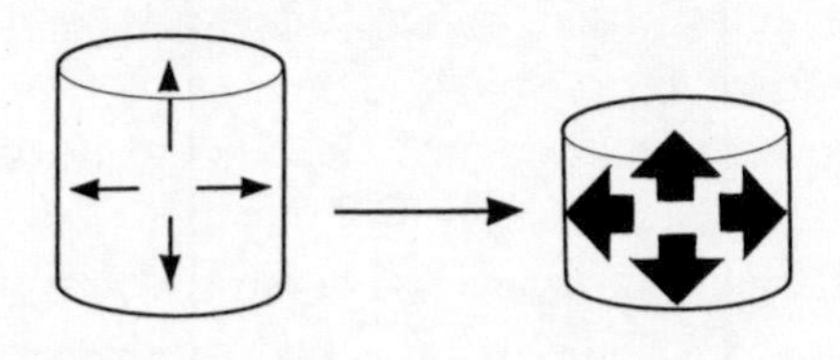

Fig. 7.18

Boyle's Law says that if the temperature stays the same, pressure × volume is constant. Halving the volume will double the pressure. Doubling the volume will halve the pressure.

Pressure Law

Imagine that the original tin of gas was heated up. The molecules get more kinetic energy, so they hit the tin harder. Eventually the tin would explode (which explains the warnings on aerosol cans not to throw them in the fire). So as the temperature increases, the pressure increases.

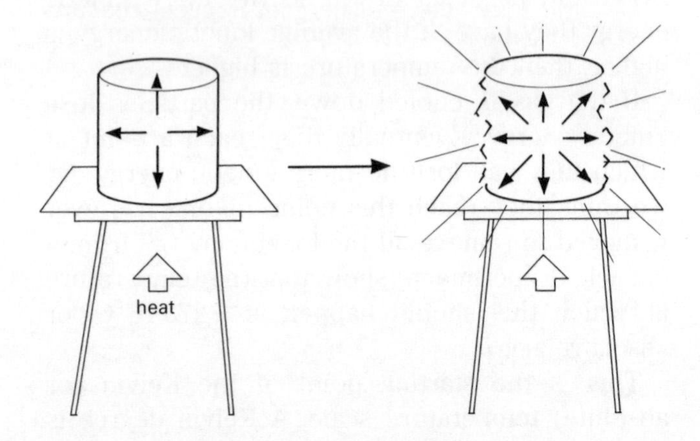

Fig. 7.19

The Pressure Law says that pressure divided by absolute temperature is constant, if the volume does not change. Doubling the absolute temperature will double the pressure.

Charles's Law

Imagine that the tin of gas is made of a special metal that expands incredibly fast as it gets hot – so fast that the pressure inside the tin does not change at all. As the molecules move faster, they will be able to take up more space. If the temperature increases and the pressure stays the same, the volume will increase.

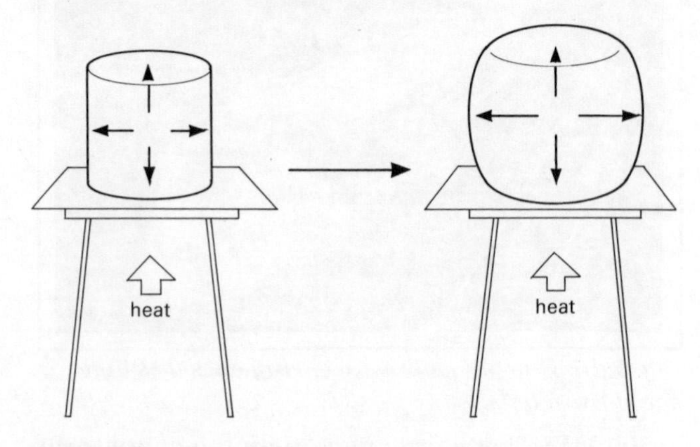

Fig. 7.20

Charles's Law says that volume divided by absolute temperature is constant, if the pressure stays the same. Doubling the absolute temperature will double the volume.

Changes in motion

WHAT YOU SHOULD KNOW

1 Objects stay as they are unless a force acts on them (Newton's First Law).
2 Force = mass × acceleration, or force is the rate of change of momentum (Newton's Second Law).
3 Any force or action is balanced by an equal and opposite force or reaction (Newton's Third Law).
4 Momentum = mass × velocity.
5 In any collision, the total momentum before the collision equals the total momentum after the collision (Law of Conservation of Momentum).
6 An object moving in a circle has a constant force on it to keep it moving in a circle. This is called centripetal force.
7 Kinetic energy = $\frac{1}{2}$ × mass × (velocity)^2
8 Increasing the pressure on a gas reduces its volume, if the temperature stays the same (Boyle's Law).
9 Increasing the temperature of a gas increases its pressure, if the volume stays the same (Pressure Law).
10 Increasing the temperature of a gas increases its volume, if the pressure does not change (Charles's Law).

QUESTIONS

1 The mass of the Ferrari in the assignment on page 67 is 750 kg.

(a) Calculate the thrust produced by the engine in the first 4 seconds.

(b) The largest possible braking force is 5250 N. Work out the deceleration produced by using this force when the car is going at $35\,m\,s^{-1}$.

(c) How long will it take to stop?

(d) How far will it travel in this time?

2 The Law of Conservation of Momentum says that the total momentum before a collision is equal to the total momentum after a collision. Explain how this works for the following collisions.

(a) A very bouncy ball is dropped, collides with the ground, and bounces back up again.

(b) A lump of Plasticine is dropped, collides with the ground, and stays there.

(c) A pellet is fired from an air rifle.

(d) A stone is fired from an elastic catapult.

3 Diesel engines do not have a spark plug. The fuel/air mixture explodes spontaneously when it is compressed.

(a) Suggest why this happens.

(b) Find out what happens to a petrol engine if the mixture is ignited by pressure instead of by a spark.

4 How could you measure the acceleration of a falling 100 g mass? You can use ordinary laboratory apparatus, but a stopwatch is *not* adequate. Explain what you would do, and what measurements you would make. How could you calculate the acceleration?

5 If you repeated the experiment in question 4 with a 200 g mass, what would the result be? Explain your answer.

6 Plastic bullets are used as a last resort to control riots.

(a) Calculate the momentum of a plastic bullet which weighs 50 g and moves at $100\,m\,s^{-1}$.

(b) A plastic bullet hits a rioter who happens to be wearing roller skates. What speed will the rioter, who weighs 50 kg, start moving at?

(c) Why are plastic bullets less dangerous than normal rifle bullets? (There are several reasons.)

(d) Plastic bullets do cause serious injuries, and may even kill people. Suggest reasons for this.

Change and Stability

KEEPING THINGS STABLE

Stability is vital at times. Engineers, chemists, animals and even you could not manage without it. What is being kept stable in each of the pictures in Fig. 7.21?

This topic covers the way in which mechanical, chemical and biological systems are kept stable.

Fig. 7.21 What is being kept stable in each of these pictures?

STRUCTURES AND STABILITY

Designing a structure like a motorway bridge may seem simple. All you have to do is to fill a gap and check that the bridge is strong enough. Or is it? Think of the problems:

How big is the span?
What materials are available?
What sort of traffic will use the bridge?
How much can be spent on it?
What will it rest on?

Building bridges

A good designer has to understand the basic physics of structures. In a simple girder bridge, the load on the bridge must be balanced by upward forces in the supports. The total downward force is always balanced by the total support force.

Fig. 7.22 The load on this bridge is balanced by upward forces in the bridge supports

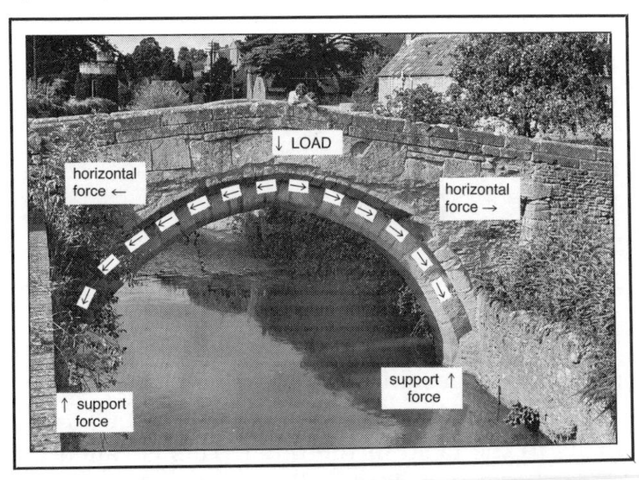

Fig. 7.23 There is an outward force as well as a downward force in this bridge

Bridge designs have to transmit the force from the load to the supports. A stone arch bridge is designed so that the stones transmit the load force outward and then downward. The downward force is supported by the bridge foundations. However, you can see from Fig. 7.23 that a horizontal force is also produced. What supports this force?

Assignment – Bridge designs

The four stamps below were issued by the British Post Office in 1968. They represent changes in bridge design through the ages.

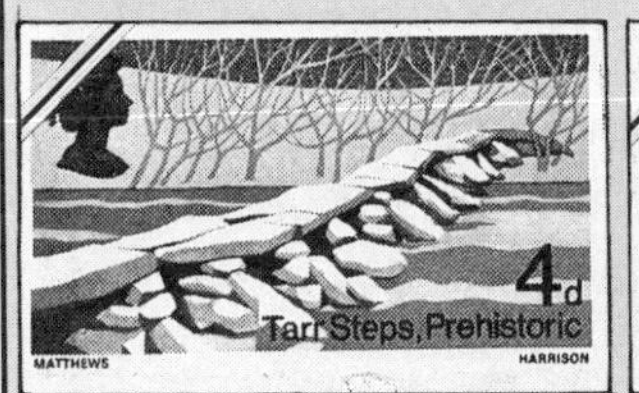

1 If the date was not on the stamps, you would still be able to tell the approximate age of each bridge. Explain how you would know.
2 For each bridge in turn, answer these questions.
   (a) Do a side-on sketch of the bridge (an elevation).
   (b) Put arrows on the sketch to show where the main support forces are.
   (c) Write down the materials that the bridge is built from.
   (d) Make a list of the advantages and disadvantages of those materials for building a bridge.
   (e) Explain how the design of the bridge depends on the materials that it is built from.
3 Which bridge is the most attractive? Explain your choice.

Investigation 7.12
Making structures

In this investigation you will use some very simple materials to make structures. It may seem simple in places, but there are some important principles involved.

Collect

40 straws
40 pins
100 g masses to test the structures

What to do (A)

Your first task is to make a bridge that will span a gap 50 cm wide. You may use 20 of the straws, and all the pins. The bridge must rest on the two supports, but it must not be fixed to them. It must not touch the ground between the supports.

When you have made the bridge, test it to destruction by hanging masses from the centre point.

Questions

1 Do a sketch of your bridge. Mark on your sketch the weakest points of the bridge.
2 Explain how you could strengthen the bridge using only straws and pins.

What to do (B)

Use the remaining straws, and the pins from the bridge to make the tallest possible structure which will support a pencil. The structure must rest on the ground; it may not be fixed to it. No outside support may be used. The pencil must be at least 15 cm long, and must rest on the structure.

Questions

3 Do a sketch of your structure.
4 Explain how the straws have to be arranged to make a stable shape.
5 Compare your structure with those made by other people. How have they made a stable tower? Has everyone used the same sort of method?

Combining forces

In real life, more than one force acts on most structures. A large tower block will have to withstand a significant sideways force due to the wind, as well as a downward force because of its weight.

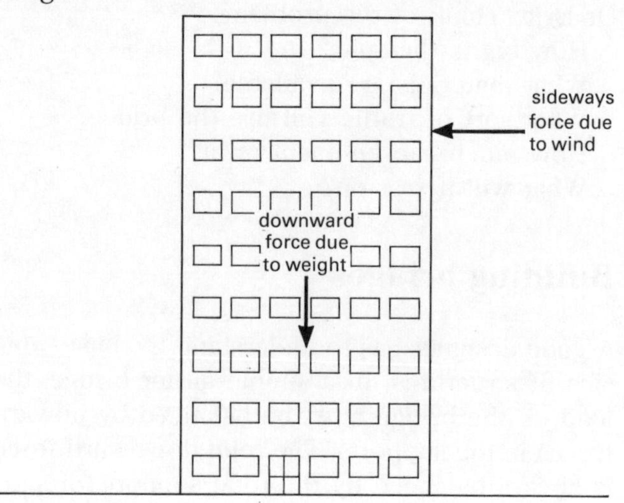

Fig. 7.24

Designers can combine these forces. Force is a vector quantity; it has size and direction. Because they are vectors, forces can be added to give a single resultant. The method is simple. Each force is represented by an arrow. The direction of the arrow is the same as the direction of the force. The size of the arrow is a scale drawing of the size of the force.

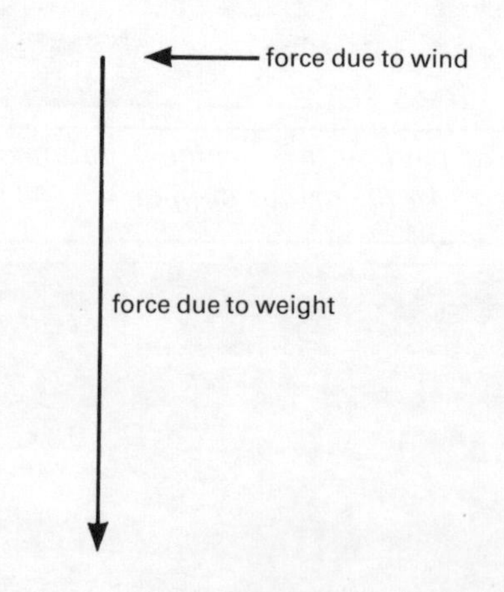

Fig. 7.25

When both forces have been represented by arrows, they are joined nose to tail. A new arrow can then be drawn from the tail of one force arrow to the head of the other. This is called the RESULTANT FORCE. Its length gives its size, and its direction shows the way in which it acts.

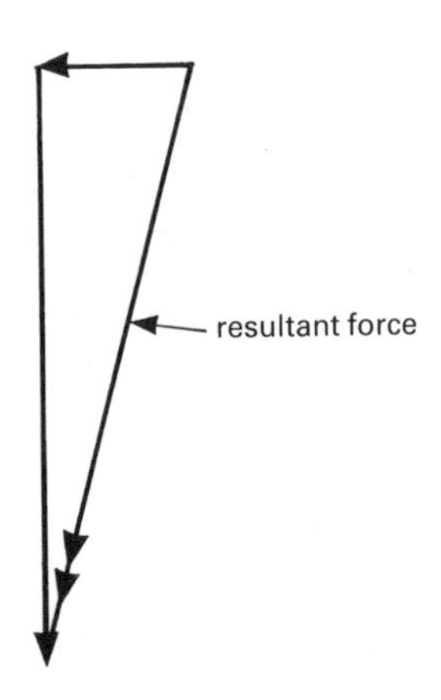

Fig. 7.26

The same method can also be used to work out the effect of two forces on a moving object. An aircraft will be moved by both the thrust from its engines, and the force of the wind.

Suppose that the thrust of an aircraft's engines is 100 kN, and it is heading due east. A sidewind with a force of 10 kN blows from the south. By drawing the forces to scale, the effect of the sidewind can be calculated. This makes it easy to correct the flight plan.

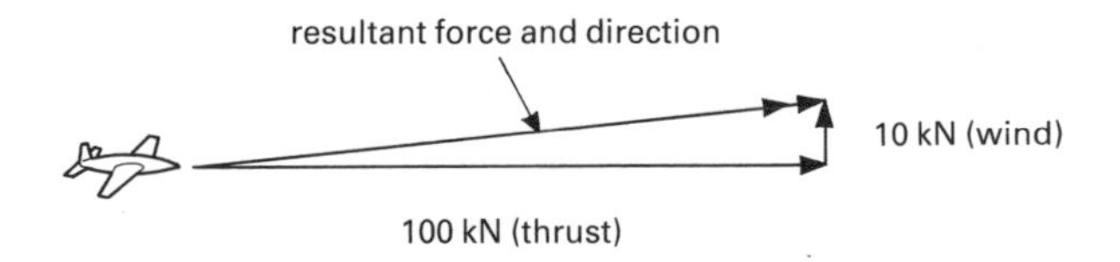

Fig. 7.27

The same principle can be used with any vector quantity. Velocities often have to be combined in this way.

CONTROL IN MECHANICAL SYSTEMS

Much of modern life is built around control devices. Some are shown in Fig. 7.28. Control devices are generally used to keep one part of a system constant. Try to work out what is being kept constant in each picture.

These control systems rely on FEEDBACK. Information about the system that is being controlled is fed back to a master control. Figure 7.29 shows an example.

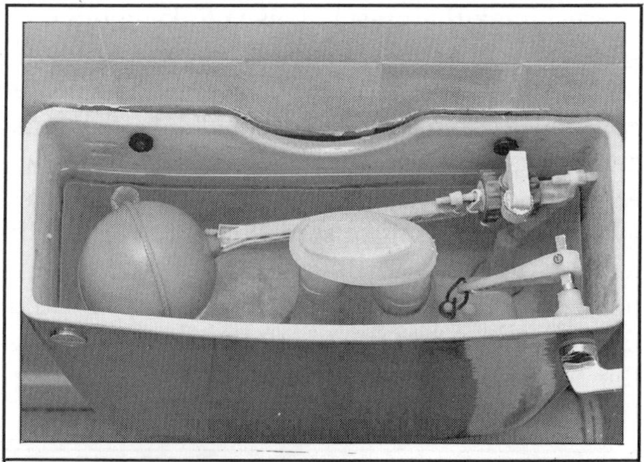

Fig. 7.28 Control devices: (a) a ballcock, (b) an aquarium thermostat

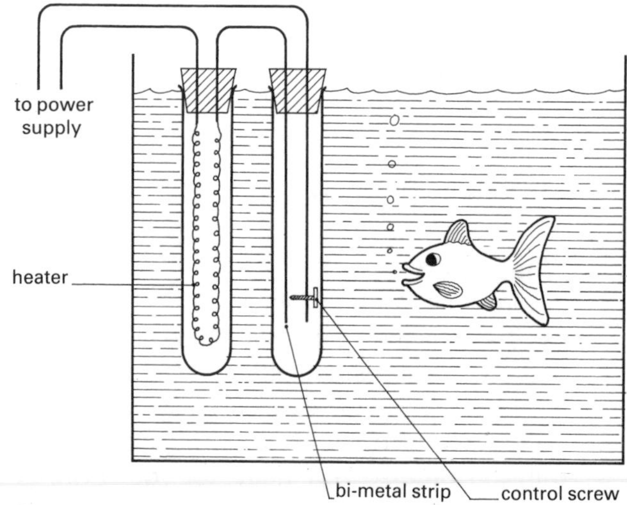

Fig. 7.29 A fish tank heater and thermostat in the off position

A fish tank thermostat is used to keep the tank water at a constant temperature. First, the temperature is set with the control screw. The heater coil and thermostat are than put into the tank, and plugged in. The water will start to warm up. As it gets hotter, it makes the bi-metal strip bend away from the control screw. Eventually, the bi-metal strip no longer touches the screw, and the heater circuit is broken. The heater stops heating the water, and it starts to cool down. As it cools, the bi-metal strip straightens. When it touches the control screw, the heater will come back on. This system will keep the water near the right temperature as long as none of the parts fail.

All control systems contain the same basic parts:

(a) something that is controlled;
(b) something that senses change;
(c) something that compares what is controlled with a set value;
(d) something that alters the thing that is controlled.

Work out for yourself which parts of the fish tank thermostat do each of these jobs.

Investigation 7.13 A model feedback system

This is a demonstration of how a feedback system works, using people instead of mechanical devices.

One person, wearing a blindfold, is 'controlled'. She has to follow a set route round some obstacles. The other person directs her, using only 'left' or 'right'.

What to do

Decide who is going to be controlled, and who will do the controlling. Set up a route to follow, around some tables or stools. The controller has to direct the controlled person as quickly and accurately as possible around the course, using only the commands 'left' and 'right'.

When you have completed the course, swap over. This time the controller may also use the words 'forward', 'back' and 'stop'. This is a more sophisticated form of feedback, and should work better.

Questions

1 What is being controlled in this system?
2 How are any changes in the system sensed?
3 What decides if the changes in the system are correct or not?
4 How is information transferred to the object that is being controlled?

Investigation 7.14 A biological feedback system

Your body contains many feedback systems. One that is easy to investigate is your own eye. Again, you will need to work with someone else.

Collect

a bright light or torch

What to do

In dim light, look at the pupils in your partner's eyes. Then carefully shine a light into one of his eyes, shielding the other eye from the light. Observe carefully what happens to *both* eyes. Repeat the experiment with the other eye. Finally, let your partner try the experiment on your eyes.

Questions

Before answering the questions, you may find it helpful to remind yourself how the eye works. You studied this in Unit 3.

1 What happens to the pupil of the eye when a light is shone into it?
2 How does the other eye respond?
3 Why do the eyes react like this to light?
4 What is being controlled in this feedback system?
5 How are changes in the amount of light entering the eye detected?
6 What decides that the pupil size needs altering?
7 How is the adjustment made?
8 Draw a picture of the eye. Label the parts which correspond to the four elements of a feedback system.

Assignment – Controlling a shower

A shower is a good example of a feedback system. Think about what you have to do when you want to have a shower. How can you adjust a shower? What happens when you do?

1 There are four parts to the feedback system.
   (a) What is being controlled?
   (b) What senses any changes?
   (c) What decides if the system needs altering?
   (d) How are any alterations made?
2 Suggest two problems with the system.
3 Do a design for a more efficient system. Label all the parts. Explain how it is better than the original shower.
4 Expensive showers contain a thermostat where the hot and cold water supplies enter. When you want a shower you just set the temperature you want, and wait for it to warm up. Once there, it stays there. What is the thermostat replacing?
5 Explain why a thermostatic shower is more likely to give an even water temperature than a simple shower.

The fish tank thermostat, and the other systems in Fig. 7.28, are all examples of NEGATIVE FEEDBACK. The information that is 'fed back' to the control device is used to keep the output at the same level.

Modern electronics provides many examples of feedback. Automatic volume control (AVC) and automatic frequency control (AFC) are examples which are found in TVs, radios or cassette recorders. Your teacher may show you some examples of simple electronic feedback systems.

CONTROL IN BIOLOGICAL SYSTEMS

In the last section, you saw how the amount of light entering the eye is controlled. Living things are packed with feedback mechanisms. Control in living organisms is so important that it is given a special name: HOMEOSTASIS. Temperature and water control are two examples of homeostatic systems found in your body.

Temperature control

Warm-blooded animals have to maintain a constant core temperature. Too hot and they dry out, too cold and their chemical processes slow down and eventually stop. Either way, the animal will die if it cannot stay at the right temperature.

Think about your own temperature control. If it gets cold, what happens?

You shiver: this means your muscles contract and produce heat.
Your skin becomes cold: the body cuts down blood flow to the less important places. That is why your nose and fingers get cold easily.
You stop sweating.

All these responses are automatic. They are controlled by your brain. Your brain will also be sending you urgent messages to get warmer. Finding more clothes, or a warm room, might be your response. In the end, if you cannot prevent heat loss, you suffer from hypothermia. This is often the cause of death in mountaineering accidents.

Your body can also cope with temperatures that are too hot.

Your blood vessels dilate (get larger). The blood flows nearer the surface of the skin so you go red, and you lose heat faster.
The rate of sweating increases. As the sweat evaporates, it cools the skin. (Why?)

Your brain also tells you to find somewhere cooler. If you cannot prevent overheating, your body core temperature starts to rise. A blood temperature over 41 °C (normal is 37 °C) will lead to collapse and death. This is a particular problem for distance runners in hot or humid conditions.

The controlling mechanism is a thermostat in your brain which monitors the temperature of the blood. When it detects a change, it tells the control area of the brain. The brain then sends messages through the nervous system and the glands to try to correct the problem. The result is the heating or cooling actions described above.

Assignment – The Challenge of the Marathon

Ron Hill was one of Britain's best marathon runners in the 1960s and 70s. He still runs very fast times. When he started running, he spent a lot of time thinking about the problems of long-distance races. Two important items are clothing and drink. Here are some extracts from his book, *The Long Hard Road*.

'One of the dangers in long distance running is overheating. Once the body overheats it becomes very inefficient and pace drops disastrously. Additionally, sweat rates can rise alarmingly and lead to dehydration which in its chronic stages can also be very dangerous physically.

'The body's natural reaction to heat is perspiration. The passage of air over sweat-wetted skin causes evaporation of the liquid, with consequent cooling. Lick the back of your hand and blow on it and you will see how it works. However, if air does not circulate over the skin, for instance if it is covered in clothing, then the cooling effect cannot work, and the body is sweating for nothing.

'The simplest thing would be to run with no vest at all, but the rules say that a vest must be worn; so a vest had to be found that gave a maximum exposure of the skin. I tried a vest with enormous armholes cut out, and thinner shoulder straps than usual. Naturally it was white to reflect the heat.'

1 Name three things about Ron Hill's vest that helped him to keep cool.
2 What does the word 'dehydration' mean?
3 What must marathon runners do during a race to avoid dehydration?
4 Suggest why dehydration can be 'very dangerous physically'.
5 Why does evaporation of sweat from the skin produce 'consequent cooling'?
6 Ron Hill later designed a vest that exposed more skin than the one described above. Find out or guess what sort of vest it was. Suggest why it was better than the one described above.
7 Athletes often have to provide a urine sample after racing to check that they have not been using drugs. Why do distance runners often find it difficult to provide a sample quickly?

Assignment – Water control

We are continually losing water. We lose it by sweating, excreting and breathing. Every day, in cool countries, an average adult loses about

1000 cm^3 as urine
1000 cm^3 as sweat
500 cm^3 in breathing

This has to be replaced by eating and drinking.

Your body has to control this carefully. If you sweat heavily, because the weather is hot or you have done some hard exercise, you lose a lot of water. The body makes up for the loss of water by passing out less water as urine. The diagram shows you how the system works.

1 All feedback systems have four elements. The volume of water in the body is obviously what is being controlled. Which parts of the body do these jobs?
   (a) Senses changes in the volume of water.
   (b) Compares the volume of water with a set value.
   (c) Alters the volume of water.
2 Describe carefully how your body would cope if you suddenly drank four pints of lemonade.
3 Describe how your body would react if you ran five kilometres on a hot day.
4 Describe the journey of a molecule of water around the body. Start with it being drunk, and finish with it being excreted.

THE HUMAN WATER CONTROL SYSTEM

The **brain** monitors the concentration of chemicals in the blood

The **pituitary gland** produces a hormone called ADH. A lot of ADH is made when the blood chemicals are concentrated. Only a little is made when they are diluted.

The heart circulates blood, containing ADH, around the body

The **kidneys** filter the blood (see Unit 2). If a lot of ADH is present, more water passes to the bladder as urine

The **bladder** stores the urine until it can be passed out

– – –► represents flow of urine

——— represents simplified blood circulation

The body's control centre

The central controller in all biological homeostatic systems is the BRAIN. The brain is divided into two halves or hemispheres. Each area of the brain is responsible for controlling a different function. The brain does the following jobs.

(a) It receives and co-ordinates signals from all parts of the body.
(b) It monitors the body's chemical levels.
(c) It controls the body's actions through electrical and chemical messages.
(d) It stores information about previous actions.

The way in which messages from the brain reach other parts of the body is ingenious. There are two routes: the nervous system for rapid messages, and the endocrine system for chemical messages.

The NERVOUS SYSTEM is like a vast and complex telephone network. Messages travel as electrical impulses to and from the brain. The 'wires' in the system are bundles of nerve fibres. The brain and the spinal cord are the 'telephone exchange' that controls all the messages. There are also local control systems which do not rely on the brain for control. An example of this is when you blink if something comes near your eye. It is a REFLEX action. Reflex actions have to take place very fast – so fast that there is not time for a message to get to the brain and back. (Fig. 7.31)

The ENDOCRINE SYSTEM is the name given to all the glands that control the body's chemical balance. Special chemicals produced by these glands actually do the controlling. They are called HORMONES. Insulin is a well known hormone that controls the body's sugar level. It is a lack of insulin that causes the illness diabetes.

The glands that produce these hormones are situated throughout the body. The main glands are shown in Fig. 7.32. As chemicals travel more slowly than electrical impulses, the glands are responsible for longer-term changes in the body than the nerves are. These include growth rate, digestion, reproduction and chemical reactions like respiration.

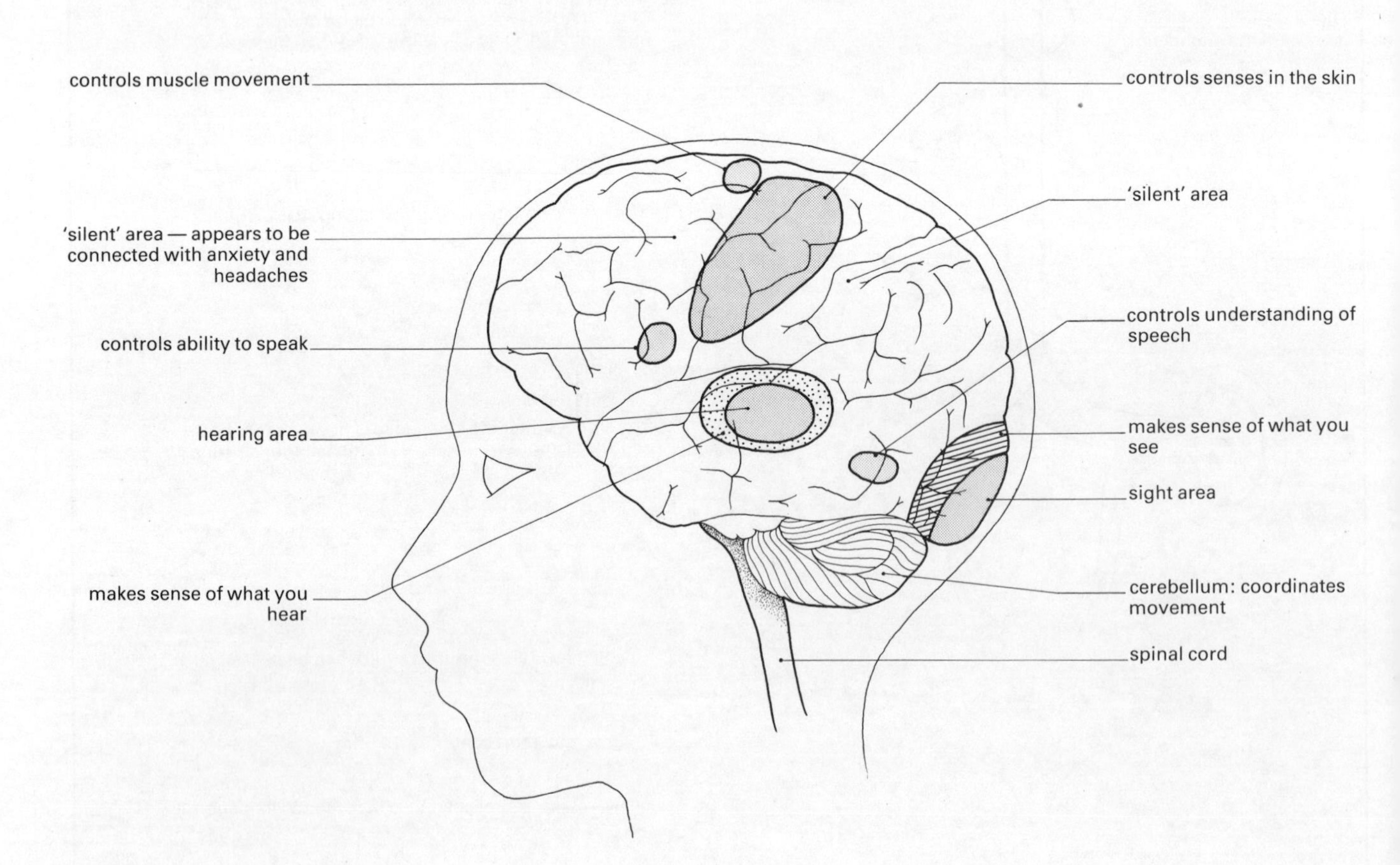

Fig. 7.30 Functions of the brain

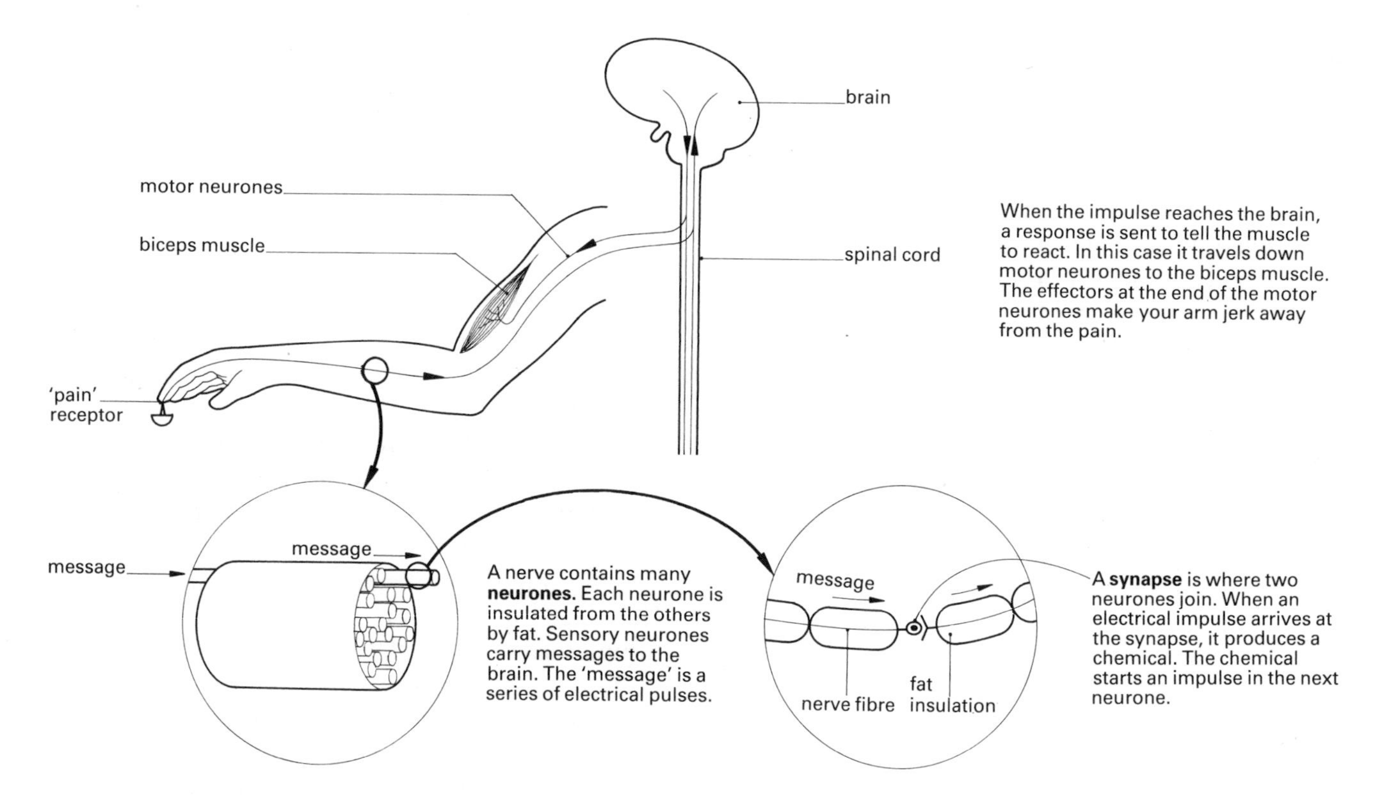

Fig. 7.31 The human nervous system

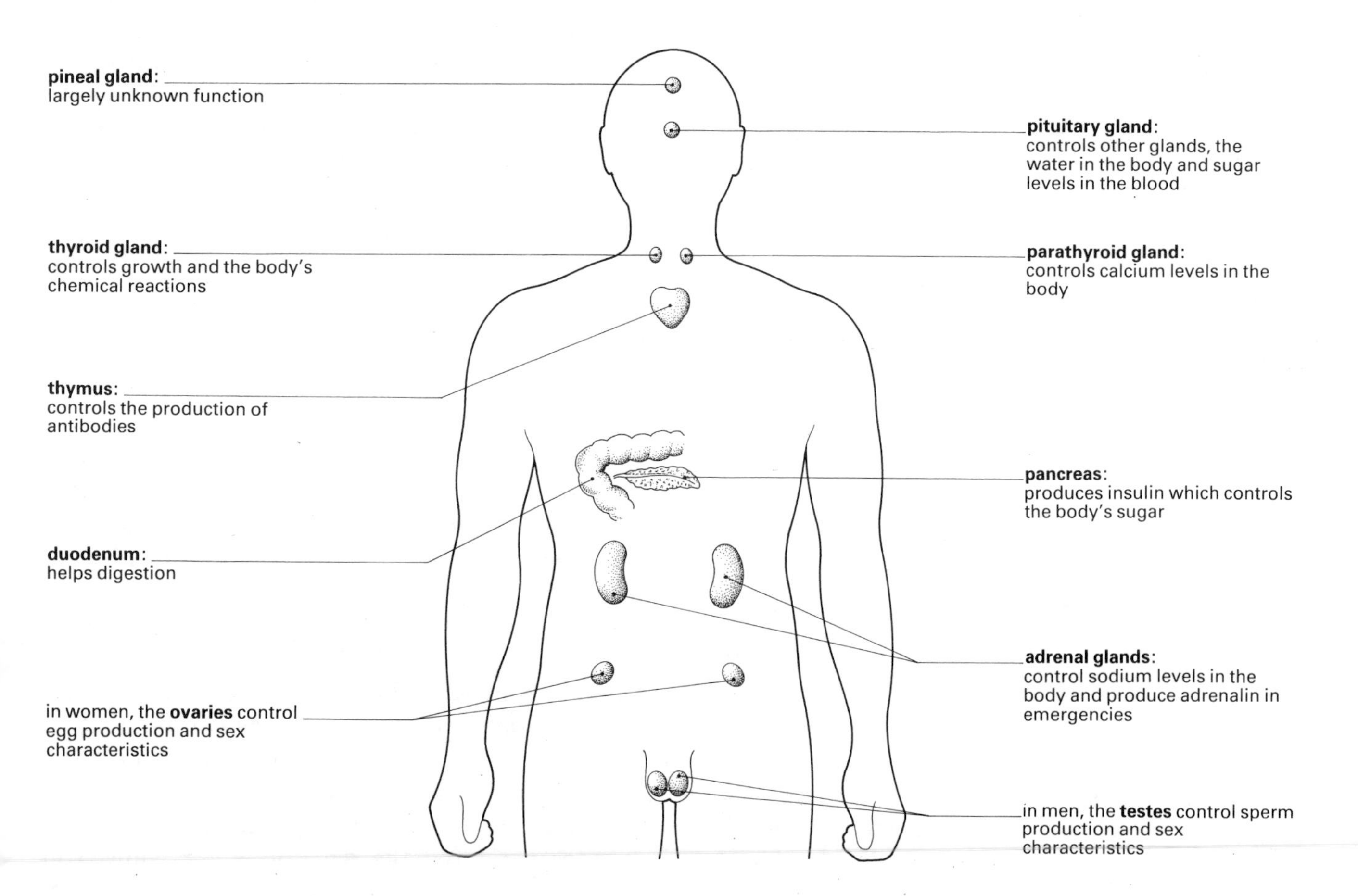

Fig. 7.32 The main glands of the human endocrine system

STABILITY IN CHEMICAL SYSTEMS

Like mechanical and biological systems, stability is also important in chemical reactions. All reactions eventually reach a stable position. This is called the EQUILIBRIUM position.

You can imagine that a chemical reaction is like a see-saw. The substances that react together are on the left of the see-saw, and the products that they form are on the right.

At the start of the reaction, the only chemicals present are reactants. There are no products.

Fig. 7.33

At the end of the reaction, which may take a long time, there are three possibilities:

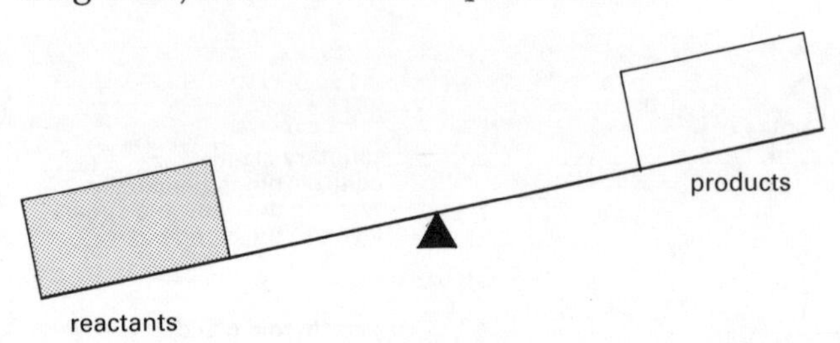

Fig. 7.34

Here there has been no change at all; no reaction. The equilibrium is 'to the left'. An example of this is the reaction between a glass vase and the water in it at room temperature.

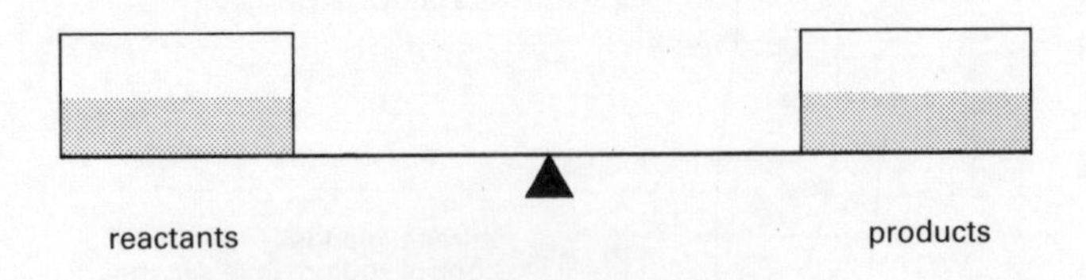

Fig. 7.35

In this case there has been a reaction. But not all of the reactants have become products. At the end of the reaction there is a mixture of both reactants and products. An example of this is the reaction between sulphur(IV) oxide and oxygen described on page 54.

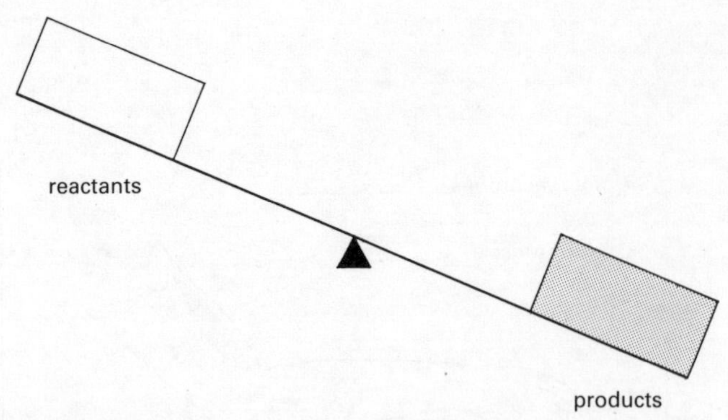

Fig. 7.36

Here all of the reactants have become products. The equilibrium position is 'to the right'. An example of this is the reaction between sodium metal and water.

Many other equilibrium positions are possible. They will range from completely to the left to completely to the right.

These ideas are very important in the chemical industry. The position of equilibrium in a reaction is vital. If it is too far to the left, it may be too expensive to manufacture the chemical that way. There is a further problem in many reactions. The equilibrium position may only be reached slowly. For an industrial process this is useless. Modern society demands plentiful, cheap chemicals. If a reaction takes a long time to reach equilibrium, the chemicals will be hard to make, and expensive. One example of this problem is in the manufacture of ammonia, an important fertiliser.

Making ammonia

The formula of ammonia is NH_3. It is made by reacting nitrogen with hydrogen. The nitrogen is obtained from the air. The hydrogen is produced in a series of reactions from natural gas (CH_4). The reaction between nitrogen and hydrogen has an equilibrium position which is well to the left. The equation is:

$$N_2(g) + 3H_2(g) \rightleftharpoons 2NH_3(g)$$

The $\rightleftharpoons$ sign means that the reaction can go either way. It is REVERSIBLE. Two things can be done to increase the amount of ammonia made. The pressure in the reaction vessel can be increased, or the temperature can be reduced. However, at low temperatures, the reaction is very slow. You can find out more about speeding reactions up on pages 57–9 of this unit.

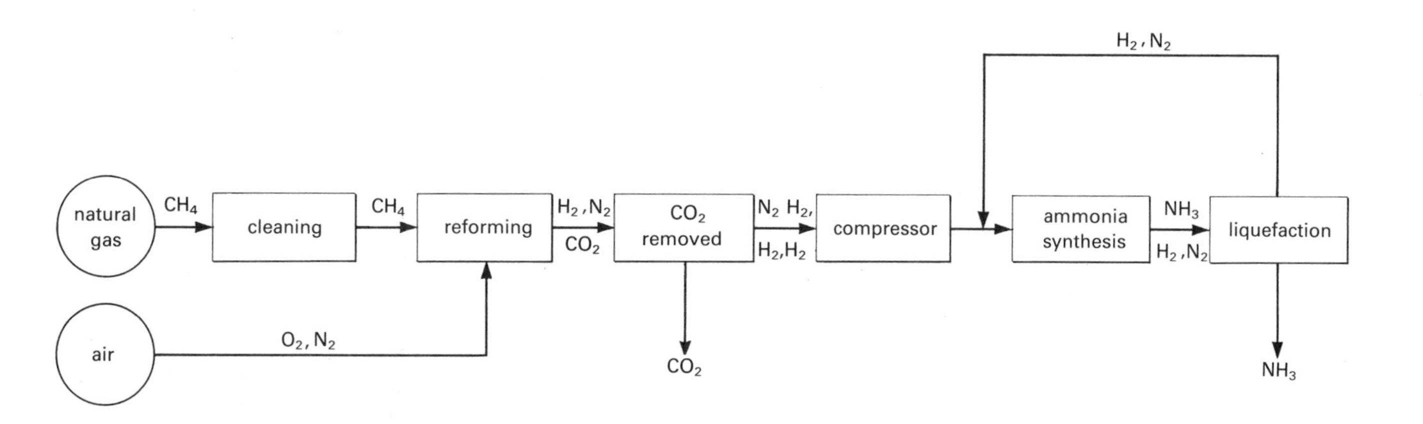

Fig. 7.37 A simplified diagram to show how ammonia is made by the Haber process

To get a reasonable yield of ammonia, the chemical engineers responsible for the plant have to be very smart. They use these techniques:

(a) The reaction is done at a high pressure (200 atmospheres or $2 \times 10^7 \, N \, m^{-2}$).
(b) An iron catalyst is used to speed the reaction up.
(c) The ammonia produced in the reaction is removed as it is made.
(d) The unused nitrogen and hydrogen are re-cycled.
(e) The reaction is not left long enough to reach equilibrium. It would take too long. Only about 15% of the possible yield of ammonia is made.
(f) The reaction is done at a 'middle' temperature: about 400 °C. This is hot enough to make the reaction work, but not so hot that the equilibrium is moved too far back to the left.

It is important, but difficult, to understand the two principles involved. Industrial chemists usually want a fast reaction, with the equilibrium to the right. The two things are *not* related. Go through the six points used in the production of ammonia. Try to work out which of them speed the reaction up, and which move the equilibrium to the right.

Some reversible reactions

There are several simple reversible reactions that can be done in the laboratory. The conversion of potassium chromate(VI) to potassium dichromate(VI) is one example.

Investigation 7.15
Reversing a reaction

Collect

test tube containing potassium chromate(VI)
test tube containing potassium dichromate(VI)
teat pipette
dilute sodium hydroxide
dilute sulphuric acid
2 empty test tubes
eye protection

What to do

1 Put a little potassium dichromate(VI) solution in one of the empty test tubes. Add drops of alkali (sodium hydroxide) until the colour changes.
2 Try adding drops of acid to the tube.
3 Try adding more alkali to the tube.
4 Write down exactly what happened.
5 Put a little potassium chromate(VI) solution in the other empty tube. Add drops of acid to it until it changes colour.
6 Try adding drops of alkali to the tube.
7 Try adding more acid to the tube.
8 Write down what you saw.
9 Repeat the two experiments, but use different acids and alkalis.

Questions

1 What is the effect of acid on chromate(VI) ions?

2 What is the effect of alkali on dichromate (VI) ions?

3 The ion equation for the reaction is:

$$Cr_2O_7^{2-}(aq) + OH^- \rightleftharpoons 2CrO_4^{2-}(aq) + H^+(aq)$$

dichromate ions + hydroxide ions (alkali) ⇌ chromate ions + hydrogen ions (acid)

How does adding alkali alter the position of equilibrium?

4 How does adding acid alter the position of equilibrium?

5 How do you know that it is the hydroxide ions that alter the equilibrium position, and not the sodium ions?

Assignment – Ammonia in the world

Why we need ammonia

Ammonia is one of the most important chemicals being manufactured today. In Britain we produce about 2 million tonnes per year. 80% of this ammonia is used to make nitrogen-containing fertilisers. Other products made from ammonia include synthetic fibres like nylon, dyes, plastics and explosives. Graph A shows how ammonia production and fertiliser production have increased in the last 70 years.

A World ammonia and fertiliser production, 1910–1985

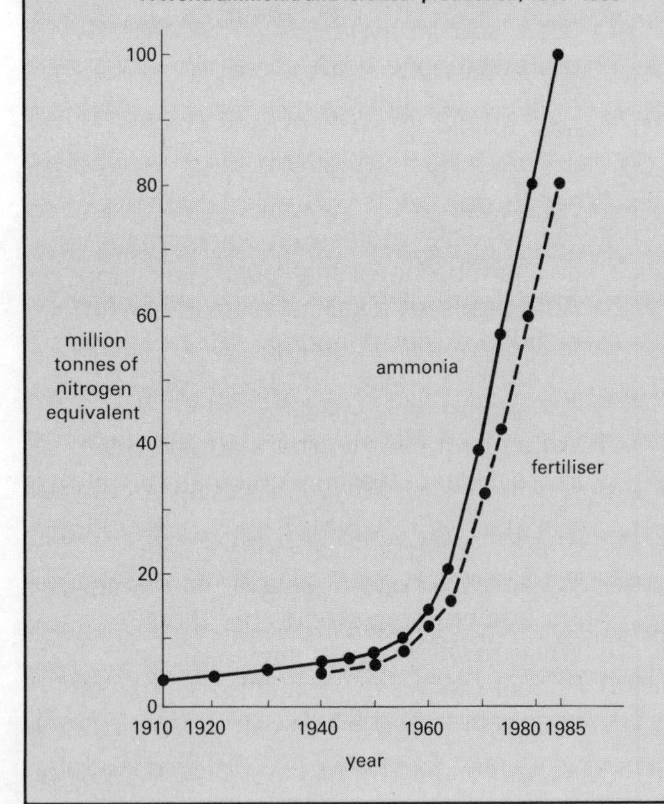

Nitrogen compounds are a major part of the protein that we need in our diet. Protein comes from meat or, more directly, from vegetables like beans. The protein in meat comes from the plants that the animals eat, such as grain. Fertilisers help all plants to grow; on average, a tonne of ammonia applied as fertiliser produces 15 tonnes of grain.

By the year 2000, the world's population will be almost twice as high as it was in 1970. The largest growth will occur in the developing countries. All these people will only be fed by improving the efficiency of farming methods and particularly by increasing the use of nitrogenous fertilisers.

Where ammonia is made

Ammonia plants have always been situated close to their sources of:

(a) energy, usually natural gas, but occasionally oil, or even coal where it is exceptionally cheap;

(b) water, usually required in fairly large quantities for the process;

(c) transport, by sea, rail, barge or road.

In Britain, the major site is Billingham on the River Tees where ICI has four large ammonia plants. It was originally chosen by the government for its proximity to the Durham coalfields and surplus electric power. Billingham is now convenient for North Sea oil and gas.

Liquid ammonia at −33°C or in the form of ammonia fertiliser is more easily transported by sea than the natural gas from which it is made so ammonia fertiliser manufacture is a profitable use of the gas from an oil field or refinery. Oil producers therefore often build plants near their oil fields to increase the value of their products.

Making a profit

A 1200 tonne per day plant, with its facilities for water purification, costs about £80 million at 1985 prices. This money must all be spent before any ammonia can be sold, or any cash can be generated. So the plant must start working on time, without technical hitches. Delays are very expensive because of the interest charges on the loan which is needed to pay for building the

B Typical movement of a bank account for the initial operating period of a 1200 tonne/day ammonia plant

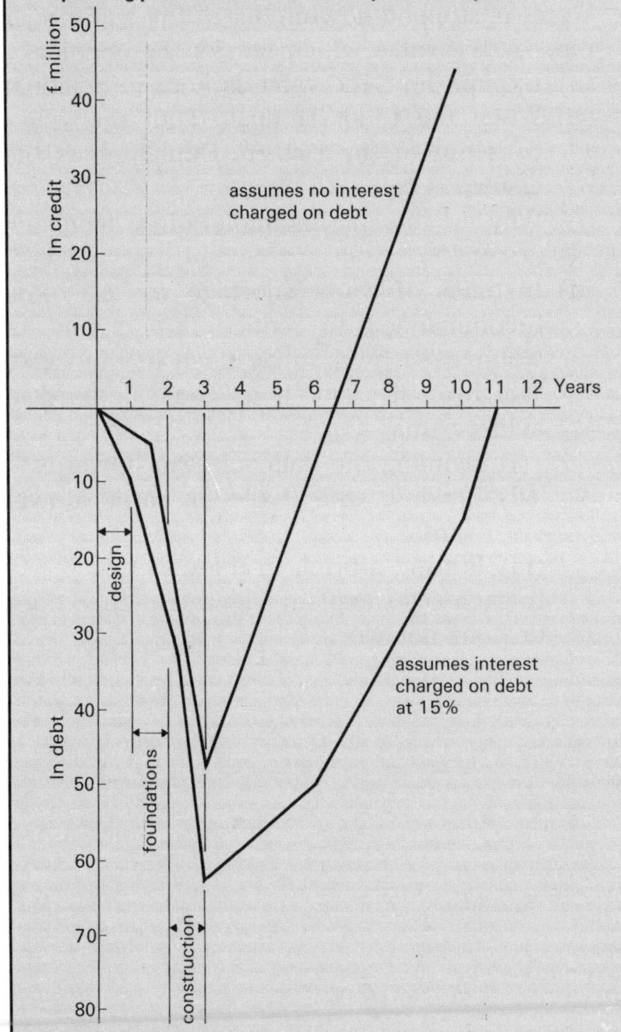

Graph B shows what it costs to build an average size ammonia plant. It takes 11 years before the sales of ammonia pay for the cost of the plant, assuming that the interest on a bank loan is 15%.

Running the plant

Large chemical plants take a long time to start up and shut down. It would take about two to three days for a 'cold' plant to make any ammonia at all. Plants are therefore operated round the clock, 365 days a year, except for breakdowns and overhauls. Catalysts last between two and eight years, so they have to be changed from time to time. Periodic overhauls are required by law so that the equipment can be inspected for defects. The plants are very highly instrumented, and are operated largely from a central control room.

(Adapted from an article by M.C.V. Cane in the ICI magazine *Steam*)

1 (a) What does ammonia contain that makes it a good fertiliser?
  (b) How does this help growing plants?
2 How do firms decide where to build an ammonia plant?
3 Only a small number of chemical companies make ammonia. They are mostly very large firms. Why is this?
4 Why is ammonia production increasing?
5 World fertiliser production is increasing much faster than the world's population. Why do you think this is?
6 You are the adviser to a developing country which wants to increase its food production. Make some suggestions about what it should do ('buy fertiliser' on its own is not an answer!). Explain each suggestion.
7 If you ignore interest charges, how long does it take a 1200 tonne/day ammonia plant to pay for itself?
8 The lower line on graph B assumes that a firm pays interest on the money needed to build an ammonia plant. Why do firms include interest in their costs? (Hint: think about how they could use the money if they did not build the plant.)

WHAT YOU SHOULD KNOW

1 In a stable structure, the force due to the load is balanced by an equal supporting force.
2 Force is a vector quantity: it has size and direction. Two or more forces can be added to give a single resultant force.
3 Control systems rely on feeding back information about the output to a controlling device.
4 In human beings, the brain and the central nervous system are the main control devices. Control is carried out by nerves and by hormones.
5 In chemical reactions, equilibrium is reached when the amounts of products and reactants remain constant.
6 A reversible reaction is one where reactants can be converted to products, or products to reactants.
7 Ammonia is an important industrial chemical. It can be manufactured only by careful adjustment of the conditions of the reaction.

QUESTIONS

1 In the manufacture of sulphuric acid, sulphur(IV) oxide is reacted with oxygen to make sulphur(VI) oxide. The equation is:

$$2SO_2(g) + O_2(g) \rightleftharpoons 2SO_3(g)$$

(a) Suggest how the rate of the reaction might be increased.
(b) Suggest how the position of equilibrium might be moved to the right.
(c) What is done in the manufacture of sulphuric acid to stop the loss of unreacted chemicals (like sulphur(IV) oxide)?

2 The diagram shows a lavatory cistern.

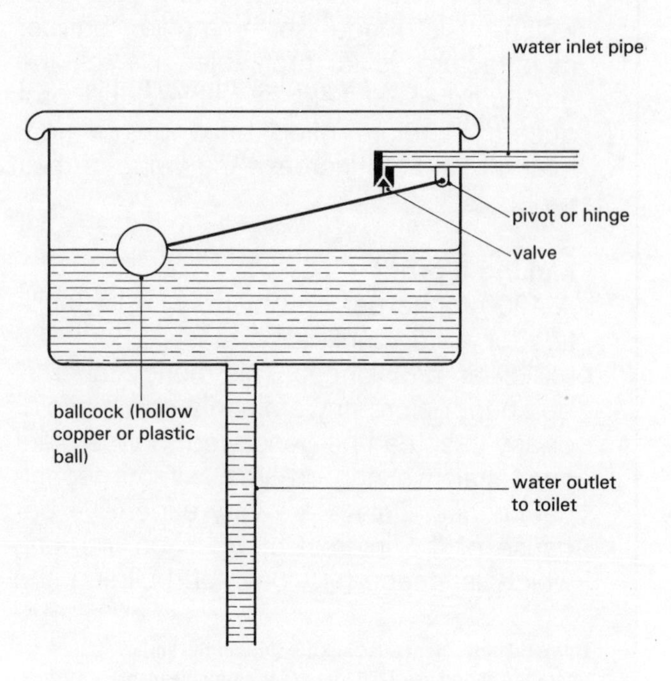

Water is stopped flowing out of the water inlet pipe by the valve which blocks the opening.
(a) Explain how the valve is kept in position.
(b) When the toilet is flushed all the water runs out of the cistern. Explain how the cistern is refilled.
(c) Why is this an example of a feedback system?
(d) In times of water shortage we are often recommended to reduce the amount of water used by a single flush of the toilet. Which of these methods could be used to achieve this?
 (i) Bending the ballcock arm upwards.
 (ii) Bending the ballcock arm downwards.
 (iii) Putting a brick in the cistern.
 In each case, explain why the method does or does not work.

3 (a) What is meant by homeostasis?

(b) A warm-blooded animal has to keep its body at a constant temperature.

(i) Give two examples of warm-blooded animals

(ii) How do these animals keep their blood warm?

(iii) How does a warm-blooded animal control the temperature of its blood?

(c) The graph below shows the relationship between the rate of heat production in the human body and the external air temperature.

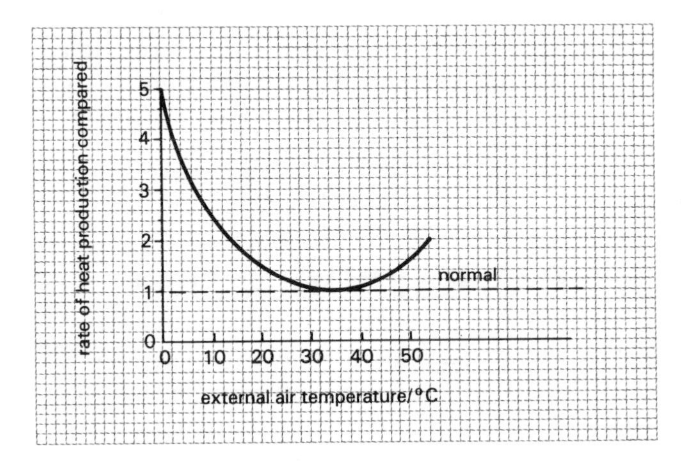

At which temperature does the response cease to be homeostatic? Explain your answer.

(d) The thermostat in an oven regulates the oven's temperature (internal environment) by means of feedback. Give *two* other examples of non-biological feedback mechanisms found in everyday life (at least one example should not involve temperature regulation).

4 Two tugs attempt to pull the QE2 out of Southampton harbour. Unfortunately, a sailor leaves one rope attaching the QE2 to the dockside.

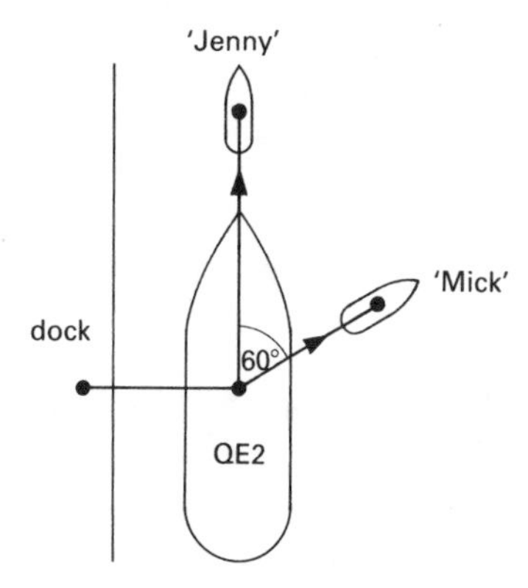

Tug 'Mick' pulls with 100 tonnes thrust at 60° to the dockside. Tug 'Jenny' pulls with 50 tonnes thrust parallel to the dock. (1 tonne thrust is 10 000 N.)

(a) If the rope from 'Mick' has a breaking strain of 2 000 000 N, will it break now? Explain your answer.

(b) By scale drawing, or otherwise, calculate the size of the force on the rope holding the QE2 to the dock.

(c) Use your drawing to find the angle that the rope makes to the dockside (marked X on the diagram).

(d) Will this rope break if it is the same type of rope as the rope from 'Mick'?

8. Reproduction, Growth and Development

REPRODUCTION AND GROWTH

No living thing lives forever. But there are always living things on the earth because all living things have the ability to create new life by reproducing.

How do living things reproduce and grow? How did you begin? Where did you come from? New plants and animals can be created by two methods – by ASEXUAL REPRODUCTION or by SEXUAL REPRODUCTION. You were created by sexual reproduction: humans cannot reproduce by the asexual method. Most living things reproduce sexually, many can reproduce asexually as well.

SEXUAL REPRODUCTION

This type of reproduction involves two parents. The young are usually different from either parent.

In sexual reproduction, special sex cells called GAMETES are produced in the sex organs of both parents. In humans the female gametes, called

Fig. 8.1 All living things reproduce themselves

egg cells, are made in the ovaries of the woman. Male gametes, called sperms, are made in the testes of the man. All gametes are formed by a special process of cell division called *meiosis*. (This will be described in detail on page 148.)

One male gamete and one female gamete fuse together in a process called FERTILISATION to form one cell called a ZYGOTE. This zygote contains genetic information from both parents. The zygote will grow and develop into a new individual. The process of sexual reproduction in both plants and animals is explained later in this unit (starting on page 99).

ASEXUAL REPRODUCTION

This type of reproduction involves only one parent and all the offspring are identical to that parent. Asexual reproduction does not involve sex cells, it involves a type of cell division called *mitosis*. (This is explained on page 146.)

The two types of reproduction are summarised below.

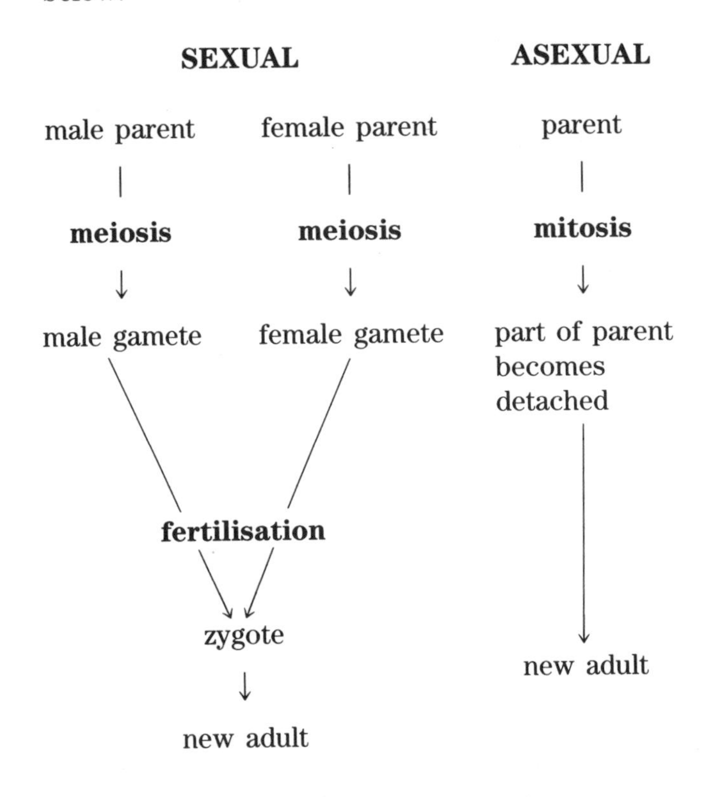

Various methods of asexual reproduction are possible and are used by different living things.

1. Binary fission

This method is used by very simple animals like *Amoeba*. The cell divides into two identical 'daughter' cells. Most bacteria also reproduce in this way.

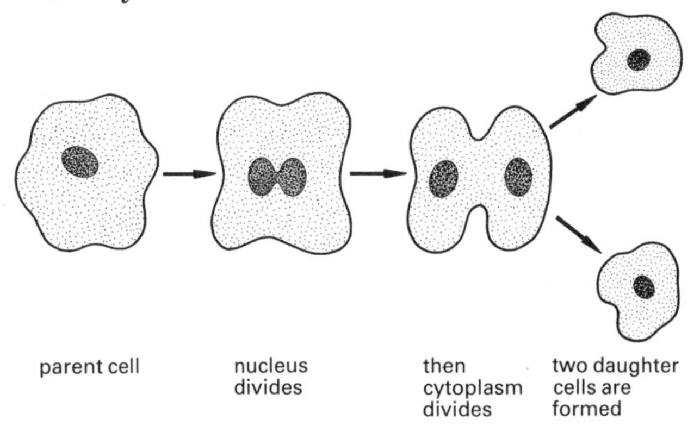

Fig. 8.2 Amoeba reproducing asexually by dividing in two

2. Budding

In some organisms, such as *Hydra*, bud-like growths develop from the parent. The bud grows into a new individual and eventually separates from the parent. Yeasts also reproduce by budding.

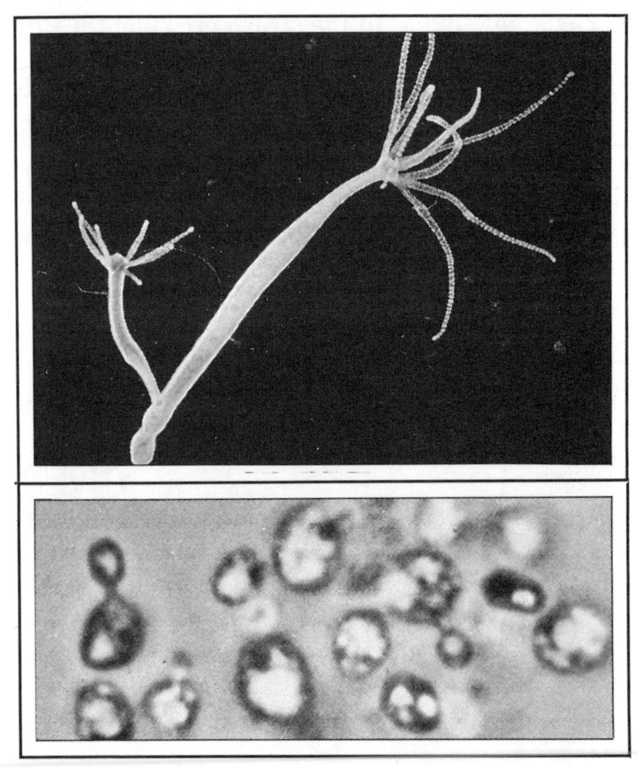

Fig. 8.3 Hydra and yeast reproducing asexually by budding

3. Spores

Many moulds reproduce asexually by producing thousands of spores in capsules on the end of fine vertical threads. The spores are formed by the division of cells in the capsule and not as a result of male and female gametes fusing together. When they have been dispersed each spore can form a new mould plant.

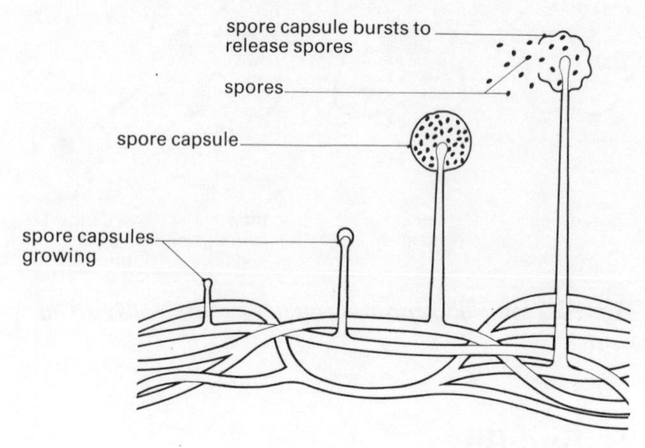

Fig. 8.4 Pin mould reproducing asexually by producing spores

4. Vegetative propagation

Runners and stolons

Many plants send out a side branch which produces a new plant. In the strawberry plant a side shoot called a *runner* grows horizontally along the surface of the soil. At intervals roots and shoots form to make a new plant. As soon as this is growing well the runner usually rots away. Some plants such as *Chlorophytum* (spider plant) send out a drooping branch called a *stolon*, at the end of which a new plant develops.

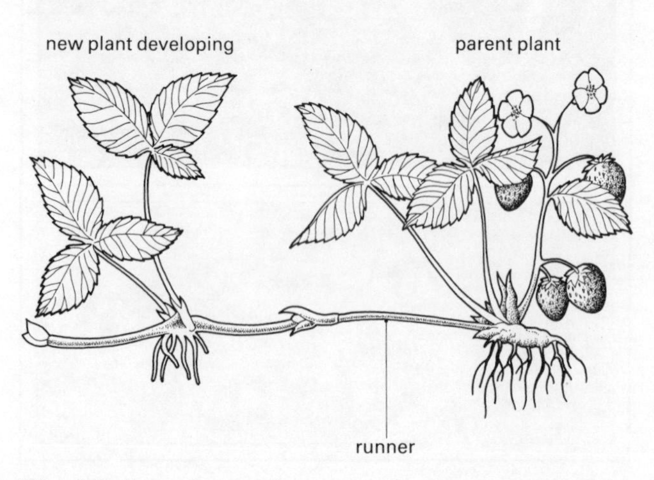

Fig. 8.5 Strawberry plant reproducing asexually by runners

Fig. 8.6 Spider plant reproducing asexually by stolons

Perennating organs

In some plants the asexual reproductive organs also act as food stores. They enable a plant to survive from one year to the next. During the winter the plant may lie dormant in the soil and then use the stored food to grow again in the spring. Survival from year to year is called *perennation*. The special asexual reproductive organs are therefore called *perennating organs*.

Some examples are shown in Table 8.1 and Fig. 8.7.

Table 8.1 Examples of perennating organs found in different plants

Perennating organ	Plant
bulb	daffodil
corm	crocus
tuber	potato
rhizome	iris

Artificial propagation

Gardeners use many methods of asexual reproduction to get new plants – they do not always use seeds. (N.B. Seeds are produced by sexual reproduction in flowering plants – see page 101 later in this unit.) The simplest and most often used method is that of taking *cuttings*. Usually, small pieces of stem with a few leaves attached are cut from the parent plant. The cutting, when put into damp soil, will produce roots and grow into a new plant.

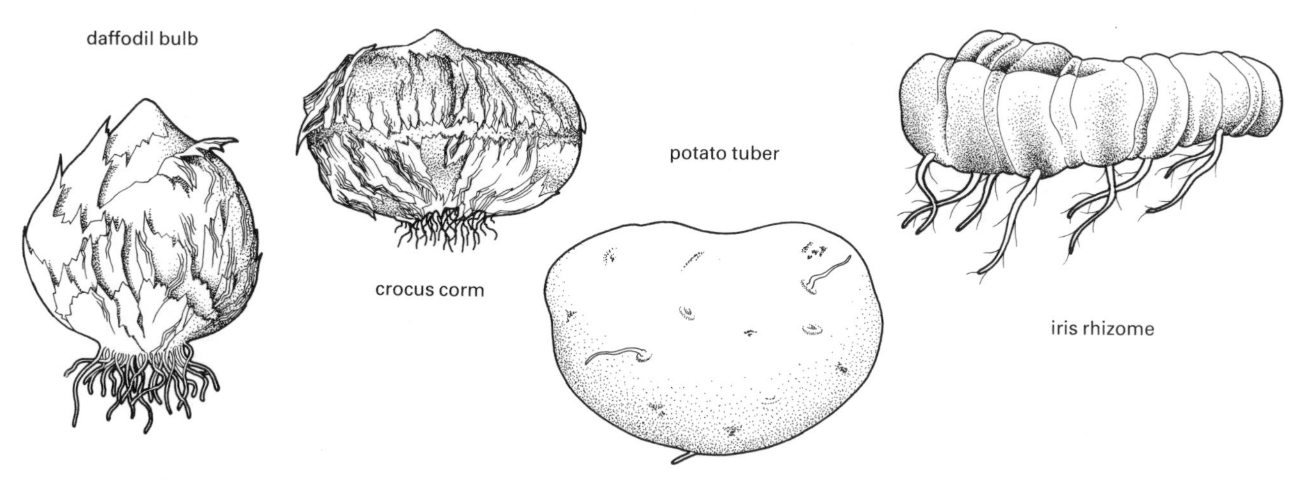

Fig. 8.7 Perennating organs

Investigation 8.1 Growing a plant from a cutting

You can grow a new plant without using seeds.

Collect

Busy Lizzie plant (*Impatiens* sp.)
sharp knife
sticky label or a chinagraph pencil
beaker half full of water

What to do

1 Cut off a side branch from the plant. It should have 6–10 leaves.

2 Remove any flowers and the bottom two or three leaves.

3 Put the cutting in the beaker of water. The leaves should all be above the water level.

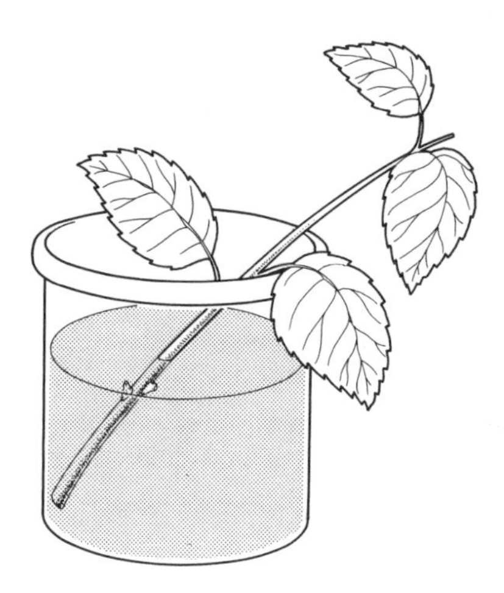

4 Label the beaker with the name of the plant and the date.
5 Put the beaker in a warm, light place.
6 Look at the cutting every few days. Add more water if the level drops.

After about a week you will probably find that your cutting has begun to grow roots. When the roots are about 2 cm long you could plant the cutting in a pot of compost. If you water it regularly and keep it in a fairly warm place (15 °C) it should grow into a large plant like the parent plant.

Assignment – The potato

A potato is a swollen underground stem called a *tuber*. The tuber is a kind of perennating organ. This means that it can rest in the soil during the winter and then produce a new plant the next summer.

When a 'seed' potato is planted shoots begin to grow from the 'eyes'. The shoot grows above the soil and leaves form. Roots also grow from the 'seed' potato. Look at the diagram.

1 The 'seed' potato withers as its food supplies are used up.
2 Food made in the leaves by photosynthesis (see page 203) is sent down to the tubers to help them grow.
3 Underground stems grow out from the base of the main stem and swell up into tubers.

Potatoes big enough to eat should be ready about two or three months after planting. The plant above the ground will begin to wither and die; the potatoes should then be dug up.

When harvested, potatoes should be stored in the dark. Light can make them go green and then they may become dangerous to eat.

1 Why are the potatoes we plant called 'seed' potatoes?
2 Are there any advantages in the potato being able to reproduce asexually? What are they?
3 Potatoes are the main part of some people's diet. Why are they such a useful food?
4 In diagram (iii) the seed potato has shrivelled up. Why is this?
5 If the potatoes were not harvested, describe what would happen to the plant and to the potatoes left in the soil.
6 Gardeners will sometimes cut a large potato in half before planting in the hope of getting more plants. Where would they make the cut if they used the 'seed' potato shown in diagram (i)?

(i) A shoot grows from the eye of a seed potato bud (called an 'eye')

(ii) The shoot forms leaves

(iii) Tubers form at the end of branches coming from the main stem. Food made by leaves is sent down to the tubers (new potatoes).

(iv) The leaves, stem and old tuber die. The new tubers remain dormant if not harvested.

5. Parthenogenesis

This is the development of a new individual from an unfertilised egg. It may happen in certain groups of invertebrates such as aphids (greenfly). During the summer, wingless female aphids may produce many more wingless females. This is a good way of increasing numbers quickly and it does not need a male.

Table 8.2 Sexual and asexual reproduction compared

	Sexual	Asexual
Parents	two	one
Offspring	not identical to parent, characteristics inherited from both parents	genetically identical to parent
Method(s)	Gametes, formed by *meiosis*, fuse at fertilisation to form a zygote	binary fission budding spores vegetative propagation parthenogenesis all new individuals formed by *mitosis*
	Both sexes needed	Only one individual is needed to start a large population
	May not increase population size (two parents may have only one or two offspring)	Always increases population
	Not very rapid	Can be very rapid
	Food reserves only in seed	Much larger food reserves in perennating organs or in parent plant

SEXUAL REPRODUCTION IN FLOWERING PLANTS

Have you ever wondered why plants have flowers? Many people enjoy the beauty and scent of flowers. But the biological importance of flowers is that they are the reproductive organs

Investigation 8.2 Looking at the structure of flowers

A flower is made up of many different parts which are arranged in rings (whorls).

Collect

wallflower and several other flowers
forceps
hand-lens
Sellotape
scissors
scalpel

What to do

1 Have a good look at your flower. Make sure you know its name.
2 Starting with the outermost layer, *sepals*, which are probably green, use the forceps to remove them all. Count how many you have.
3 Do the same as in step 2 with the *petals*.
4 Carefully remove all the *stamens* and count them.
5 Remove the *carpels* and count.
6 Set out a page in your exercise book like the one below.

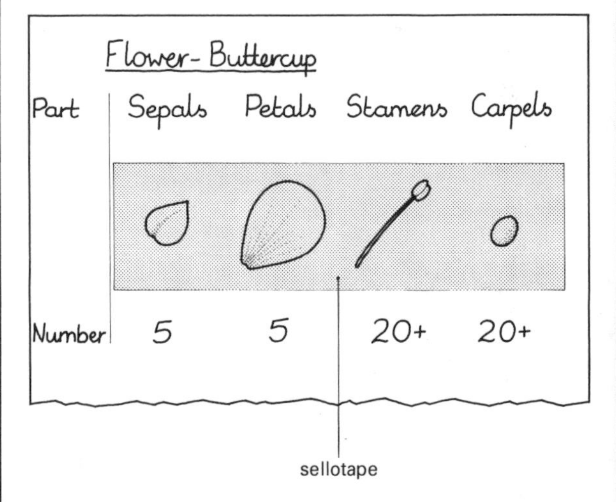

Stick one sepal, one petal, one stamen and one carpel into your exercise book with Sellotape. Record the numbers of each part.
7 Cut open a carpel. Can you see an ovule inside it? Use the hand-lens to look.
8 Repeat 1–7 with other flowers.

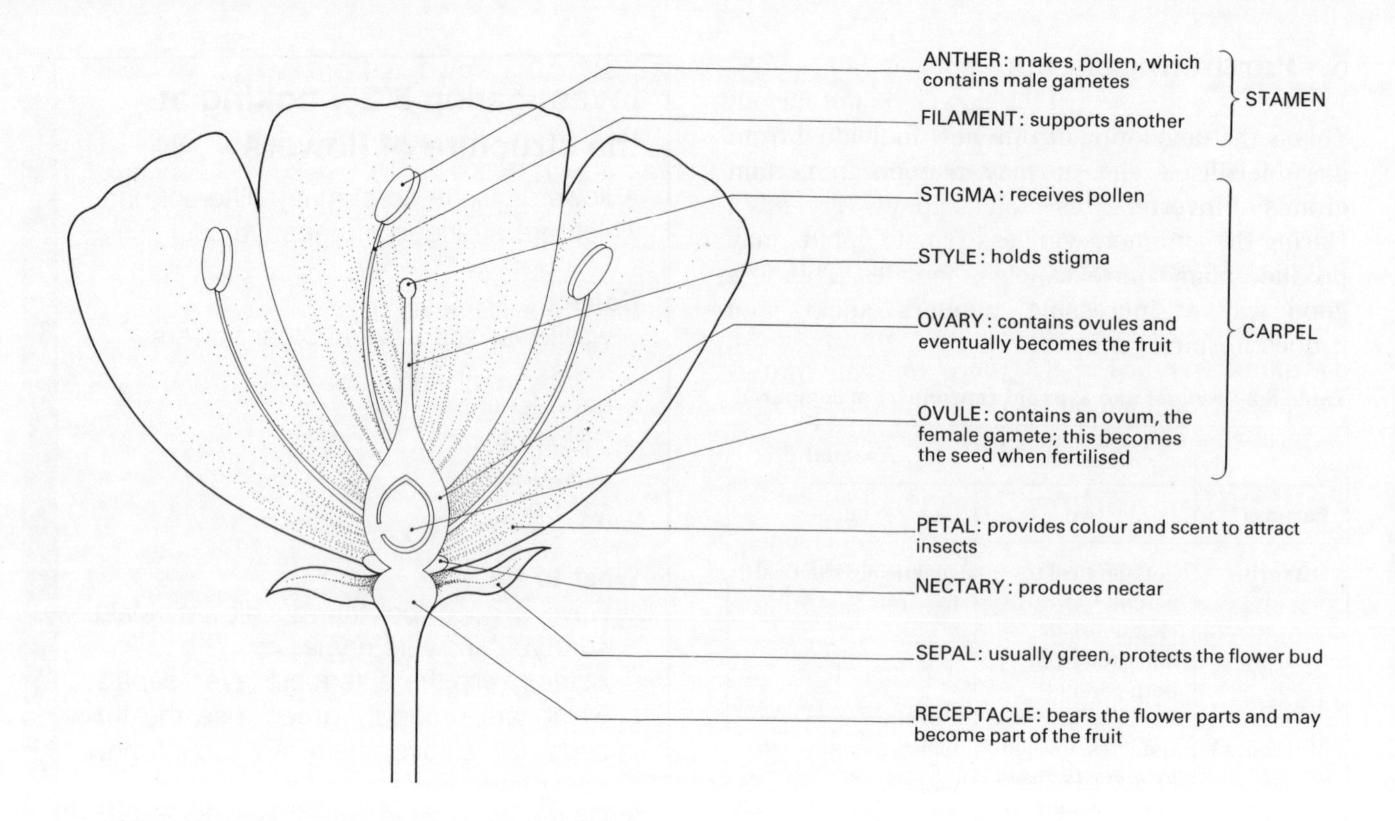

Fig. 8.8 Structure and function of an insect-pollinated flower

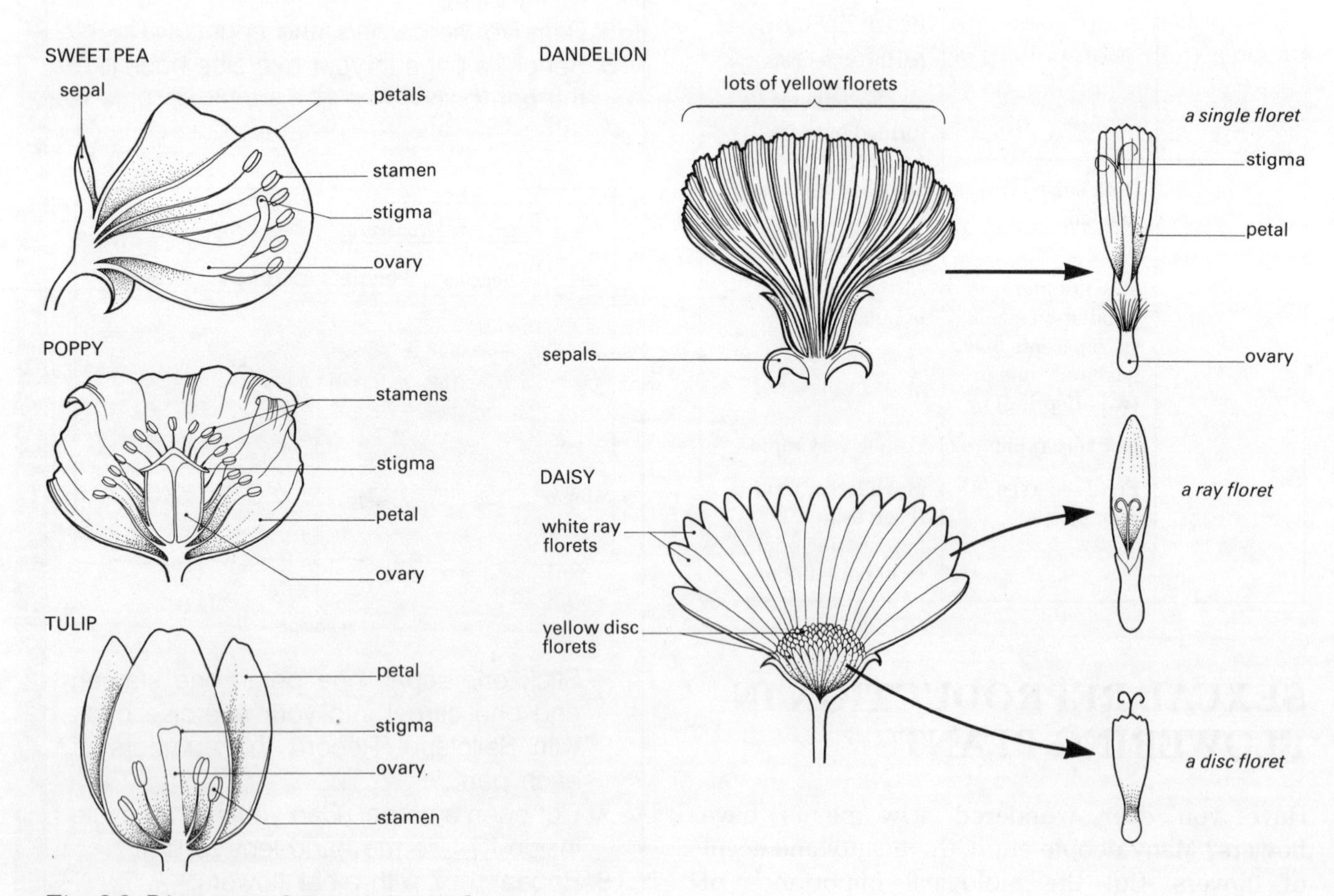

Fig. 8.9 Diagrams of some common flowers

of the plant and allow it to reproduce sexually. The colour and scent of the flowers is not for our benefit but to attract insects to the flower. Insects help in the process of reproduction in many flowers.

Flower structure

Although most flowers have the same parts as the one shown in Fig. 8.8, there are many different arrangements and numbers of parts. Also there are many different shapes of flowers.

Pollination and fertilisation

There are two main stages in sexual reproduction in plants. The first is POLLINATION and the second FERTILISATION.

Pollination is the transfer of pollen from the stamens to the stigmas. Pollen grains are very small. They may be carried in the air by wind from the stamen to the stigma or they may be carried on insects. When a pollen grain lands on a stigma it will begin to grow a pollen tube. This pollen tube grows down the style to the ovary.

The pollen tube goes into the ovule and the nucleus from the pollen grain (male gamete) fuses with the nucleus of the ovum (female gamete). The fusion of the male and female gametes inside the ovule is called *fertilisation.*

Although pollination must occur before fertilisation, pollination does not always result in fertilisation.

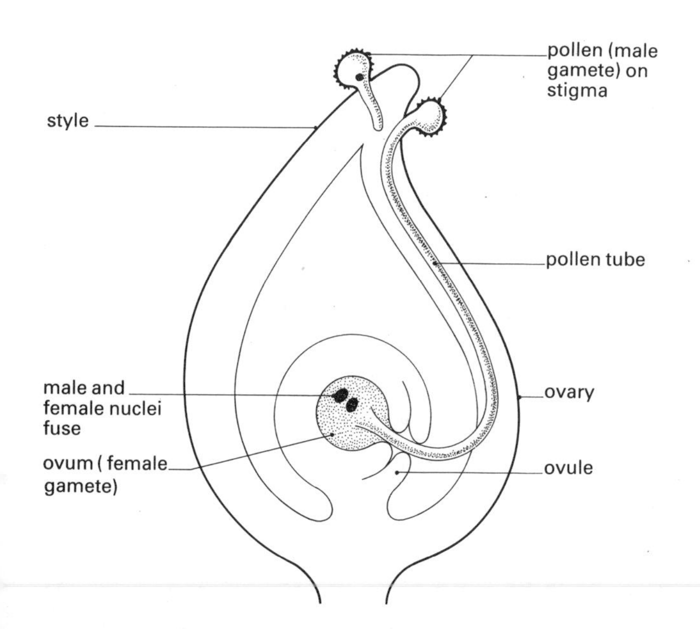

Fig. 8.10 Fertilisation in a flowering plant

Fruit and seed formation

After fertilisation, when the pollen nucleus has fused with the ovule nucleus, the resulting cell is called a ZYGOTE. The zygote will develop into a SEED. First the zygote divides many times and forms a miniature plant called an EMBRYO. The embryo consists of a tiny root and shoot with one or two special leaves called *cotyledons*. The cotyledons swell with stored food and enclose the embryo. The outer wall of the ovule becomes thicker and forms the seed coat or *testa*.

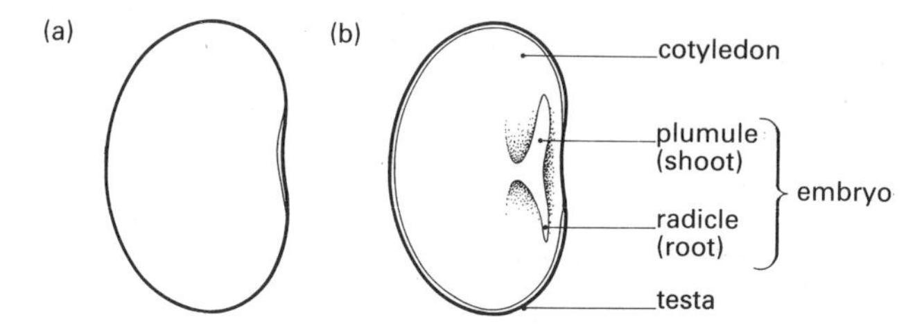

Fig. 8.11 A broad bean seed (a) whole, (b) cut in half

As the seed grows the ovary also becomes larger. Most of the other flower parts drop off: that is, the sepals, petals, stamens, stigma and style. The ovary is now called a FRUIT. The biological meaning of a fruit is a fertilised ovary, so not all fruits are edible. For example, in peas the pod is the fruit and the peas are the seeds (see Fig. 8.12).

Fig. 8.12 The pod is the fruit of the pea plant: the peas are the seeds

Assignment – Pollination

Pollination is usually carried out either by wind or by insects. It may also be carried out artificially by people.

Insect-pollinated plants include roses, wallflowers, buttercups and many others. Insects such as bees visit the flowers to feed on nectar. As an insect does this, its hairy body or legs get covered with pollen and it transfers the pollen to the stigmas of other flowers it visits.

Wind-pollinated plants include grasses and some trees. The petals of the flowers are usually small and are often green, but they produce very large numbers of small, light pollen grains. The flowers often hang down so that pollen is easily shaken out. The feathery stigmas also hang out of the flower so that any pollen in the air may be caught.

The diagram shows a typical insect-pollinated and a typical wind-pollinated flower.

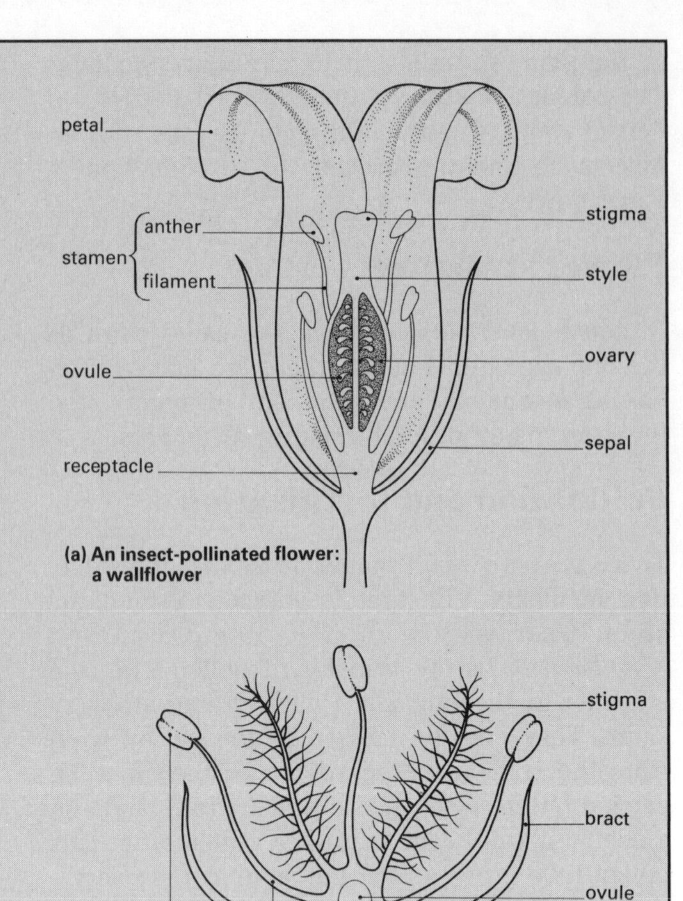

(a) An insect-pollinated flower: a wallflower

(b) A wind-pollinated flower: one grass flower

1 Using the information above, and looking at the diagrams, copy the table below and fill in the spaces.
2 What do you think the terms *self-pollination* and *cross-pollination* mean?
3 How could you pollinate a plant artificially? Describe what you would actually do.

	Wind pollination	**Insect pollination**
Petals	Small – stamens and carpels exposed Dull colour Do not produce nectar or scent	------------------ ------------------ ------------------ ------------------
Stamens	---------------------- ----------------------	Stiff filaments and anthers inside the petals
Pollen	Large quantities ---------------------- ----------------------	------------------ Rough pollen grains which 'stick' to insects.
Stigmas	Large Exposed to the wind ---------------------- ----------------------	------------------ ------------------ Often like a pinhead May be sticky

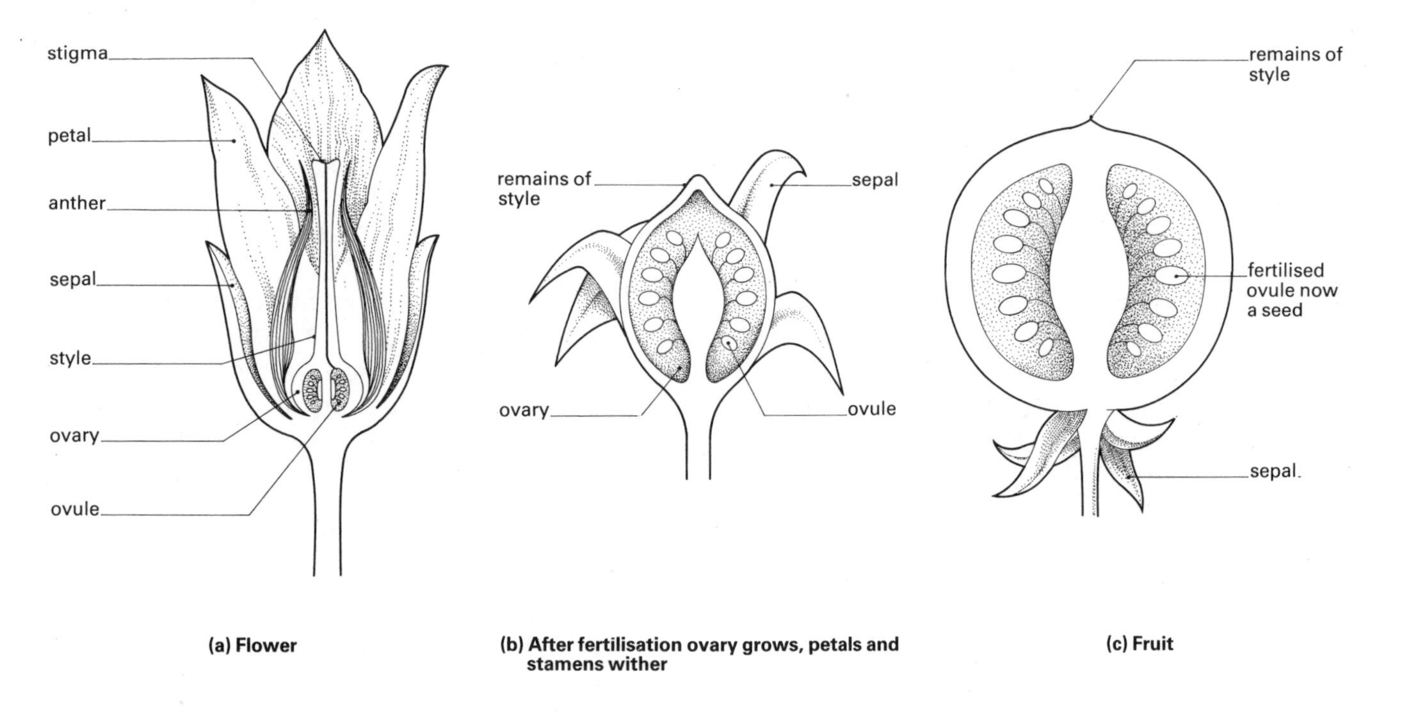

Fig. 8.13 Fruit formation in a tomato

There are many types of fruits. Some, such as cherries and plums, have one seed. Others have many seeds, such as tomatoes and gooseberries. Some, such as blackberries and raspberries, are formed by many small fruits clustered together. Yet another type is that of the strawberry, where the fruits are the pips and the fleshy part is formed from the receptacle of the flower.

Fruits protect the developing seeds and also help to disperse the seeds away from the parent plant. This reduces the competition for light and water between plants of the same kind. Some fruits have special adaptations to get them a long way.

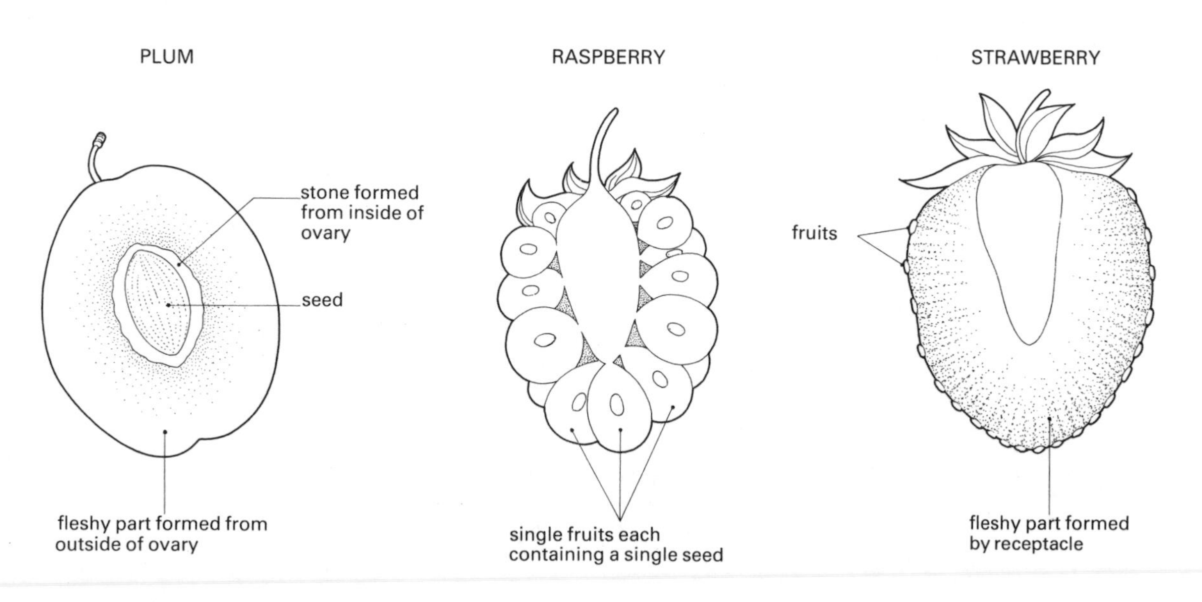

Fig. 8.14 Fruits

Assignment – Seed dispersal

There are various different methods of dispersal. Some of these are described below.

(a) *By 'explosion'.* As certain fruits dry they shrivel and this puts tension on the fruit. The fruit will suddenly burst open, throwing out the seeds. Peas are dispersed in this way.

(b) *By wind.* Some fruits have a large surface area to catch the wind. Two common examples are the dandelion which has a 'parachute' of hairs and the sycamore which has 'wings'.

(c) *By water.* Fruits may contain air or oils which makes them buoyant. If they drop into water they will float and may be carried long distances by rivers or the sea. Seeds from coconuts are dispersed in this way.

(d) *By animals.* There are two main ways in which animals can help to disperse seeds.

(i) Some plants, such as cleavers, develop fruits which have small hooks on them. The hooks may cling to an animal and be carried away from the parent plant.

(ii) Many fruits are juicy and fleshy and are eaten by various animals and birds. Many of the seeds contained in the fruits have a hard covering which protects the seed as it passes through the animal's digestive system. The seeds are passed out in the animal's faeces: this will probably happen at some distance from the parent plant.

1 Look at the diagram. Make a table to show the method by which each fruit or seed shown is dispersed.

2 By what methods do humans disperse seeds? Explain with reference to particular examples.

3 Why is it an advantage for a plant to have an efficient seed dispersal mechanism?

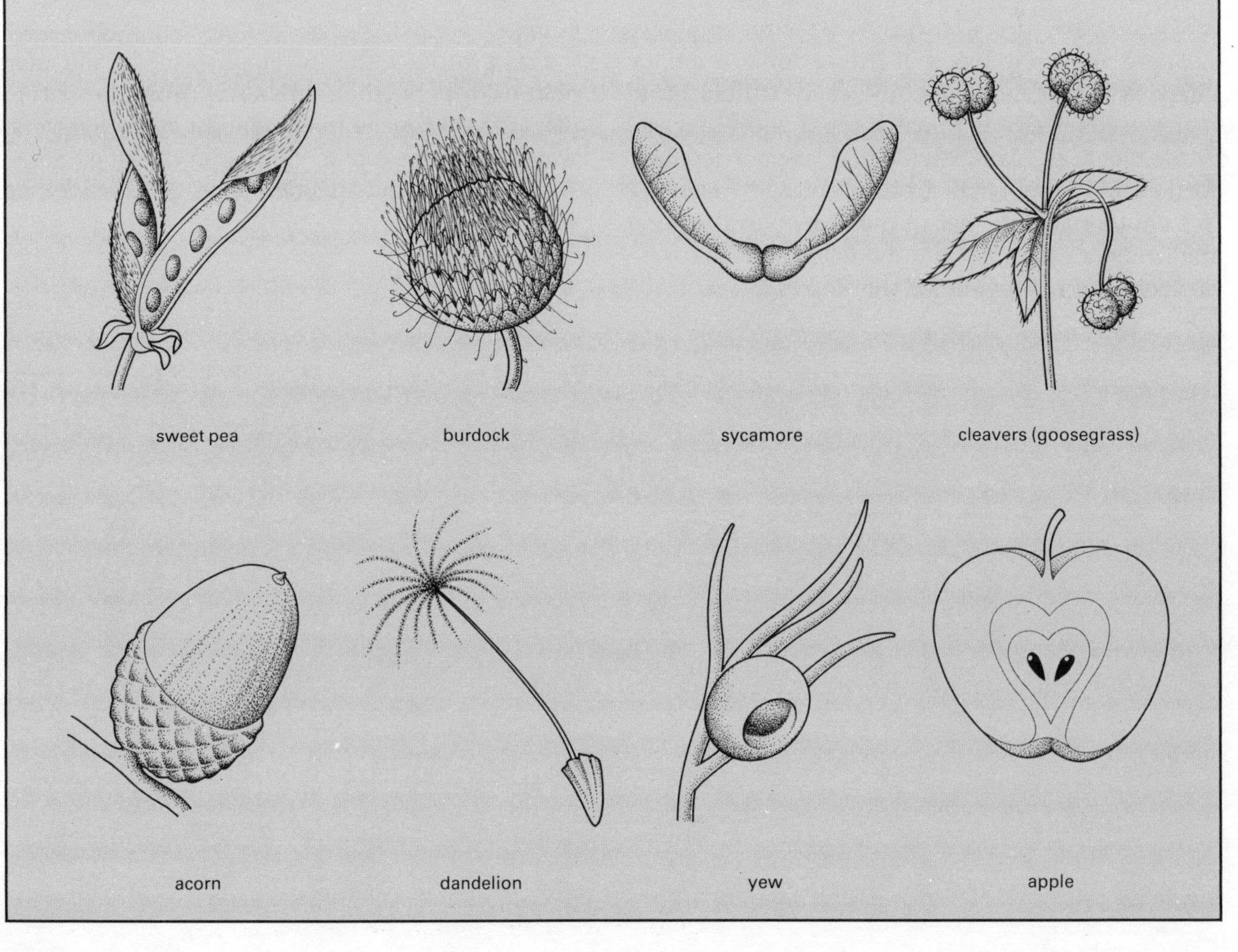

Germination and growth

When the seeds have been dispersed, if they land in favourable surroundings, they may GERMINATE; that is, grow into a new plant.

Investigation 8.3 To find out what factors are needed for germination

New plants grow from small seeds, but to start growing – this is called germination – they need certain conditions around them.

Collect

5 boiling tubes/large test tubes
cotton wool
cress seeds
cardboard box or aluminium foil
oil
5 labels or a chinagraph pencil

What to do

1 Label the test tubes A, B, C, D and E.
2 Put a piece of cotton wool into the bottom of each test tube. Then set up the test tubes as shown in the diagram.
3 Pour some water into B, C and D to make the cotton wool very damp.
4 Put some cress seeds (10–20) on top of the cotton wool in each test tube.
5 Pour water carefully into D to a depth of about 2 cm, then add a thin layer of oil.
6 Cover E with a cardboard box or with foil. Make sure no light can get in.
7 Put C in a refrigerator.
Put A, B, D and E in a warm place.
8 After 3 or 4 days look at the test tubes again. In which test tubes have the seeds germinated?

Results

Draw a table like the one below.
In it record the conditions in each test tube, and whether the seeds germinated. The first two have been started for you.

Test tube	Conditions	Germination
A	warm, oxygen, no water, light	
B	warm, oxygen, water, light	
C		
D		
E		

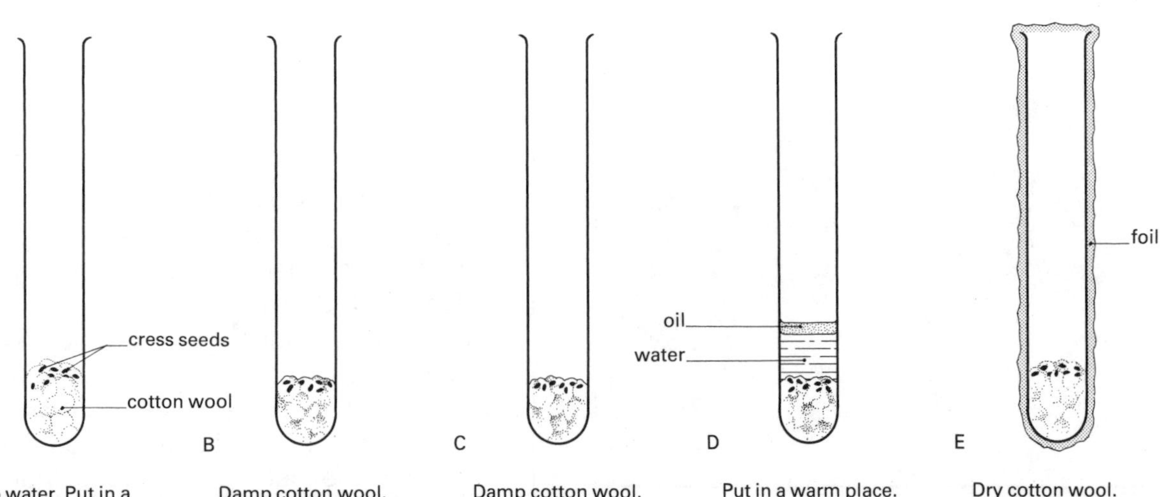

From your results for Investigation 8.3 you will see that certain conditions are needed for cress seeds to germinate. They need (1) water, (2) oxygen, (3) warmth. The cress seeds do not need light to germinate.

Some seeds do need light for germination, some seeds will germinate only in the dark, but most will germinate in either the light or the dark. The temperature needed for germination also varies according to the type of seed and the part of the world where it normally grows.

Even under the best conditions not all seeds germinate; it seems that some seeds are not able to grow at all. This is obviously a nuisance to farmers and gardeners. One method of improving the growth of seeds, on which research is still being done, is to use artificial seed coverings. Each seed is covered with a compound so that all seeds are the same size and shape to allow even sowing in the field. In each of the seed 'balls' there are chemicals to help the seed grow and others to protect it against insects and disease. In some experiments more than 90% of the coated seeds have been found to grow whereas with uncoated seeds grown under the same conditions only about 50% of them grow.

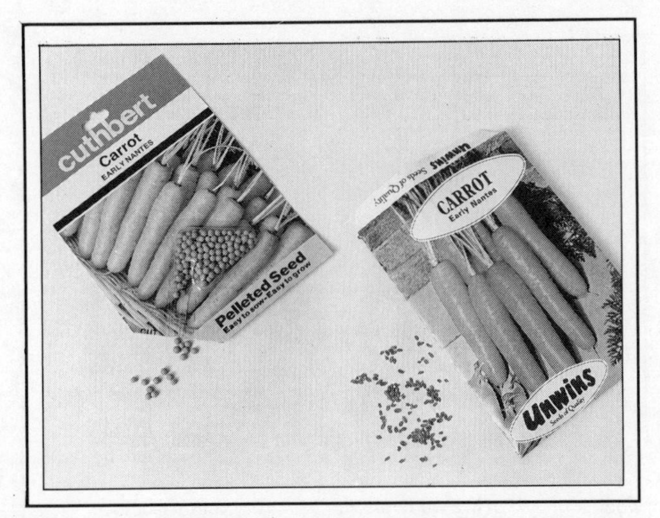

Fig. 8.15 Seeds may be sold in their natural form (right) or as 'seed balls' (left)

Investigation 8.4
Finding out germination success rates in different seeds

The percentage of seeds that germinate may not be the same for all plants.

Collect

3 petri dishes with lids
3 pieces of filter paper
3 different kinds of seeds – 50 of each type
scissors
3 labels or a chinagraph pencil

What to do

1. Cut the pieces of filter paper so that they fit the petri dish.
2. Draw grid lines 1 cm apart on the paper, with a pencil, as shown in the diagram. You should have at least 50 squares.

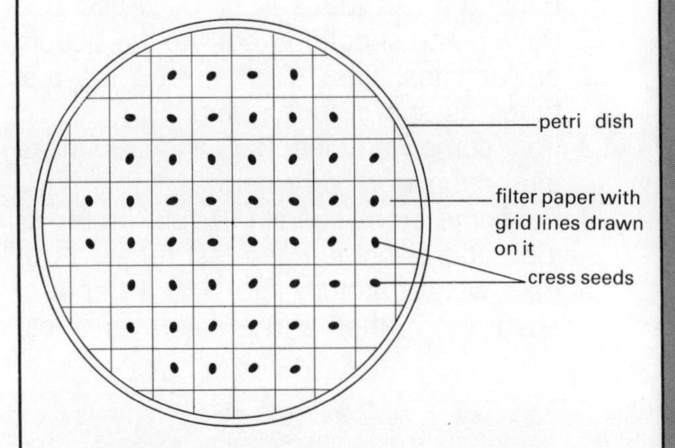

3. Put the filter paper in the petri dishes and make it very wet.
4. Take 50 seeds of one type and put one seed in each square. Put the lid on and label the dish.

   Repeat using two other types of seed in the other two dishes.
5. Leave the dishes in a warm place for 4–7 days. Make sure the filter paper does not get dry.
6. Count the number of seeds which did not germinate and then count or work out the number that did germinate. To find the percentage of seeds which germinated, multiply the number of seeds that germinated by 2.

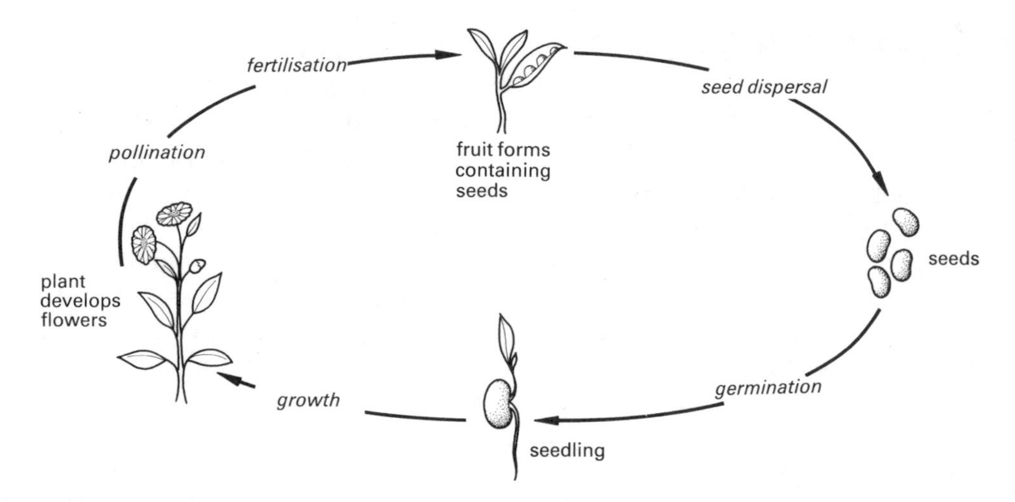

Fig. 8.16 The life cycle of a flowering plant

Assignment – Germination conditions needed by different types of seeds

Go to a shop or a garden centre that sells seeds in packets. On the back of the packet you should find a description of the plant, instructions for sowing (when, where and how to plant, with the recommended germination temperature if this is important) and the best situation for growth, together with other helpful information.

1 Make a table like the one below and then fill it in for three types of flower seeds and three types of vegetable seeds. The first example has been done for you.

Name	Germination temperature	Depth	Germination time	When to sow
Pansy	10–15 °C	3 mm	10–21 days	Jan–Mar

2 Seeds which are planted too deep may not germinate.
   (a) Suggest possible reasons for this.
   (b) Describe an experiment which you could carry out to test one of your suggestions.
3 Why is it useful to know the length of time it usually takes a seed to germinate?
4 Which of the seeds in your table would need to be germinated in a heated greenhouse or propagator?

Assignment – Annuals, biennials and perennials

How long do flowering plants live? The life cycle varies in different plants. In its life cycle a plant grows from a seed, produces more seeds and then dies. Some plants get through their life cycle in one year. They are called *annuals*. Many garden flowers such as petunias and forget-me-nots are annuals and must be grown from seed each year.

Some plants take two years to complete their life cycles. They are called *biennials*. In the first year they send up a leafy shoot but produce no flowers, in the second year they produce flowers and seeds. A common example is the wallflower.

Some plants live for years and years. These are called *perennials*. They may be herbaceous, such as geraniums, or woody, such as shrubs and trees.

1 Visit your local garden centre or gardening shop. Look at the instructions on packets of seeds and make lists of plants that are annuals, biennials and perennials.
2 If you moved into a new house which had a bare garden what would you plant if
   (a) you wanted lots of flowers quickly;
   (b) you were not too bothered about flowers but wanted plants that would not need much looking after from year to year?
   Give reasons for your selections.

SEXUAL REPRODUCTION IN ANIMALS

Although some animals can reproduce asexually, (see page 95) most reproduce sexually. In this book we are going to explain how reproduction occurs in humans.

To understand how reproduction works you first need to know about the structure of the reproductive system in both men and women. Fig. 8.17 shows the reproductive system of a human male. The diagram has been annotated to help explain what happens in some of the parts.

The male gametes, *sperms*, are made in the testes which are positioned outside the main part of the body. This is important because the body makes sperms more rapidly and effectively in cool conditions and the testes have a lower temperature than the rest of the body. Inside the testes are many sperm tubules where the sperms are made by the process of cell division called meiosis (see page 148). The sperms move from the tubules to the epididymis where they are stored. The Cowper's gland, prostate gland and seminal vesicles all produce fluids which are added to the sperm to make *semen*.

From Fig. 8.17 you can see that both the

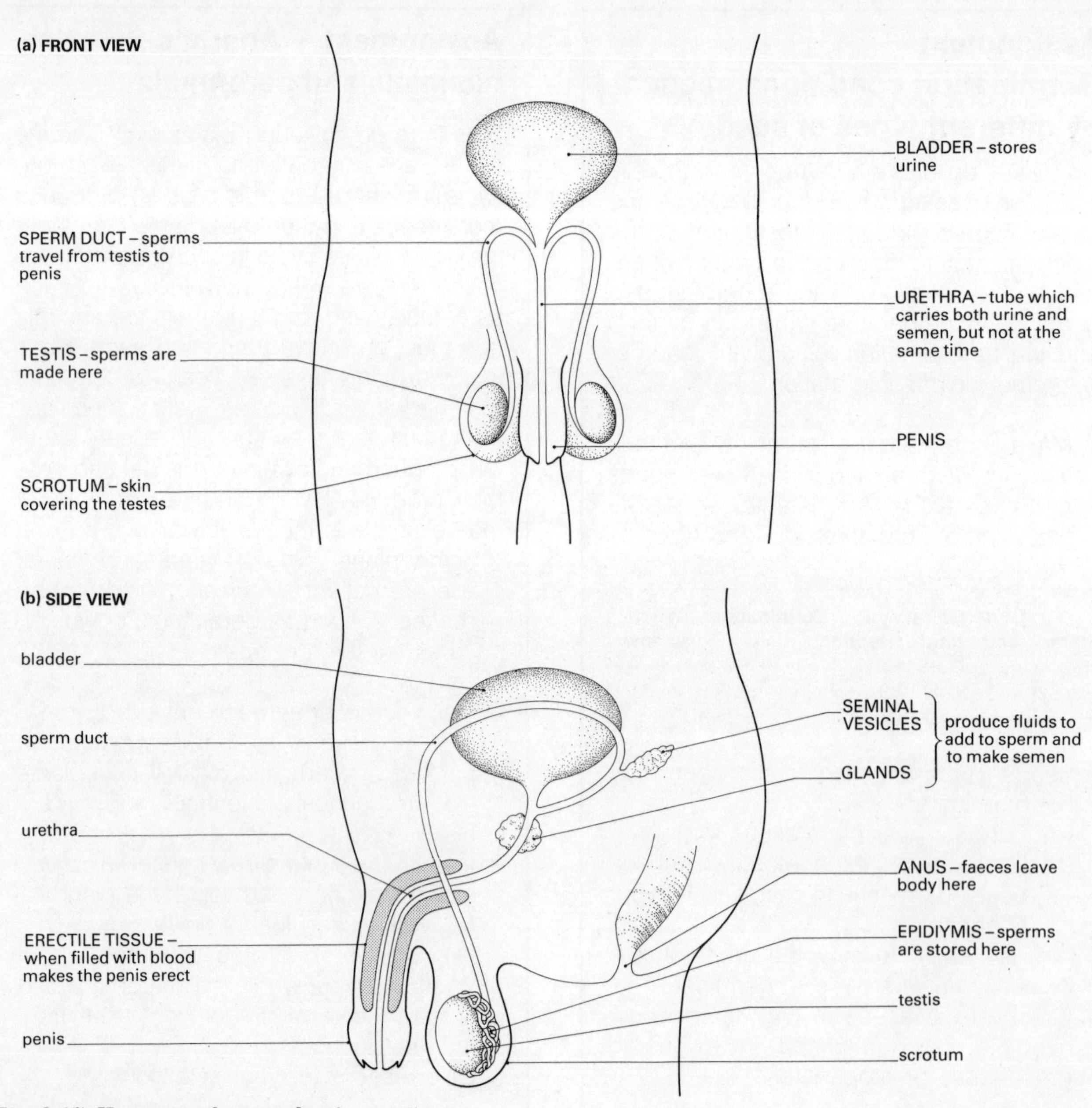

Fig. 8.17 Human male reproductive organs

bladder and the sperm ducts open into the urethra. However urine from the bladder and semen never pass through the urethra at the same time. It is difficult for a man to urinate when his penis is erect and it is impossible for him to ejaculate, that is release semen, when his penis is not erect.

The arrows on the diagram in Fig. 8.17 show the route the sperms take when leaving the body.

Figure 8.18 shows the reproductive system of a human female.

The female gametes, *eggs*, are made in the ovaries. As with sperms, the eggs are made by the process of cell division called meiosis. Eggs are already present in the ovaries when a baby girl is born. When a girl first starts to have periods one egg ripens each month. About every 28 days, one or other of the ovaries releases an egg into the oviduct. If the egg is fertilised, it will develop in the uterus.

The vagina is a passage from the uterus to the outside, from Fig. 8.18(c) you can see that it is separate from the urethra. Close to the entrance to the urethra is a small 'lump', the clitoris. The clitoris is the female's equivalent of a penis and this can become erect during sexual intercourse when it is stimulated.

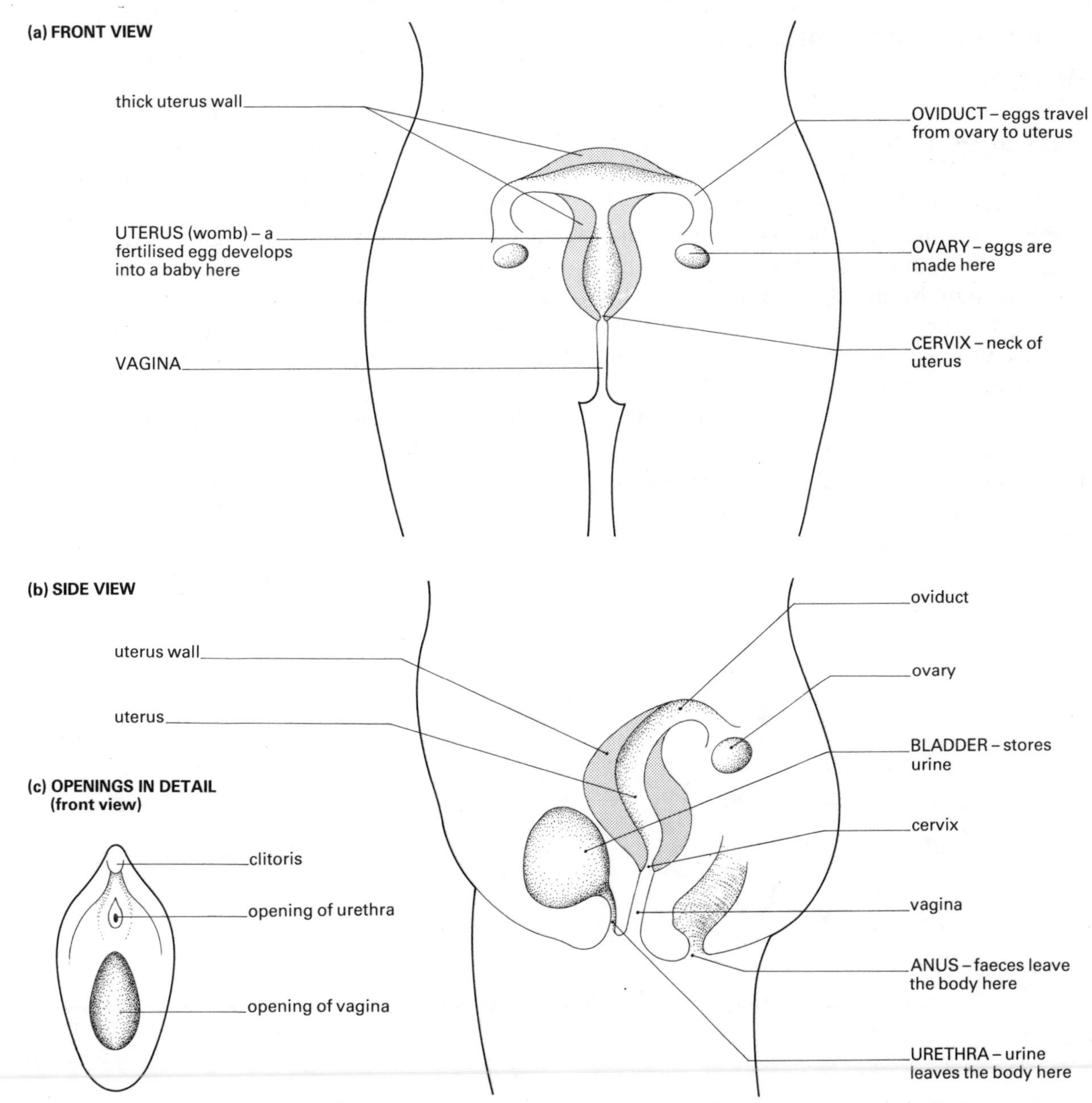

Fig. 8.18 Human female reproductive organs

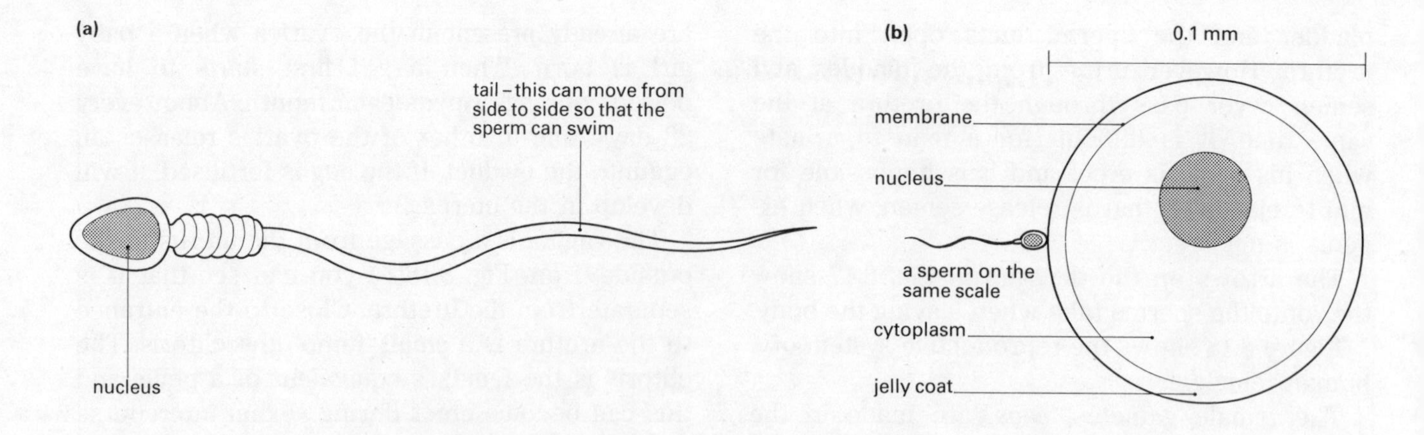

Fig. 8.19 (a) A human sperm. (b) A human egg, with a sperm alongside

Intercourse (copulation) and fertilisation

For sperms to get to the egg, intercourse must take place. The penis of the male becomes stiff and hard. This is called an *erection* and is caused by more blood coming into the penis. When the penis is erect the man can put his penis into the vagina of a woman. Natural lubricants, mucus in the vagina and some fluid from the man's glands which comes through the tip of the penis, help the insertion of the penis into the vagina.

The man then moves his penis rhythmically backwards and forwards inside the vagina. The tip of the penis is very sensitive and nerve endings in this are stimulated by contact with the vagina. This sets in motion a reflex action which brings about *ejaculation*. At ejaculation the *semen* (sperms plus fluid) is expelled from the penis in a sudden spurt. At each ejaculation about 4 cm^3 of semen, containing about 100 million sperms, are deposited in the vagina of the woman, near the cervix. In men ejaculation is accompanied by a very pleasant feeling called an *orgasm*. Women may also have an orgasm but it usually takes longer for a woman and often needs stimulation of the clitoris. A woman will not necessarily have an orgasm each time a man ejaculates during intercourse.

The sperms swim up through the uterus and into the oviducts. If there is an egg in one of the oviducts fertilisation may take place. One sperm may penetrate the egg and the sperm nucleus and the egg nucleus may fuse to form a zygote.

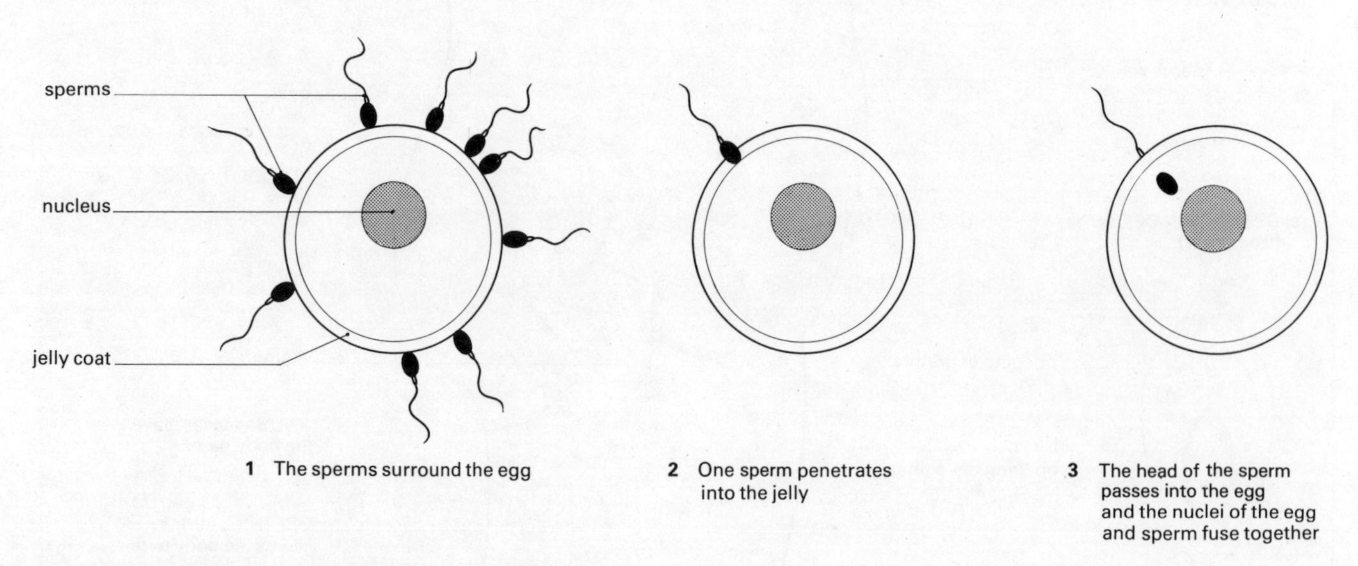

Fig. 8.20 The stages of fertilisation in a human

Assignment – The menstrual cycle

Study the diagram and then answer the questions.

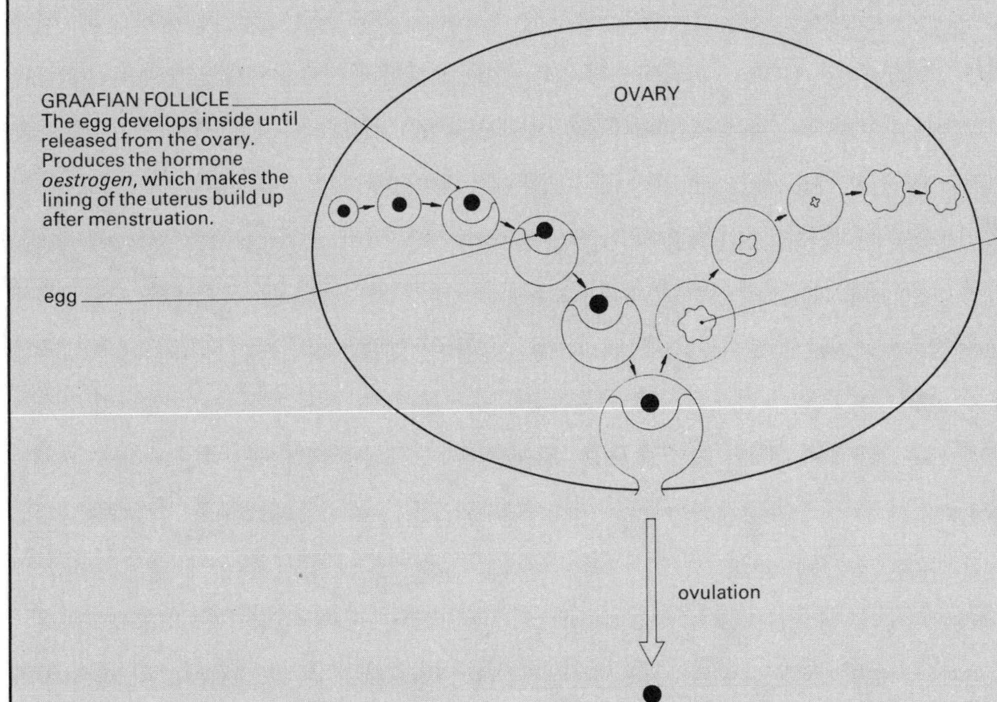

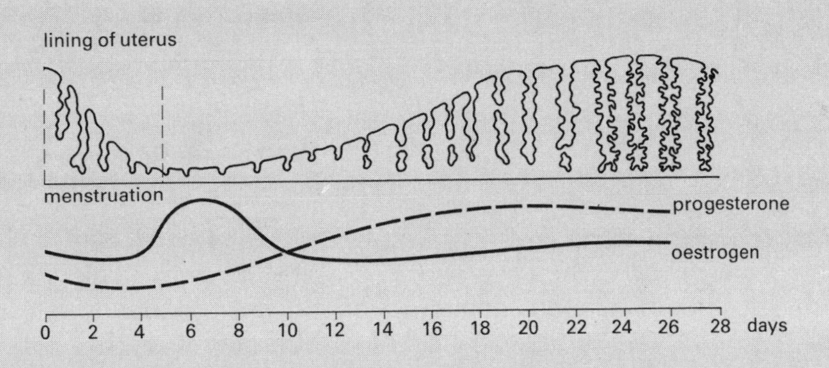

1 On which days of the menstrual cycle will a woman have her periods?
2 On which day of the menstrual cycle will ovulation (the release of an egg) usually occur?
3 How many days after the beginning of a period will ovulation occur?
4 What happens to the lining of the uterus when a Graafian follicle is developing in the ovary? Why is this necessary?
5 On which days of the menstrual cycle is fertilisation most likely to occur?
6 When ovulation has occurred, what happens to the Graafian follicle?
7 What is the function of the corpus luteum (yellow body)?
8 If fertilisation does not occur, what happens to the corpus luteum?
9 What is usually the first sign of a woman being pregnant?
10 What causes menstruation to start?

Assignment – Contraception

Contraception means the prevention of fertilisation and/or implantation of an egg. There are various methods available, many of which are very reliable. These are shown in the diagram.

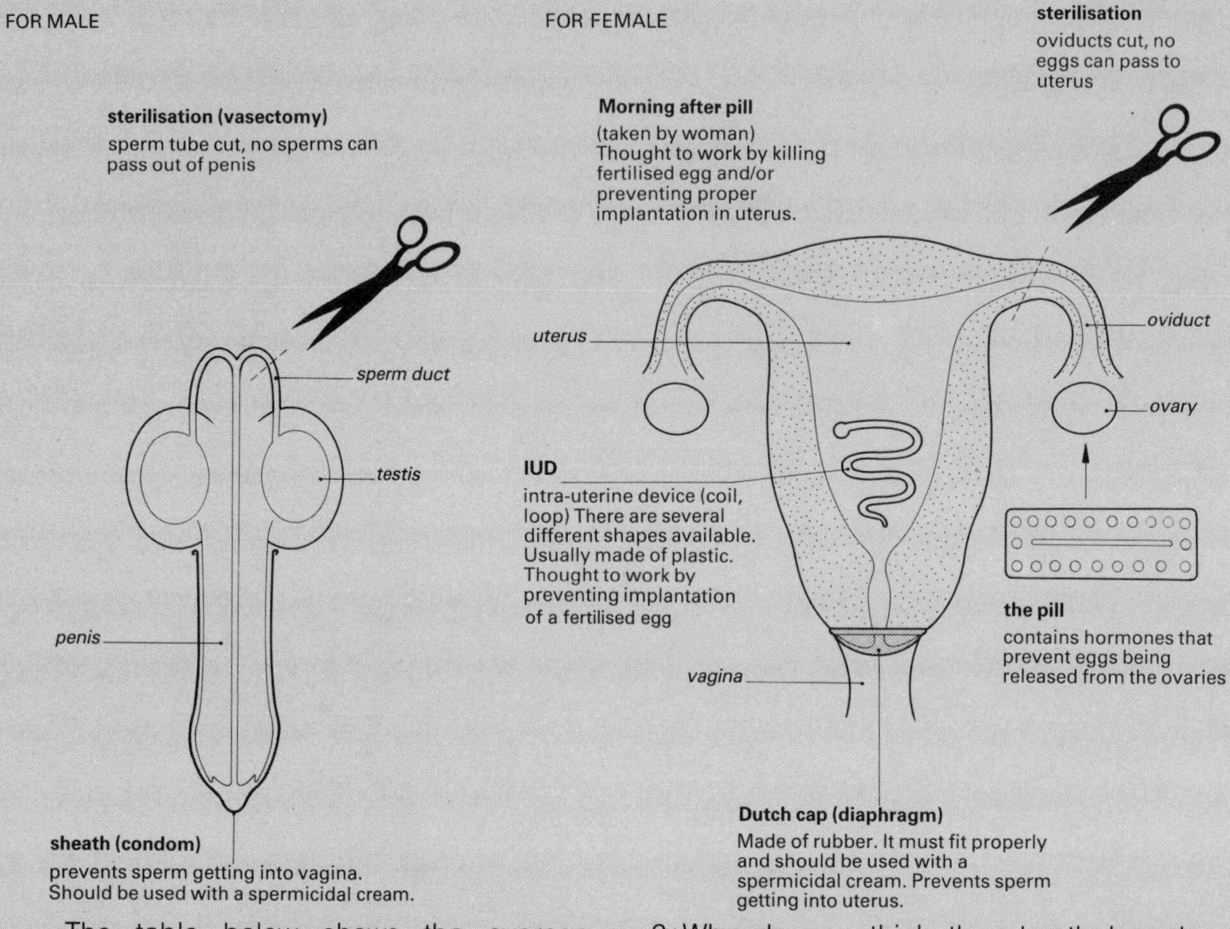

The table below shows the average pregnancy rate for some different contraceptives.

Contraceptive method	Pregnancy rate
Contraceptive pill	0.2
IUD	2.0
Sheath	12.5
Cap	2.2
Sterilisation	0.0

1 Which methods are the most reliable?
2 Describe how sterilisation operations in (a) a man, (b) a woman, work to prevent pregnancy.
3 Why do you think the sheath has the highest pregnancy rate as shown in the table above?
4 Do you know of other methods of contraception? If so describe them and how you think they work. Comment on their reliability.
5 Make a table like the one below to show where you can get the various different contraceptives (e.g. chemist, doctor, family planning clinic).

Type of contraceptive	Available from

Development of the embryo

A fertilised egg takes about a week to pass down the oviduct and reach the uterus. During this journey it divides many times, into two cells, four cells, eight cells etc. (see page 146, mitosis). When it gets to the uterus the zygote has developed into a ball of cells, and is now called an EMBRYO. The embryo sinks into the lining of the uterus and the woman is then pregnant.

The embryo begins to get nourishment from the mother as soon as it has been implanted in the uterus. It needs food and oxygen in order to grow. At first these come from the uterus wall, but gradually a special structure called the PLACENTA grows from the tissues of the embryo and the mother. This is shown in Fig 8.21.

The placenta is connected to the embryo by the *umbilical cord*. Blood vessels in the umbilical cord take oxygen and dissolved food to the embryo and remove carbon dioxide and other waste products. The embryo's blood does not mix with the mother's blood. The capillaries in the placenta are very close to those of the uterus wall, so substances can diffuse from one to the other. This can be seen in Fig. 8.21.

When the embryo has developed into a recognisable little human it is then called a FOETUS. It is surrounded by watery fluid called amniotic fluid which is enclosed in a bag called the *amnion*. This protects the baby from any damaging blows. The baby develops inside the mother for 40 weeks. This is called the GESTATION period. Other animals may have very different gestation periods; that is, the length of time between fertilisation and birth. For example, in a mouse or rat it is about 3 weeks, in a dog about 8 weeks, in a horse 48 weeks and in an elephant 96 weeks.

The diagrams in Fig. 8.22 show some stages of development of the foetus.

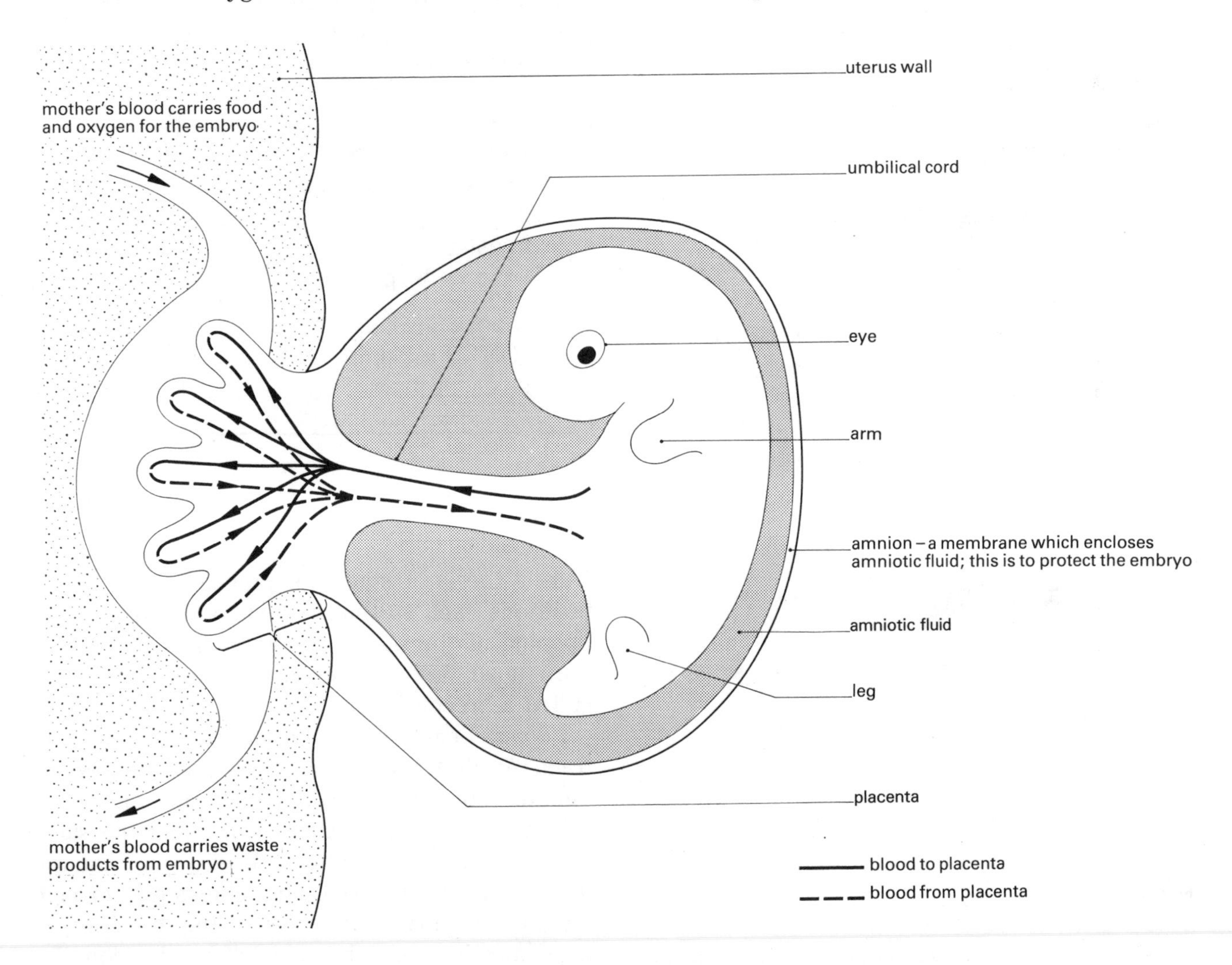

Fig. 8.21 An embryo and placenta in the uterus

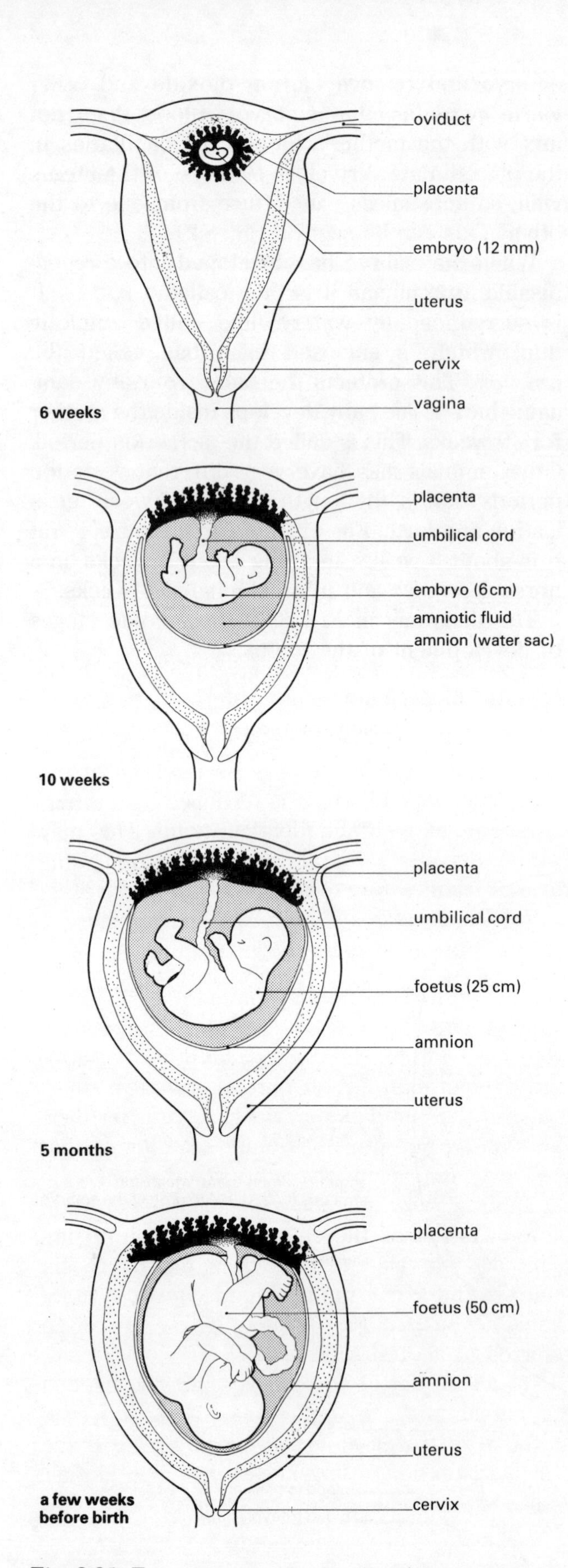

Fig. 8.22 Four stages in the development of a human in the uterus

Assignment – Do's and don'ts in pregnancy

Pregnancy is not an illness but mothers should take care of themselves during pregnancy for the sake of their own health and that of the baby they are expecting.

1 Make a table headed DO'S and DONT'S. Put each of the statements below into the appropriate column.

go for regular check-ups
eat enough food for two
sleep a lot more
listen to the doctor or midwife's advice
drink a lot of alcohol
take iron tablets if prescribed
wear tight clothing
take any pills not prescribed by a doctor
have vaccinations
have a sensible balanced diet
smoke
have sexual intercourse
attend ante-natal classes
eat more sugary foods
come into contact with rubella (German measles)
avoid constipation

2 Choose two of the statements in the DONT'S column and explain why you have put them there.

Birth

During the later stages of pregnancy the foetus positions itself so that its head is down near the opening of the uterus. About 95% of babies are born head first. If the feet are born first it is called a breech birth.

Labour begins when the uterus begins to contract. At first the contractions may be several minutes apart but they gradually get closer together and also get stronger. At about the same time as the contractions begin, the amnion breaks and the amniotic fluid comes out through the vagina. This is often called the 'breaking of the waters'; it may happen before the contractions start.

The contractions of the uterus gradually push the baby out of the vagina. As soon as the baby is born it starts to breathe, the umbilical cord is cut and has a clip attached, or it is tied. (The scar becomes a person's 'belly button' or navel.)

Soon after the birth the placenta comes away from the wall of the uterus, and it comes out through the vagina. This is often called the *afterbirth.* The average mass of a newborn baby is 3 kg and the placenta will probably have a mass of about 0.6 kg.

Twins

Usually a woman releases one egg from one of her ovaries each month and if this is fertilised it will start to develop into a baby. Sometimes two babies may develop at the same time. These are called *twins* and there are two types.

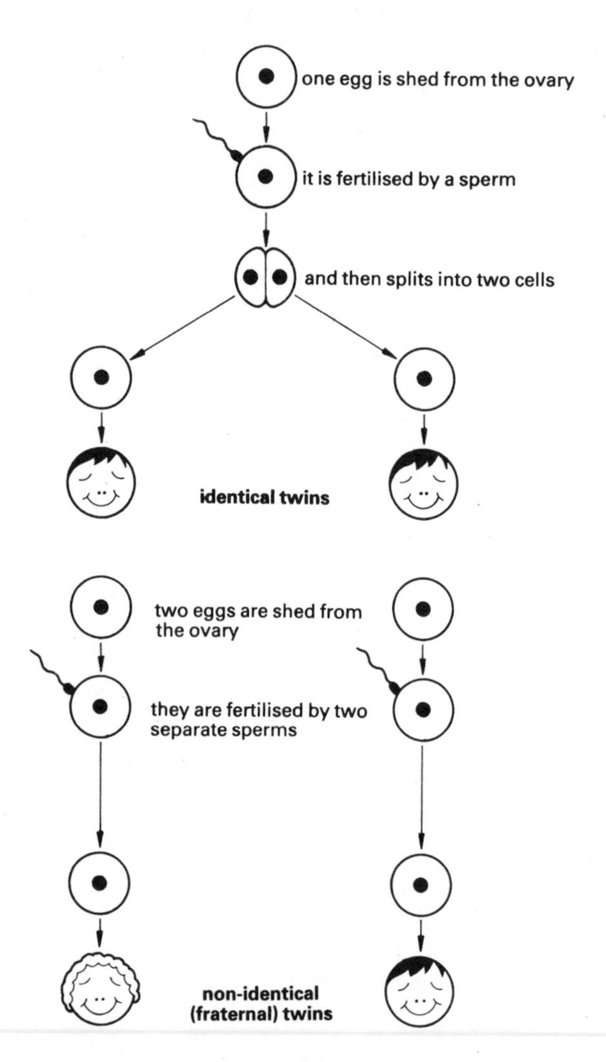

Fig. 8.23 How twins are formed

Identical twins. One egg is fertilised by one sperm in the usual way. The fertilised egg then splits into two cells which separate. Each of these cells develops into an embryo. The two cells both have the same genetic information (see page 150) so both will be the same sex and will develop into babies which look exactly alike (see Fig. 8.23).

Non-identical twins. Two eggs are sometimes shed from the ovaries at the same time and each may be fertilised by a different sperm. The babies may be of different sexes and will be no more alike than brothers and sisters normally are. They are sometimes called fraternal twins (see Fig. 8.23).

Multiple births

Sometimes a woman may give birth to three, four, five or even six babies. There are two main reasons why this may happen.

(a) Three or more eggs may be released at the same time, each of which is fertilised by a different sperm, as with non-identical twins. This may happen because a woman has been taking some form of fertility drug to help her to become pregnant. The drug may stimulate the ovaries to release more than one egg per month.

(b) If a woman has not been able to get pregnant some eggs may be removed, in a small operation, from her ovaries. These eggs may then be fertilised by sperms from her partner in a 'test tube'. Some of the embryos that develop after fertilisation will be put into the woman's uterus so that they may implant in the uterus and develop in the normal way for the rest of the gestation period. As not all the embryos are likely to implant and survive, some doctors put up to six embryos into the mother; consequently more than one baby might develop. These are often referred to as 'test tube babies'.

Not all the fertilised embryos may be needed for implantation, so some may be frozen and stored. The frozen embryos may be used if the first implantation is unsuccessful or if a mother wants another child a few years later.

Assignment – Ways to have a baby

Most mothers would prefer to have a natural birth if possible. However, giving birth can be quite painful and many mothers do use pain-killers to make the birth less painful. It is also possible to have a 'high-tech' birth. Monitors are used for measuring the contractions of the uterus and to check the baby's heartbeat; an episiotomy may be given; forceps may be used and in some cases a Caesarean section may be carried out.

An episiotomy is a cut made at the opening of the vagina birth canal to make more room for the baby's head. This will be done if it looks as if the mother's skin is going to tear or if forceps are used. The cut is stitched up after the birth.

Sometimes the delivery of the baby is helped by using forceps. These may be used if the baby is being born early and its head is very soft; they will protect the head from too much pressure. They may also be used if the last part of labour is taking a long time and the mother is very tired. Forceps are rather like large salad servers. They are put around the baby's head to protect it and are pulled gently as the mother tries to push the baby out.

A baby may be delivered through the abdomen instead of through the vagina. This is called a Caesarian section. This method is used if a vaginal delivery would be difficult or dangerous, for example, if the baby's head is too wide to go through the pelvis, if the baby is the wrong way up in the uterus (this is called the breech position) or if the baby needs to be delivered very quickly because its oxygen supply is running out. Usually a horizontal cut is made across the lower part of the tummy so that the scar will be hidden even when wearing a bikini.

A mixture of gas and air may be given as a form of pain relief. This is breathed in through a mask which the mother holds herself so she can control how much she uses. It may make her feel drowsy or sick.

Pethidine is the most commonly used pain-killer which is given by injection. This relaxes the muscles and may also make the mother feel drowsy. The pethidine may affect the baby so that it is very sleepy when it is born.

With an epidural anaesthetic it is possible to take away virtually all the pain. A doctor injects a local anaesthetic into the back very close to the nerves which affect the lower part of the body. The mother may not be aware of the contractions and may find it difficult to push at the right time.

A Caesarian may be carried out using an epidural so that the mother is awake while the operation is being carried out although she will feel no pain. Some hospitals encourage husbands to be present during the operation, and both husband and wife are protected from the sight of blood and cutting by a screen across her middle. However, many Caesarians are carried out using a general anaesthetic. This is the type used for other operations and if this is done the mother will be unconscious when the baby is delivered.

1 (a) What do you understand by the term 'natural birth'?
(b) Give three situations when it may be unwise for a woman to have a completelv natural birth.

2 Make a table with headings as shown below. Fill it in using the information above and anything else you can find out.

Painkiller	Advantages	Disadvantages

3 Give at least one reason why each of the following may be carried out:
(a) episiotomy;
(b) forceps delivery;
(c) Caesarean section.

4 What is the difference between a local anaesthetic and a general anaesthetic?

LIFE CYCLES

The sequence of changes that a living thing goes through from the moment of fertilisation until its death is termed its LIFE CYCLE.

During their life cycle most living things usually produce new individuals by asexual or sexual reproduction. During the life cycle both types of cell division, mitosis and meiosis, may occur (see page 146).

The life cycle of humans and most other animals is summarised in Fig. 8.24.

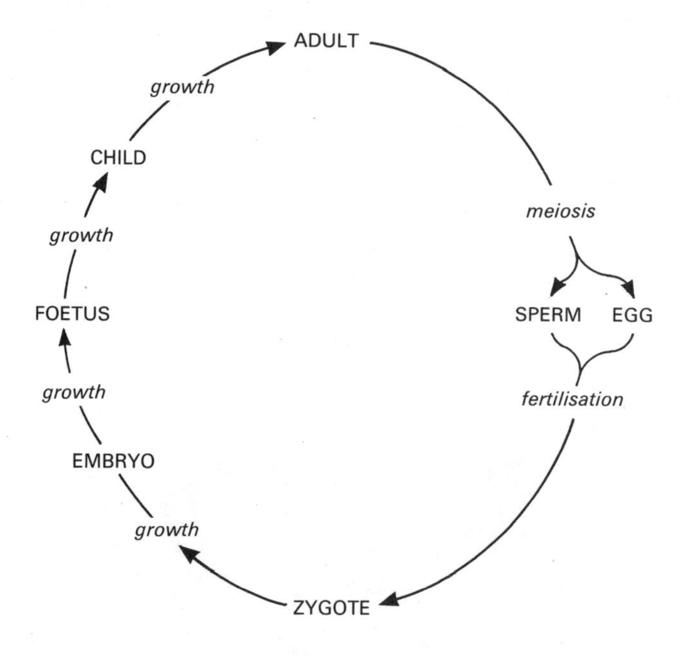

Fig. 8.24 Life cycle of human

The life cycle of flowering plants is summarised in Fig. 8.25.

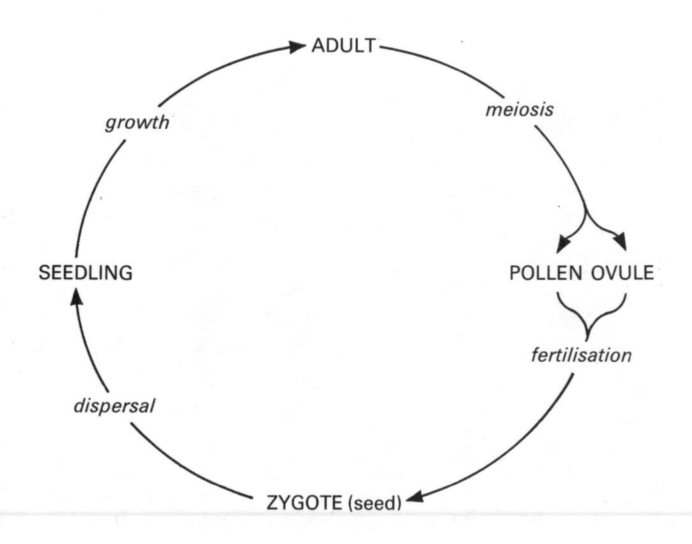

Fig. 8.25 Life cycle of a flowering plant

Some animals have a different type of life cycle. In some insects, and in amphibia, there are larval stages between the egg and the adult. The changes which take place in the body between the egg and the adult are termed *metamorphosis*. Examples are shown in Fig. 8.26.

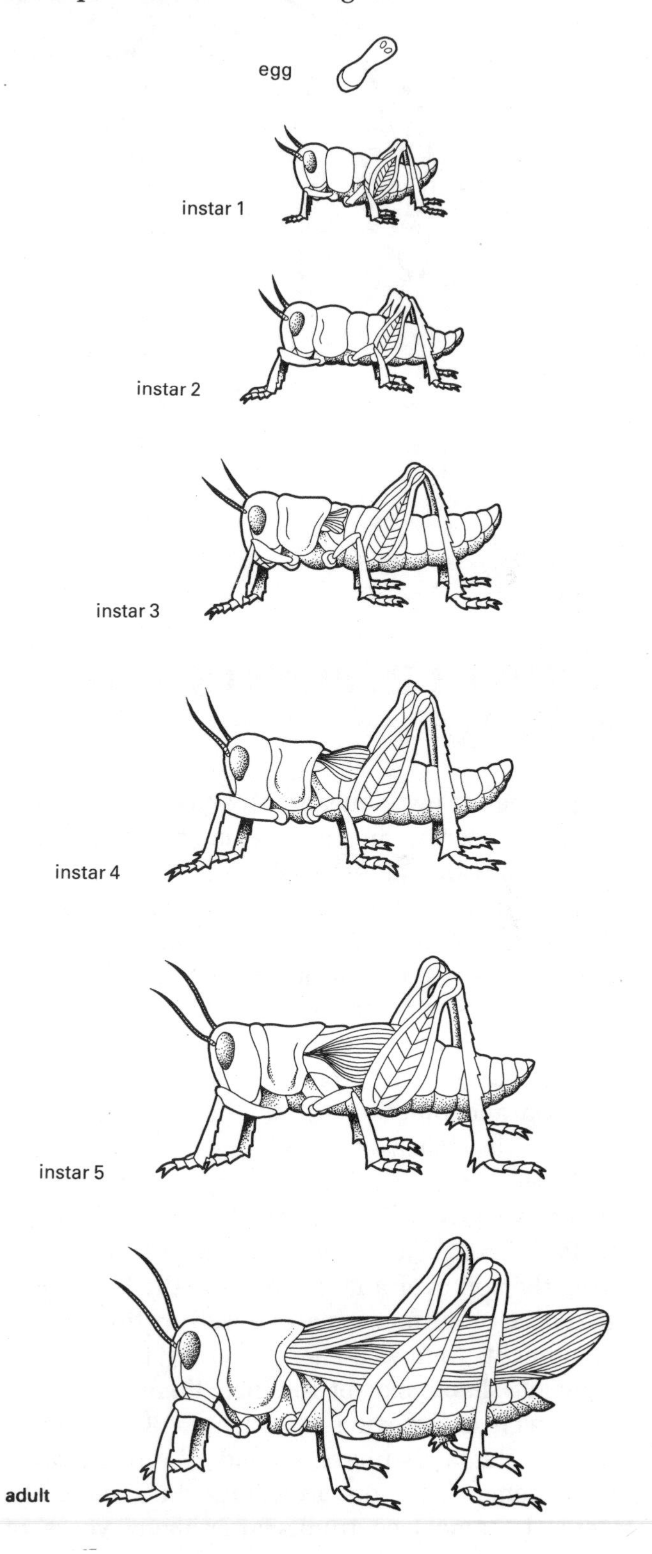

Fig. 8.26(a) Life cycle of a locust

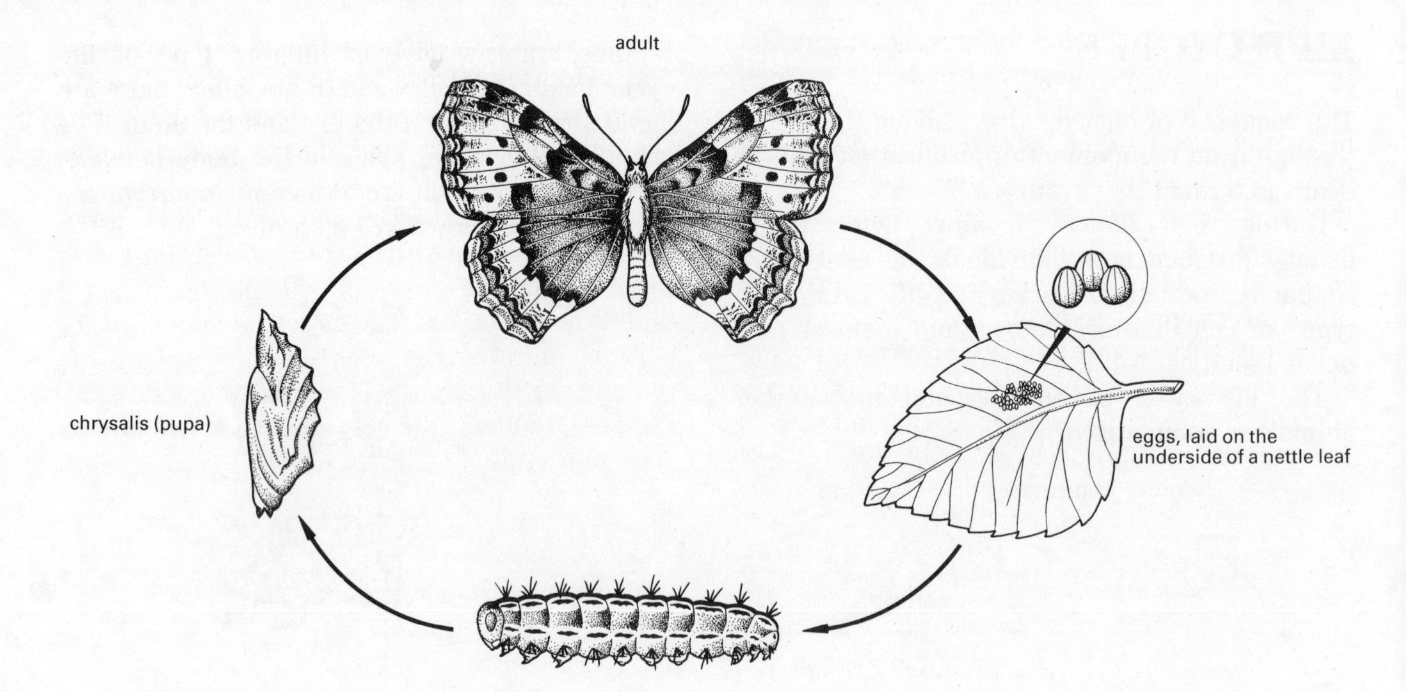

Fig. 8.26(b) Life cycle of a small tortoise shell butterfly

GROWTH IN PLANTS AND ANIMALS

What does growth mean? Most people use the term to describe an increase in the weight and size of an animal or a plant. All animals and plants can grow but the way they grow may be different, the resulting shapes being suited to their different types of nutrition.

Growth results from (a) cell division and (b) the growth of cells. The cells may then differentiate into different types according to their position and function within the living thing, as you saw in Unit 2.

Green plants grow at the tips of roots and shoots to give a branching shape with a large surface area. The roots can absorb nutrients from the soil over a large area and the leaves can give a large surface area exposed to the air and to sunlight. Animal bodies generally grow into a compact shape, except for their limbs.

As well as growing at the tips of roots and shoots, plants also have secondary growth when both stems and roots grow in girth. Eventually, stems become tree trunks after many years of growth.

Fig. 8.27 These boys are the same height – but have they both grown by the same amount?

Growth in cells, tissues, organs, organisms and populations can be measured in terms of fresh weight, dry weight, length or numbers and it can be plotted against time. Dry weight is usually the best measurement to take when measuring the growth rate of individuals, as it measures the exact amount of cell material present. Length is easier to measure but may disregard growth in other directions – see Fig. 8.27.

Growth curves to show the normal growth patterns can be drawn for various types of animal and plant. Some examples are shown in Fig. 8.28.

From Fig. 8.28 you can see that the growth pattern shown by mammals such as humans is continuous. They grow gradually in both mass and size. There are a couple of growth 'spurts' when growth is more rapid. These occur during the first few years and during puberty/adolescence.

In some insects, such as locusts, a very different pattern of growth can be seen. The insect increases in mass inside its cuticle (outer layer) until it no longer fits. The cuticle is then shed for a larger one that has been forming beneath the old one. Although the mass of the locust increases gradually the size does not, as can be seen in the graph (Fig. 8.28d).

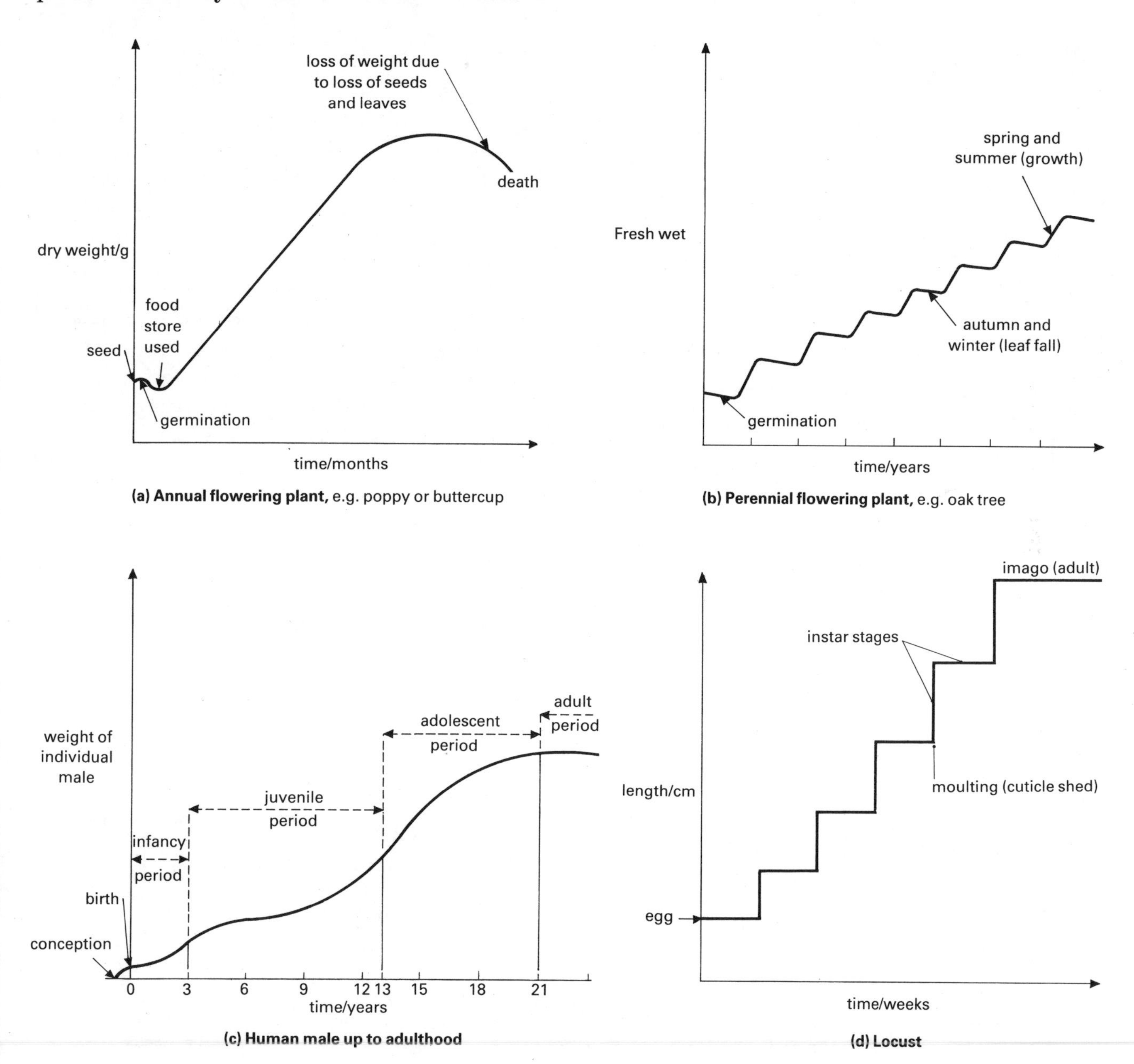

Fig. 8.28 Growth curves

Another difference in growth rates in humans is seen between different parts of the body. The body in general grows in spurts as mentioned above. The nervous system (brain, spinal cord, eyes, etc.) grows very rapidly from birth to six years when it reaches 90% of adult size. The reproductive organs remain small until puberty when they grow rapidly to adult size and become functional. These differences are illustrated in Fig. 8.29.

In humans the body proportions also change. This can be seen in Fig. 8.30.

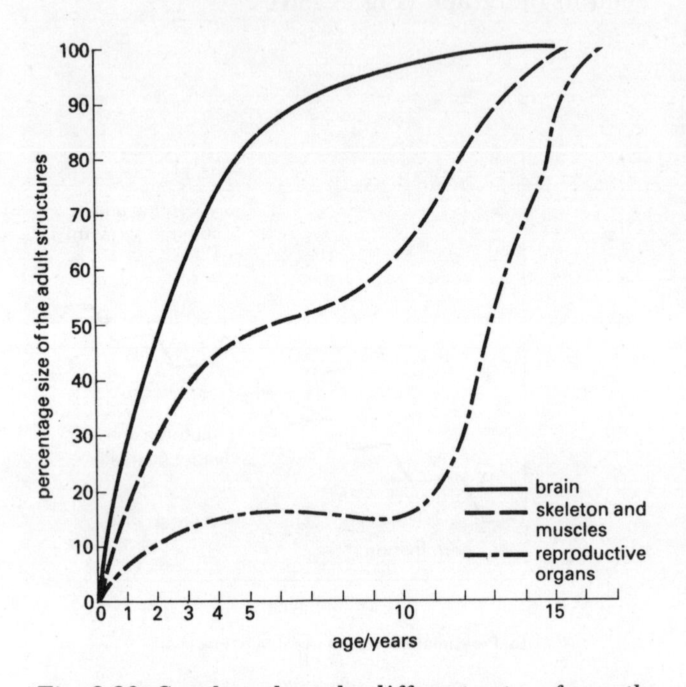

Fig. 8.29 Graph to show the different rates of growth of various organs in humans

Assignment – Growth

The table below shows the body mass of a boy and of a girl over a number of years.

Age	Body mass (kg) boy	Body mass (kg) girl
birth	4	4
5 years	23	18
10 years	39	32
11 years	42	42
12 years	47	50
15 years	68	56

1 Using the horizontal axis for age, plot graphs of these figures to show the changes in body mass for both the boy and the girl from birth to the age of 15 years.
2 How do the growth rates of the boy and girl differ?
3 What other methods, other than finding an increase in body mass, may be used to measure the growth of a young person? Suggest a disadvantage or inaccuracy which could be present in the methods you describe.

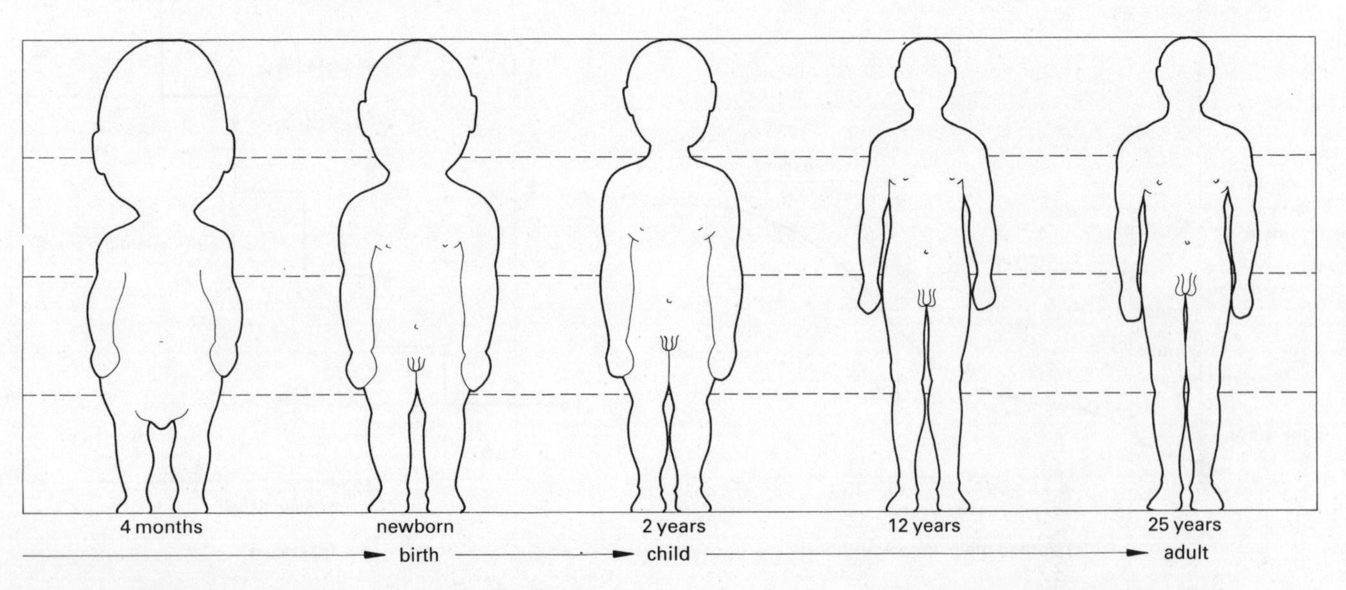

Fig. 8.30 Proportions of the male human body, from when it is a foetus before birth until it is an adult

Ageing in humans

Ageing occurs very slowly at first but speeds up later in life. Many body cells live only for a certain length of time, then they die and must be replaced with new cells (for example, skin cells and blood cells). As a person gets older the renewal of cells becomes slower and some cells, such as nerve cells, cannot be replaced at all.

An old person may not see as well, movements may be slower and the memory may be less accurate. The skin becomes thin, wrinkled and less elastic with age. Muscle power is at its maximum at about 20 years of age and can stay like that until about 40 when it begins to decline. The decline gets more rapid after the age of 70 years of age. Fractured bones mend more slowly, kidneys do not work so well. Old people should not eat too much fat or carbohydrate, as obesity reduces life expectancy. But a good balanced diet with plenty of vitamins and minerals is essential for the elderly to remain healthy.

Assignment – Cancer

Cells usually divide at a suitable rate for growth and for the repair of damaged or worn-out tissues. The process by which they divide is called mitosis (see page 146).

However some cells divide too quickly. These may be cancer cells. When cells divide too quickly they often produce a lump, called a *tumour*. There are two types of tumour.

(a) Benign tumours – these are slow growing lumps which do not destroy other tissues and which are non-cancerous.

(b) Malignant tumours – these grow fast and destroy other tissues. They are cancer tumours.

Normal cells may change into cancer cells. These changes may be caused by:

1. Chemicals – e.g. cigarette smoke and asbestos particles.
2. Radiation – X-rays, ultraviolet light and radiation from radioactive materials.
3. Viruses – some types have been found to cause cancer.

Many scientists are involved in research all over the world to try and find the exact way that some of the above cause the changes within cells. Until they find out exactly *how* cancer is caused they will not be able to prevent people from getting this illness, which is often fatal if not detected soon enough.

Cancers can start in various areas of the body. Some are easier to spot than others. For example, if a tumour, such as a lump in a woman's breast, is found early enough it can often be removed by surgery. Cancer in some parts of the body can be treated by giving the patient certain chemicals to take, and in other parts of the body where surgery is not possible it may be treated by giving a type of radiation treatment. Chemical treatment and radiation treatment are designed to try and kill all the cancerous cells so that they will not spread to other parts of the body.

If cancer is not detected soon enough, some of the cancer cells may break away from the main tumour and travel in the bloodstream to other parts of the body. If this happens, new cancerous growths may occur in other parts of the body and it is usually very difficult to treat all these secondary cancers.

1 How do cancer cells differ from normal cells?

2 (a) (i) If a man finds a lump, for example under his arm, what should he do?
(ii) Why is it important to do something quickly?

(b) If a small part of the lump is removed and it is examined under a microscope the technicians are able to see whether the cells are normal or not. The doctor will then tell the patient whether the tumour is malignant or benign. (i) What do these two terms mean? (ii) If the tumour is malignant what treatment would you expect the doctor to advise?

3 Find out what a mastectomy is.

Reproduction and growth

WHAT YOU SHOULD KNOW

1 New animals and plants may be produced by two methods, asexual reproduction and sexual reproduction.
2 Sexual reproduction involves the male and female sex cells (gametes) joining together.
3 Asexual reproduction does not involve gametes. There are various types of asexual reproduction: binary fission, budding, spores, vegetative propagation.
4 Flowers are the organs of sexual reproduction in plants. The stamen is the male organ; this produces pollen grains which are the male gametes. The ovary is the female organ; this produces ovules which contain the female gametes.
5 Pollination is the transfer of pollen from the anthers of one flower to the stigma of another. Pollination may be done by insects or by wind.
6 Fertilisation (in flowering plants) occurs when a pollen nucleus fuses with an ovule nucleus. After fertilisation the ovary grows to form a fruit and the ovules become seeds.
7 Seeds contain a food store which the embryo uses when it starts to grow, that is, to germinate.
8 Germination is influenced by temperature and by the amount of water and oxygen available.
9 In humans and most other animals the male gametes (sperms) are produced in the testes. The female gametes (eggs) are produced in the ovaries. (Gametes are formed by a special type of cell division called meiosis).
10 The male and female gametes are brought together during sexual intercourse.
11 Fertilisation in humans happens when a sperm enters an egg and the sperm and egg nuclei fuse.
12 The fertilised egg (zygote) divides into many cells and becomes embedded in the uterus lining where it grows into an embryo.
13 The embryo gets food and oxygen from the mother, and gets rid of waste products via the mother.
14 When animals and plants grow they increase their size and weight and their structures become more complicated.
15 Most living things grow from a single cell. Growth takes place as a result of cell division (mitosis), cell enlargement and cell differentiation.
16 Not all parts of living things grow at the same rate. Some parts of the body may change their proportions during growth.
17 Growth patterns vary in different types of animal. For example, in most mammals growth is gradual but in insects such as locusts there are distinct stages.

QUESTIONS

1 Explain what is meant by the following terms: (a) gamete, (b) zygote, (c) fertilisation.
2 Make a table using the headings below

Method of asexual reproduction	Plant or animal that reproduces in this way

Fill in the table with six methods of asexual reproduction and at least two examples for each.
3 Why are perennating organs also described as storage organs?
4 What are the advantages to a gardener of propagating a plant from cuttings?
5 Match up the part of the flower in column A with what it does in column B. Write your answers as complete sentences.

A	B
Sepals	is the part of the carpel which contains the ovule
Petals	supports the anther
The ovary	holds up the stigma
The stigma	is the part of the stamen which makes the pollen
The style	may help to attract insects
The filament	protect the buds
The anther	is the part of the carpel which receives the pollen

6 Explain the difference between pollination and fertilisation in a flowering plant.

7 What conditions are needed for most seeds to germinate?

8 Explain the difference between each of the following pairs:
   (a) egg and sperm;
   (b) testes and ovary;
   (c) sperm duct and urethra (in males);
   (d) penis and clitoris;
   (e) fertilisation and sexual intercourse;
   (f) erection and ejaculation.

9 What methods of contraception would you advise for the following people?
   (a) A couple who have four children and do not want any more.
   (b) A young couple who want to wait a few years before starting a family.
   (c) A young woman who has not got a steady lover but wants to be safe in case she meets someone and wants to make love.

   Give reasons for your answers in each case.

10 Find out what the following terms mean: abortion, miscarriage, menstruation, after-birth, growth spurt, puberty, menopause, surrogate mother.

11 A married woman finds she is unable to get pregnant although she and her husband are having intercourse regularly and are not using any method of contraception. Suggest at least THREE reasons for their lack of success and in each case suggest how they may be helped.

12 A scientist sows a large number of seeds. He wants to measure the rate of growth of the seedlings. There are three methods which he could use.
   (a) Measure the height of 50 plants every day and take an average.
   (b) Dig up 10 plants every day and find the mass.
   (c) Dig up 10 plants every day and dry them by putting in a hot oven to remove the water content. Then find the dry mass.

   What are the advantages and disadvantages off the three methods?

13 Make a table to compare the changes that take place in adolescence and in old age. Try to think of at least five differences: one has been done for you.

Adolescence	Old age
A considerable increase in height	A slight decrease in height

Reproduction, Growth and Development

POPULATIONS AND COMMUNITIES

In the first part of this unit we have looked at the way that individual animals and plants grow and reproduce.

What happens when more than one animal or plant of a particular type/species are together? Do they grow and reproduce at the same rate as they would if alone? Do they affect each other? Do other things like the weather, or their surroundings, affect the rate at which they grow and reproduce?

POPULATIONS

Before we look at the way in which a population grows, it is important to know exactly what the word population means.

A POPULATION is the number of a particular species of plant or animal found in a well defined area. Often when people talk about populations they are referring to humans. However we can use the term for other animals and for plants too. Some examples are the population of rabbits in a wood or of daisies in a lawn or of goldfish in a pond.

Daisies (left) and dandelions (right) – see Investigation 8.5 opposite

Investigation 8.5
Finding the populations of daisies and dandelions in a lawn

Daisies and dandelions are both very common weed plants which are often found in lawns. Look at the photographs on p. 124).

Collect

1 metre ruler or a tape measure marked in metres
large ball of string
wooden stakes or meat skewers

What to do

1 Find a suitable lawn in the school grounds or at home. A part of the school playing fields may be suitable: check with your teacher. The total area should not be too big (about half the size of a tennis court or less) and it should be square or rectangular.
2 The whole class should divide the lawn into strips, each 1 metre wide. Use a pair of stakes or skewers with string between to divide the lawn into strips.
3 The class should divide into pairs. Each pair should count the number of daisy plants in their strip.
4 The totals for each strip must be added together. This will give the population of daisies for the whole lawn.
5 Repeat steps 3 and 4, but this time count the number of dandelion plants.
6 You could also find the total area of your lawn. Multiply the length by the breadth (both in metres) to find the area (m^2).

Questions

1 What was the population of daisies and dandelions in the lawn you studied?
2 Would you expect the populations to be the same in all lawns? If not, explain why.

Investigation 8.6
Finding the population of woodlice and snails in a given area

It is possible to find the population of animals in an area by a method called the capture, recapture method.

Collect

containers for putting animals in – specimen tubes, plastic boxes or jam jars
green, water-based, non-toxic powder paint
small paintbrush

What to do

1 Choose a suitable area, e.g. a greenhouse, a small garden, a patio.
2 Collect all the woodlice you can find in the whole area. Put them carefully into a container until you can find no more. Look under pots, leaves and stones to find as many as possible.
3 Count the woodlice and write down the total.
4 Mark each woodlouse on its back with a *small* painted dot.
5 Release all the marked woodlice in the area where you collected them.
6 The next day catch and recount as many woodlice as you can. Write down
(a) the total number caught,
(b) the number of spotted woodlice.
You could make a table for your results like the one in the next assignment.
Release all the woodlice.
7 Work out the total population using the following formula:

$$\frac{\text{total caught day 1} \times \text{total caught day 2}}{\text{spotted animals caught on day 2}}$$

e.g. $\frac{60 \times 40}{24} = \frac{2400}{24} = 100$ (total population)

8 Repeat 2 to 7, but this time look for snails.

You may be surprised at how few spotted woodlice you found on the second day in Investigation 8.6. This is because woodlice like to live in damp dark places and are also well camouflaged. This should show you that just counting the numbers of animals on one day would not always be a good method of finding the population of animals. Of course it may work for large animals which are easily seen, such as elephants or even humans, but often estimates are made by taking samples from certain areas. Population surveys and censuses, which are concerned with finding the human population totals in a country, rely on the honesty of householders filling in forms and so are not 100% reliable, even though there are fines in Great Britain for giving false information. From Investigations 8.5 and 8.6 you will probably have realised that it is easier to find the population of plants than of animals, because they do not move!!

Assignment – The capture, recapture method of finding animal populations

Below is part of a page from a girl's notebook showing her results for part of Investigation 8.6. Have a look at it and then answer the questions.

Species	Day 1	Day 2	Spotty Day 2
Centipedes	𝍸 𝍸 𝍸	𝍸 𝍸	𝍸
Millipedes	𝍸 𝍸 I	𝍸 𝍸 𝍸 II	𝍸 II
Woodlice (type A)	𝍸 I	IIII	I
Woodlice (type B)	𝍸 𝍸 III	𝍸 III	𝍸 I
Spiders	IIII	𝍸 I	II
TOTALS	49	45	21

1 From the results calculate the total population of centipedes, and of spiders.
2 The last line gives *totals*. Does this show the total population of all the animals in the area? Explain your answer.

The growth of populations

How do populations grow in size? Are there any patterns of growth such as we saw with individuals (page 119)? Obviously this will depend on such things as the type of animal or plant we are talking about, its type of life cycle and how many offspring it usually has. External factors such as the weather and the food available are also important.

Populations grow by increasing in numbers. This happens when there are more births than deaths. Populations can also increase by immigration, which means organisms arriving from another area.

Populations of some small living things, such as bacteria, grow very quickly. In some species each bacterium can divide and form two bacteria in an hour (Fig. 8.31).

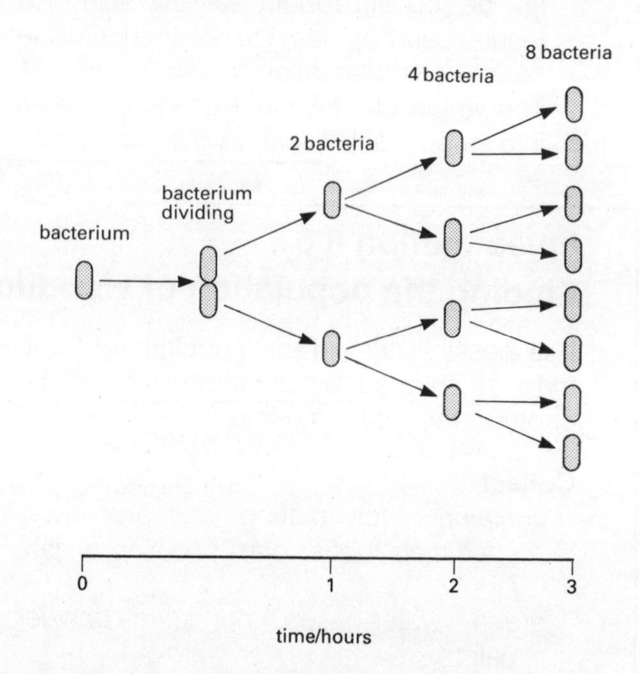

Fig. 8.31 If each bacterium divides every hour, after 3 hours 1 bacterium will have 8 offspring

If you started with a population of 100 bacteria in a closed container where they have plenty of food and all other conditions are ideal, after one hour you would have 200 bacteria, after two hours you would have 400, after three hours there would be 800 and so on. The total number doubles at regular intervals of time. This type of increase is called EXPONENTIAL. All populations grow in this way. Whether we consider bacteria, flies, mice, rabbits or humans, they all show exponential growth of their populations.

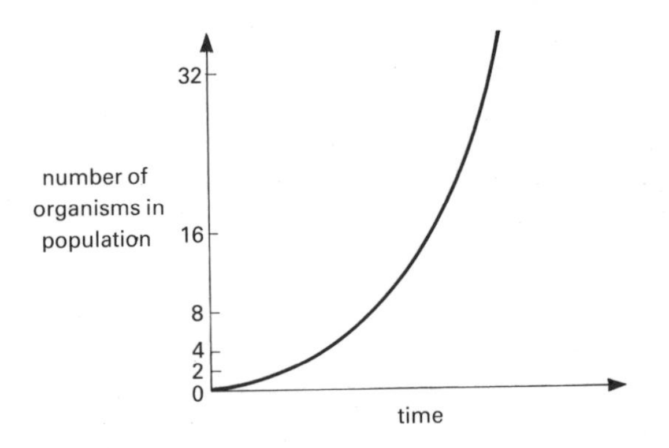

Fig. 8.32 Exponential growth

If you started a population of mice with one pair, one male and one female, you would get similar results. A pair of mice on average produces a litter of six young every three weeks. The litter of six would probably have on average three males and three females. Mice are usually about four weeks old before they can mate and have young. Most mice live for about three years.

Table 8.3 shows the results you might expect to get during one year, if all conditions for the mice were ideal.

Table 8.3

Time/weeks	0	7	14	21	28	35	42	49
Population of mice	2	6	18	48	132	360	900	2400

Assignment – Breeding mice

1. Plot a graph using the data from Table 8.3.
2. Compare your graph with those in Fig. 8.32. Does it show a similar pattern? Does it differ at all?
3. If you started with one pair of healthy mice do you think that you would really end up with 2400 after a year? Can you think of anything that might limit the growth of the population? Would it make a difference if the mice were kept in the laboratory as opposed to being in the wild?
4. Make a list of things you think may affect the growth of the population of mice
   (a) in a laboratory, e.g. at your school;
   (b) in the wild, e.g. in your house and garden.

Limiting factors

From the data in Table 8.3 and the graph in Fig. 8.32 you should be able to notice that the more living things there are the more quickly the population grows. If conditions are good and nothing happens to stop the growth of a population, the exponential growth pattern would continue. This would result in a huge and continuous increase in numbers. But in real populations many factors usually check the rate of population growth so that it does not go on rising so quickly.

The following may all limit the size of a population:

(a) food and water supplies;
(b) oxygen supplies;
(c) space;
(d) build-up of waste products which produce toxic (poisonous) substances;
(e) competition between members of the same species for food, space and light, and competition between members of different species for these resources;
(f) predators;
(g) shelter: this may be from predators or from harsh effects of the environment, e.g. heat, cold, waves;
(h) disease.

Assignment – Factors affecting populations

Below is a list of factors that affect the rate of growth of a population:

good food supply	plentiful pure water
diseases	little disease
lack of food	only a few predators
many predators	lack of space
no shelter	

1. Which factors would make it impossible for a population to grow?
2. Which factors may make the population decrease?
3. What other factors may affect the survival of plants and animals? Give specific examples to illustrate your answer.
4. Look at Table 8.3. How may each of the factors listed above change the pattern of population growth for the mice?

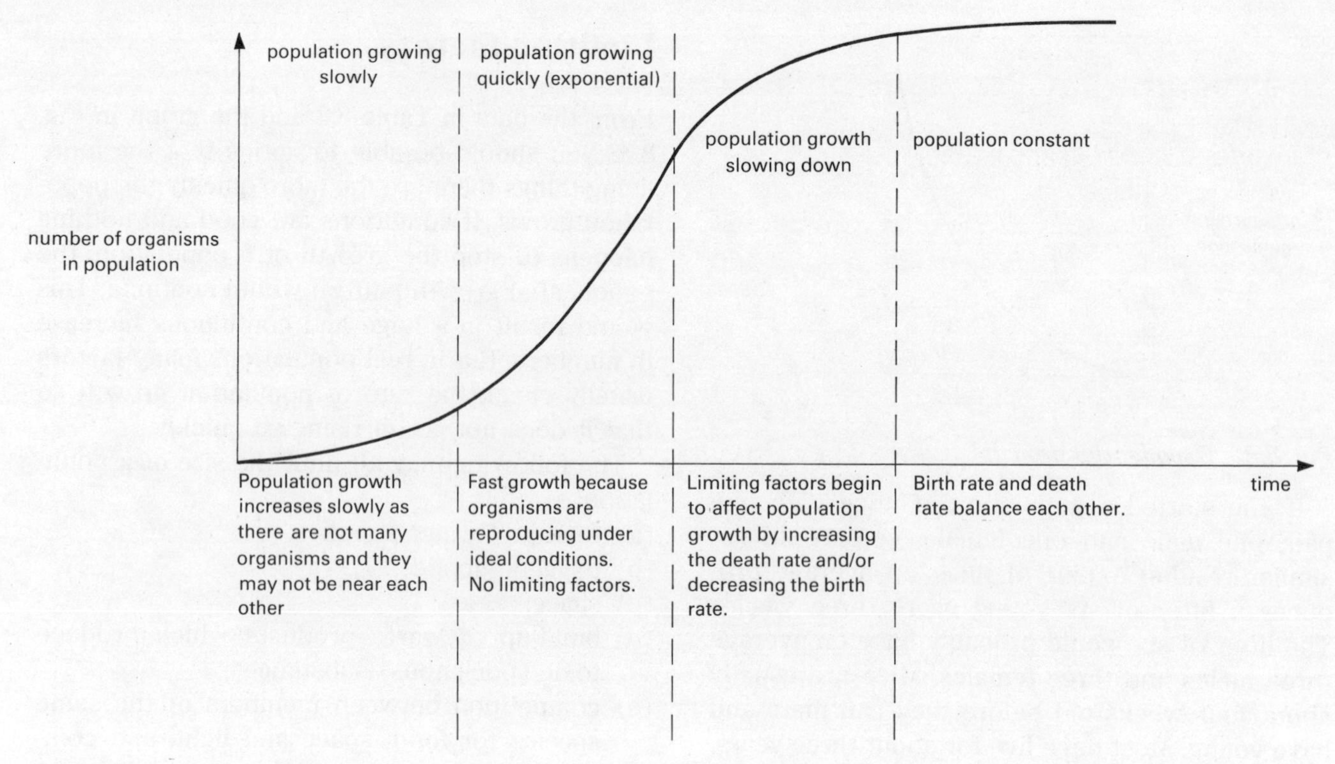

Fig. 8.33 Population growth

Figure 8.33 shows a generalised graph of population growth that takes these factors into account.

> **Assignment – Populations**
>
> Think of a list of factors that would affect the following:
> (a) the growth of a population of bacteria in a petri dish;
> (b) the population of daisies found in a lawn;
> (c) the population of humans in your town.

Changes to patterns of population growth

We have seen that if there are no limiting factors a population will grow exponentially. This rarely happens for long, as the growth rate will slow down as one or more limiting factors begin to affect the population. Some of these factors are slow to act, for example lack of space, but occasionally they may alter the pattern of population growth very rapidly; for example, drought, outbreaks of serious diseases, wars and natural disasters such as earthquakes and hurricanes.

Eventually most populations reach a stage where they remain fairly stable in numbers. For this to happen the birth rate of the population per year must balance the death rate per year. But even when a population reaches a stable state it may fluctuate slightly due to the effect of a particular limiting factor. This is often seen in predator/prey relationships and is also sometimes caused by seasonal effects.

Predators usually prevent populations getting too large but competition is also important. The main things that animals compete for are food, territory or a place to make a home, such as a bird's nest. Plants compete for light and also for water and nutrients from the soil. Without light and water they cannot photosynthesise and make new tissues (see page 203) so they will not grow and reproduce.

Control of populations by man

Humans can often control the size of a population if they want to. For example, we kill many pests with chemicals called *pesticides*. As many of the pests are insects, the chemicals are often referred to as *insecticides*. Pesticides may kill

Assignment – Predator–prey relationships

The graph below shows the interaction between populations of a greenhouse whitefly and a minute wasp which is a parasite on the whitefly.

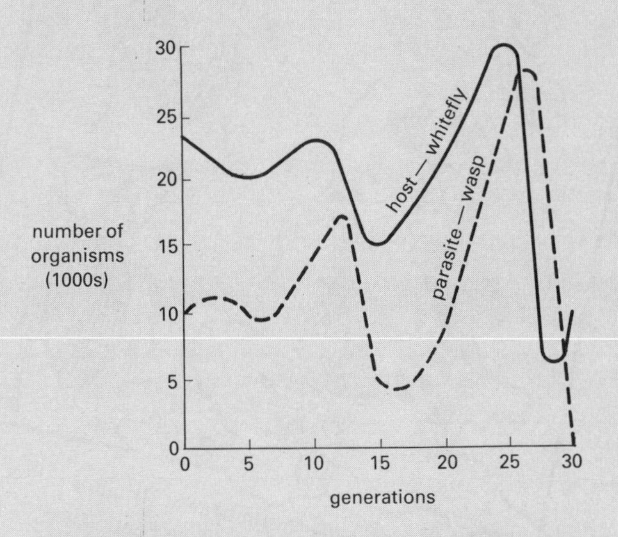

1 How many whitefly are there after (a)5 (b)15 (c)25 (d)30 generations?
2 How many wasps are there after (a)5 (b)15 (c)25 (d)30 generations?
3 What is the food of (a)the whitefly (b)the wasp?
4 When the number of whiteflies increases the number of wasps increases soon afterwards. Why does this happen?
5 Between 10 and 15 generations the number of whiteflies drops rapidly. What is the cause of this fall?
6 When the number of whiteflies falls so does the number of wasps. Explain why this happens.
7 This interaction has been used in the biological control of whiteflies which are pests in greenhouses where crops such as tomatoes and cucumbers are grown.
   (a) Find out how the whitefly harms these plants.
   (b) What advantages are there in using the parasitic wasps to kill whiteflies instead of using chemicals?
   (c) What would happen to the numbers of whiteflies and wasps in an empty greenhouse?
8 (a) What happens to the wasps at 30 generations?
   (b) Suggest three causes for this.

The pattern of cycles of abundance and rarity like those shown by the whitefly and wasp occur only when a predator or a parasite feeds on only one prey or host species. The cycles will not occur if the predator or parasite changes to another food source when its usual prey or host species becomes scarce.

9 If, after 25 generations when the population of whiteflies began to fall, the wasps started to parasitise another type of insect, what would you expect to happen to the number of wasps?

harmless organisms, including the pests' natural enemies, so it is not always the best method of getting rid of pests. Many weeds which compete with crop plants are killed by using chemical *herbicides* and many diseases which affect crops are prevented by spraying the crops with *fungicides* and other chemicals.

Apart from using chemicals, we can also reduce the population of some pests by introducing an organism which will kill the pest. This is called BIOLOGICAL CONTROL. Two well known examples of this have been used in Australia. The prickly pear cactus became a nuisance, thriving too well on farmland. Moth larvae (caterpillars) which feed on the cactus were introduced. As the moth larvae had plenty of food, they rapidly increased in numbers and the spread of the prickly pear was quickly brought under control. Another example was the introduction of the disease myxomatosis to kill the rabbits which had become so numerous that they had become a real pest. There were not enough predators to keep the population of rabbits under control. Myxomatosis killed many rabbits but it does not affect other animals.

Another example of biological control used by gardeners and horticulturalists in this country is the use of ladybirds and other predators in greenhouses to eat greenflies (aphids) and other similar plant pests. (See the assignment above.)

Assignment – Biological control

Read the newspaper article below and then answer the questions.

1 What species of tree is being attacked by the harmful beetles?
2 (a) Which of the beetles mentioned is harmful to the trees?
   (b) Explain how it harms the trees.
   (c) How did this harmful beetle get to Britain?
3 Describe how the scientists hope that this example of biological control will work to get rid of the pests.
4 The Forestry Commission has stopped the movement of pines, including large Christmas trees. How may this help to halt the damage being done to the trees?

Can the Belgian beetle (bottom) save Britain's spruce forests (shaded areas) from the bark beetle (top)?

Pines fight for life

by ROBIN McKIE
Science Correspondent

DEEP in the woods today, two microscopic armies are locked in mortal combat. Their struggle is being watched anxiously by scientists: the outcome will determine the future of our pine forests.

One one side of the woodland battlefield are marauding hordes of the Great Spruce Bark beetles, *Dendroctonus micans*. These have been destroying spruce trees across Britain for 11 years after wreaking widespread destruction in north European forests. Already acres of woodland in Shropshire and Wales have been infected.

On the other side are the battalions of Belgian beetles, specially imported in a bid to save the spruce.

Biologists are hoping the Belgian visitor, *Rhizophagus grandis*, will be the victor in this 'battle of the bugs.' If not, there could be serious destruction of Britain's main source of softwood, the Norway and Sitka spruce, which are vulnerable to bark beetle attacks.

'We are very worried about the spruce bark beetle. It has already done a great deal of damage and must be halted as quickly as possible.' said Dr Gordon Campbell, plant health officer for the Forestry Commission.

He added that the commission had brought in other measures – such as orders stopping movements of pines, including large Christmas trees.

The introduction here of the Belgian beetle is a more unusual measure.

'We have been doing this very carefully, but have already released 32 000 Belgian beetles at 960 different sites,' explained Mr Colin King, of the Forestry Commission's research centre at Ludlow, Shropshire.

The Belgian beetles, the natural predators of Great Spruce Bark beetles, were originally bred from samples brought from Belgium. 'We have high hopes but will have to wait to see how well they work,' added Dr Campbell.

Great Spruce Bark beetles were probably accidentally introduced into Britain in cheap timber imports.

They lay eggs under the bark of a spruce tree. On hatching, their larvae slowly move round the trunk eating the layers of cells that carry nutrients between roots and leaves. If left unchecked, the tree will die.

The Belgian beetle also lays eggs under spruce bark. However, its larvae eat only those of the bark beetle. As it does no other harm, it safely removes the danger ot the pest.

'It sounds very promising, but really it is too early to say if this is definitely going to work,' said Dr Campbell.

'For one thing, we do not know how well the Belgian beetles will react to a cooler climate in Britain.'

Much depends on the outcome of the trials: a fifth of all British timber is provided by Norway and Sitka spruce trees. As Britain is trying to cut down on foreign timber imports, widespread destruction of these two trees would be a serious blow.

Observer, 11 Nov. 1984

Meanings of words and population change

BIRTH RATE – number born per 1000 per year
DEATH RATE – number who die per 1000 per year
INFANT MORTALITY RATE – number of young children who die per 1000 per year
IMMIGRATION RATE – number who come into a country per 1000 per year
EMIGRATION RATE – number who leave a country per 1000 per year

If the birth rate is the same as the death rate the population remains unchanged.
If the birth rate is greater than the death rate the population increases.
If the death rate is greater than the birth rate the population decreases.

Within a country there are also other factors such as large scale immigration and emigration which may affect the total population.

birth rate – death rate = population change

(birth rate + immigration) – (death rate + emigration) = population change

Humans seem to have more control than many other organisms over their own population size. Many of the limiting factors that apply to other populations are not so relevant. There are no predators, the problems of infectious diseases have been considerably reduced in the last century and although food supply is a problem in some parts of the world it is not in others.

The average life span of humans has increased in recent years with more control over the death rate. Better hygiene, good diet and various forms of medical treatment have enabled people in many parts of the world to live longer. As the birth rate has also been increasing at the same time (possibly because there are more healthy women able to reproduce) and the infant mortality rate has decreased, the world's population has been rising dramatically.

Population growth varies from country to country. In the world as a whole, total population growth is still exponential. Many more people are born each year than die (see Fig. 8.34.)

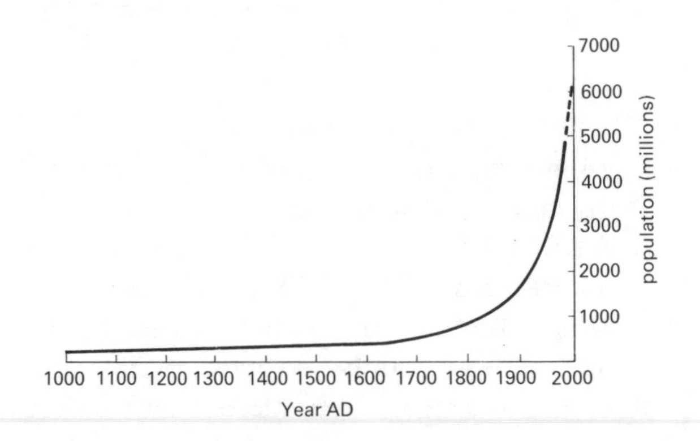

Fig. 8.34 World population from 10 000 years ago

The human population may soon run out of food, raw materials and space if it continues to increase at the present rate. The problem is more acute in some countries, such as China, than in Britain. Many predictions about this have been made, starting with Malthus in the early nineteenth century. (See Fig. 8.37.)

In some less developed countries life expectancy is still only about 40–45 years. In many countries diseases of malnutrition such as kwashiorkor cause the deaths of thousands of children. Other diseases such as cholera, typhoid, sleeping sickness and malaria cause many early deaths.

Fig. 8.35 This child is suffering from kwashiorkor

In Europe, the USA and some other developed countries, vaccinations against fatal diseases, good hygiene of water supplies and disposal of sewage together with advances in medicine have saved many lives, so that the death rate has fallen and life expectancy has risen. See Fig. 8.36.

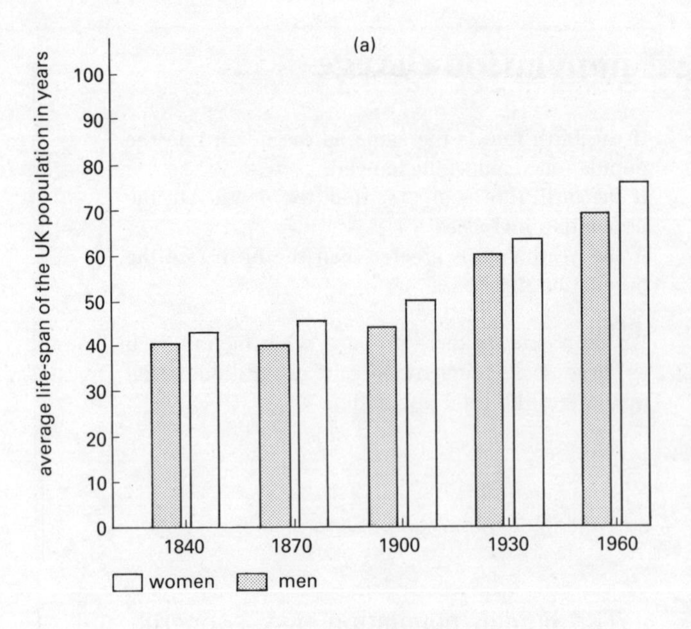

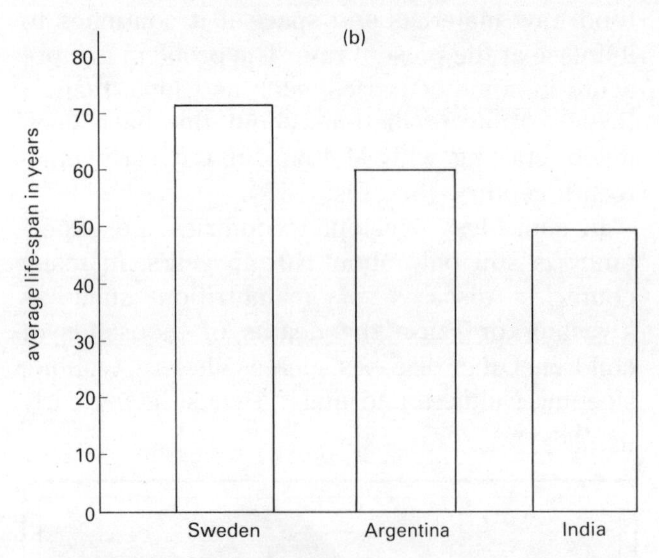

Fig. 8.36 Average life spans in (a) the UK at different times, (b) Sweden, Argentina and India in 1979

Human population control

Humans have continually increased the amount of food produced but are unable to keep pace with the birth rate. There are unlikely to be any natural processes that will limit the growth of the human population and so it can only be checked by people's own efforts.

What is the solution? Many people think that the only solution is to prevent the birth of too many babies. The introduction of reliable methods of birth control (see page 112) has meant that the birth rate can now be reduced.

Assignment – Birth rate and death rate

Study the graphs, which show birth rates and death rates in two countries: A, a Western European country and B, a South American country.

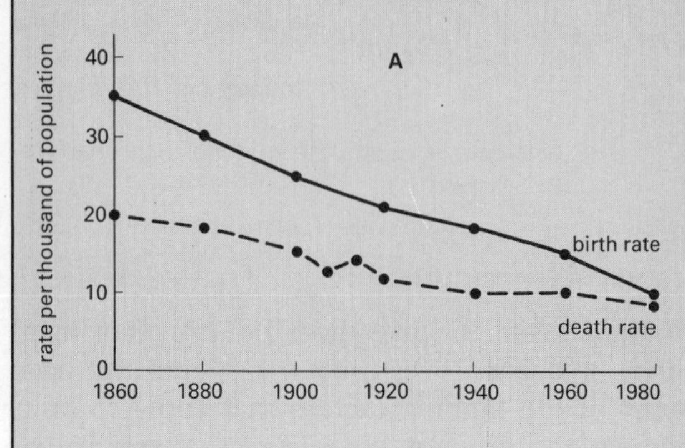

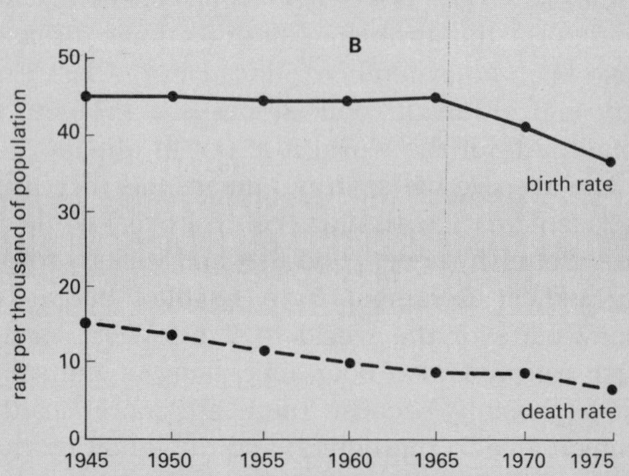

1 What was the death rate in A (a) in 1900, (b) in 1960?
2 What was the birth rate in A (a) in 1900, (b) in 1960?
3 Was the population of A increasing or decreasing (a) in 1900, (b) in 1960?
4 What was the death rate in B (a) in 1900, (b) in 1960?
5 What was the birth rate in B (a) in 1900, (b) in 1960?
6 Was the population of B increasing or decreasing (a) in 1900, (b) in 1960?
7 What changes have there been in the birth rate in country A since 1960?
8 What changes have there been in the death rate in country B since 1945?

Assignment – Composition of population and life expectancy

The structure or composition of a population may be shown in an age–sex pyramid like the one shown here.

1 How many people are under 20 years old in A and in B?
2 How many people are over 60 years of age in A and in B?
3 How do the numbers of males and females in the populations differ from each other in A and in B?
4 Which population do you think has the greater life expectancy? Give reasons for your answer.
5 Suggest a country which may have a population structure such as that shown in A.
6 Suggest a country which may have a population structure such as that shown in B.

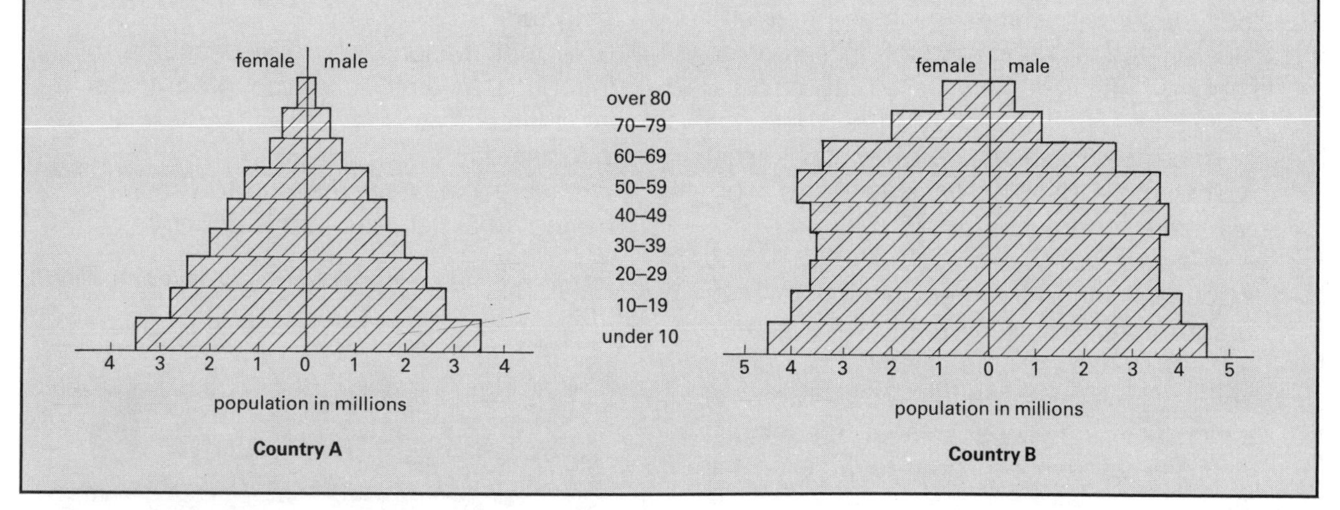

In Europe, the birth rate has fallen so that in some countries such as Sweden the population is not increasing. But will this happen all over the world? In some countries governments have been trying to educate the people about birth control and have made contraceptives and vasectomies freely available. But there is often resistance from people on cultural or religious grounds. Many uneducated people do not want to have just one or two children and do not understand the problems of the future which an ever increasing population will bring.

In the early nineteenth century Malthus predicted that unchecked population growth would outstrip food production. His predictions are shown in Fig. 8.37.

Overpopulation also causes more pollution and natural resources such as coal and oil will be exhausted more quickly. (See Unit 10.)

Look at the graph in Fig. 8.37. What do you think would happen at point X?

So far this topic has dealt with how populations of organisms grow and the factors which affect their growth. The next part of the topic will deal in more detail with how populations interact with each other when they live together and also how they interact with their environment.

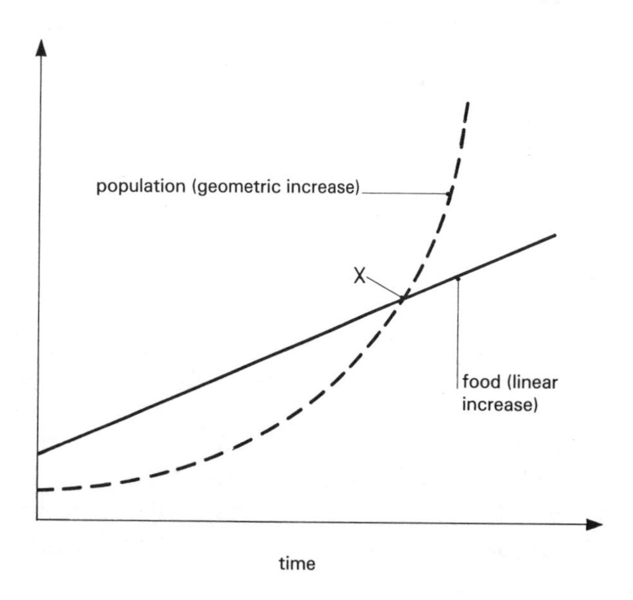

Assignment – Family planning and birth control

Attitudes to birth control by people living in Britain may not be the same as those in less developed countries. In the last century, neither the Church of England nor the Roman Catholic Church was in favour of any form of contraception. It is only during the last 50 years that contraceptives have been used more widely and with some religious approval.

Apart from religious views, other factors influence attitudes in less developed countries. In some countries, such as Brazil and Algeria, the governments are not keen to support family planning programmes. People believe that a growing population is a good thing, because they think that having more people available to do work will make the country richer.

In some countries, such as India and Indonesia, families expect a high level of infant mortality so they tend to have large families to allow for the loss of some children. In countries where most forms of agriculture are done without machinery, families may need children to work on the farm.

Some less developed countries are very poor and cannot afford to introduce medically supervised family planning programmes. Contraceptives would have to be provided free if the people were to use them and in many areas there are other medical problems which take priority, such as curing diseases and dealing with malnutrition.

Many less developed countries *have* introduced family planning programmes and are making an effort to explain to the people why such measures are necessary. They believe that family planning is the only way to decrease unemployment, raise the standards of health and education and stimulate savings.

In towns where contraceptives, clinics, schools and employment for all are available, the people are willing to accept the suggested programmes. In China the slogan used by the government is 'Late, Long, Few' – marry late, leave long spaces between children and have only a few. In many areas, couples are persuaded to have only one child and if a woman gets pregnant for a second or third time she may be forced to have an abortion.

1 Give four reasons why less developed countries may not want to restrict their birth rate.
2 Give four reasons why it is desirable to introduce a family planning programme in countries which are already overpopulated.
3 What are the main reasons for couples having small families in this country?

Governments may use advertising to encourage birth control

Assignment – Depo-Provera

Depo-Provera sounds ideal because it's a contraceptive that is more or less as effective as the pill, but you can't possibly forget to take it. One dose given as an injection lasts for at least three months.

Depo-Provera is based on a synthetic progesterone hormone, medroxyprogesterone acetate. A dose of 150 mg is usually injected to give protection for three months. A larger dose will give protection against pregnancy for a longer period. The drug prevents pregnancy in three ways.

(a) It suppresses ovulation (the monthly release of the egg).

(b) It makes the lining of the womb more likely to reject an egg that has been fertilised.

(c) It makes cervical mucus (the sticky plug at the entrance to the womb) more sticky so that it is a more effective barrier to sperms.

Depo-Provera is being used on thousands of woman in the third world. Many governments who are trying to control their population growth are in favour of this method of contraception because of its reliability. The women do not have to remember when to take pills or to use the various barrier methods such as the cap or the sheath when they have intercourse.

This drug has also been given to British and American women; however, it was banned in America in 1978 as it has not been proved to be totally safe. Some tests carried out on animals over 10 years ago show that the drug may cause some serious long-term health risks. Breast cancer was found to be more common in a group of dogs given the drug than in a control group. In monkeys, cancer of the lining of the womb developed in those given a high dose. In American women, cervical cancer was found much more often in women taking Depo-Provera than in women not taking the drug.

Apart from long-term risks, women have also suffered unpleasant side-effects. These include: disruption of the menstrual cycle, continued bleeding or no bleeding at all, headaches, weight gain, nausea, cramps, irritability, loss of hair and depression.

If Depo-Provera is not considered safe for women in America, many people feel that it should not be used for any women, including those in the third world. Apart from the safety aspect, many people also think that the drug may be misused. In Britain, some doctors have prescribed the drug for women who are very poor or who are non-English speaking as a quick solution for women they feel are social problems. In the third world, the drug is often given to women without much explanation, just to stop them having more babies.

1. What advantages has Depo-Provera over other contraceptives?
2. Why is Depo-Provera particularly suitable for use in inaccessible parts of the world, such as the foothills of the Himalayas, where there are no permanent doctors?
3. Why do many people want Depo-Provera banned from use throughout the world?
4. If this drug is banned, what effects do you think this will have in parts of the world where it is the most commonly used contraceptive?
5. Imagine that you are a mother of four children and that a doctor has just suggested that you are given a Depo-Provera injection. What would you say to the doctor in reply?

COMMUNITIES

A COMMUNITY is a group of organisms, animals and plants which live together in an area. Within a community there will be populations of many different types of plants and animals.

The type of community will depend on the HABITAT where it is found. A habitat is any place where a group of organisms can live. Habitats may be large, like woods or ponds, they may be much smaller, like rock pools or an individual plant. Because there are many different habitats there will be different types of community. (For example, the organisms found in a wood would be called a woodland community.)

All the organisms in the community depend upon their habitat and upon each other for their survival. Many different types of interaction are found between the organisms of a community.

An ECOSYSTEM is the term used for the habitat and the community living there. For example, the water, mud and rocks of a pond together with all the organisms that live in it make up a pond ecosystem.

The scientific study of ecosystems is called ECOLOGY. Ecologists may study any type of ecosystem, such as a rocky shore, beech wood, pine forest, sand dunes, chalk grassland and so on.

Fig. 8.38 (above right) A woodland habitat, (below left) a pond habitat, (below right) a seashore habitat

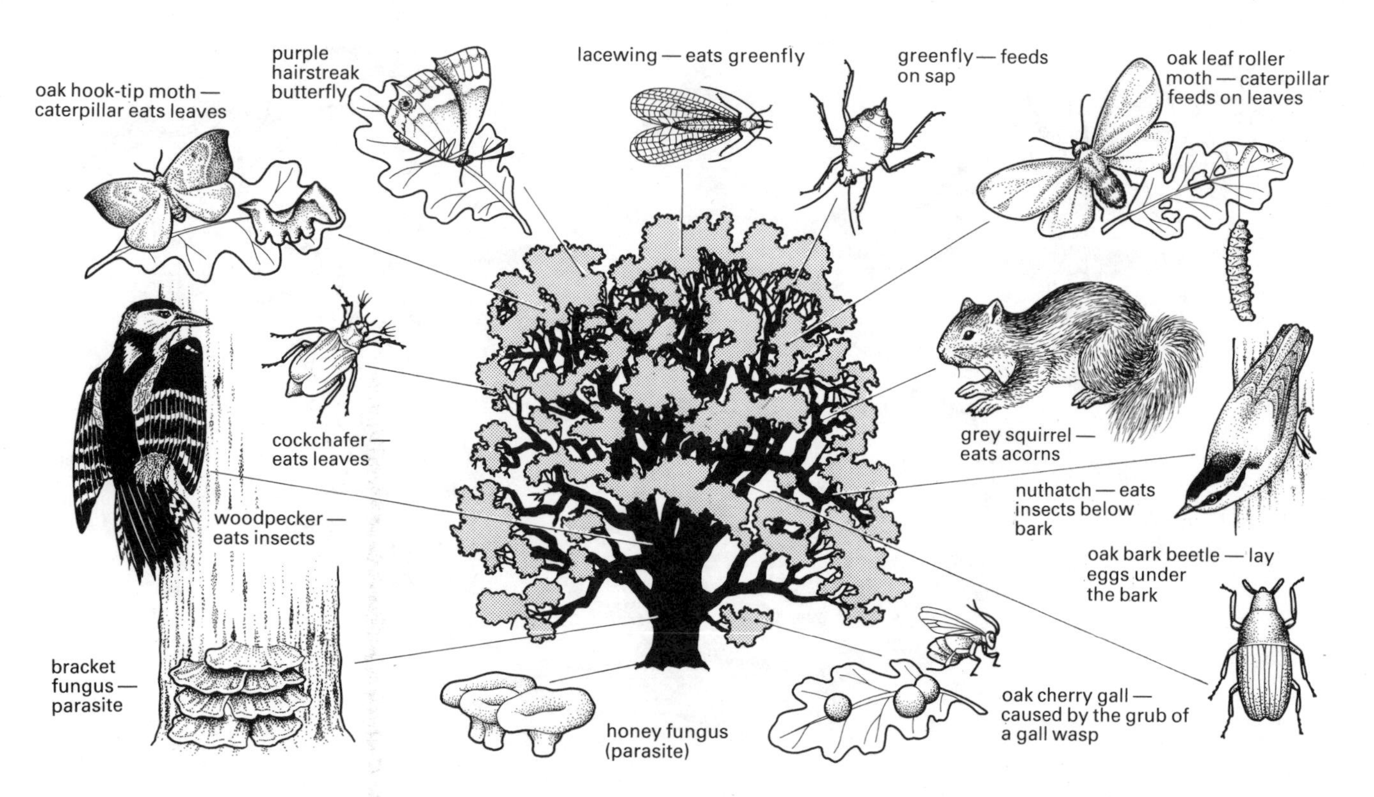

Fig. 8.39 An oak tree community: the diagram shows just some of the many species that may be found

Investigation 8.7 Studying a tree as a habitat

On one tree you will find many different types of animal and probably some plants too. All these obtain food or shelter from the tree and form a community.

Collect

umbrella
pooter
containers to put small animals in
hand-lens
identification books or charts

What to do

1 Find an oak tree near to your school or home.
2 Hold the opened umbrella upside down under a low branch.
3 Shake or beat the branch vigorously so that animals fall into the umbrella. Be careful not to break the branch.
4 Collect all the animals into your containers. Use a pooter if necessary.
5 Identify all the animals and then release them near the tree.
6 Have a good look at some of the oak leaves. See if there is anything attached to them. If there is, try to identify it.
7 Look at the bark of the tree. Is there anything attached to this?
8 Make a table to record what species you have found and the numbers of each.
9 Find a different type of tree and repeat 1 to 8. Compare what you have found on the two trees.

Questions

You will probably have found many more animals than you expected living on the trees. These all make up a community; they interact with the each other and with the tree.

1 How do they interact with each other?
2 How do they benefit from the tree? Do they harm it?

Fig. 8.39 may help you answer these questions.

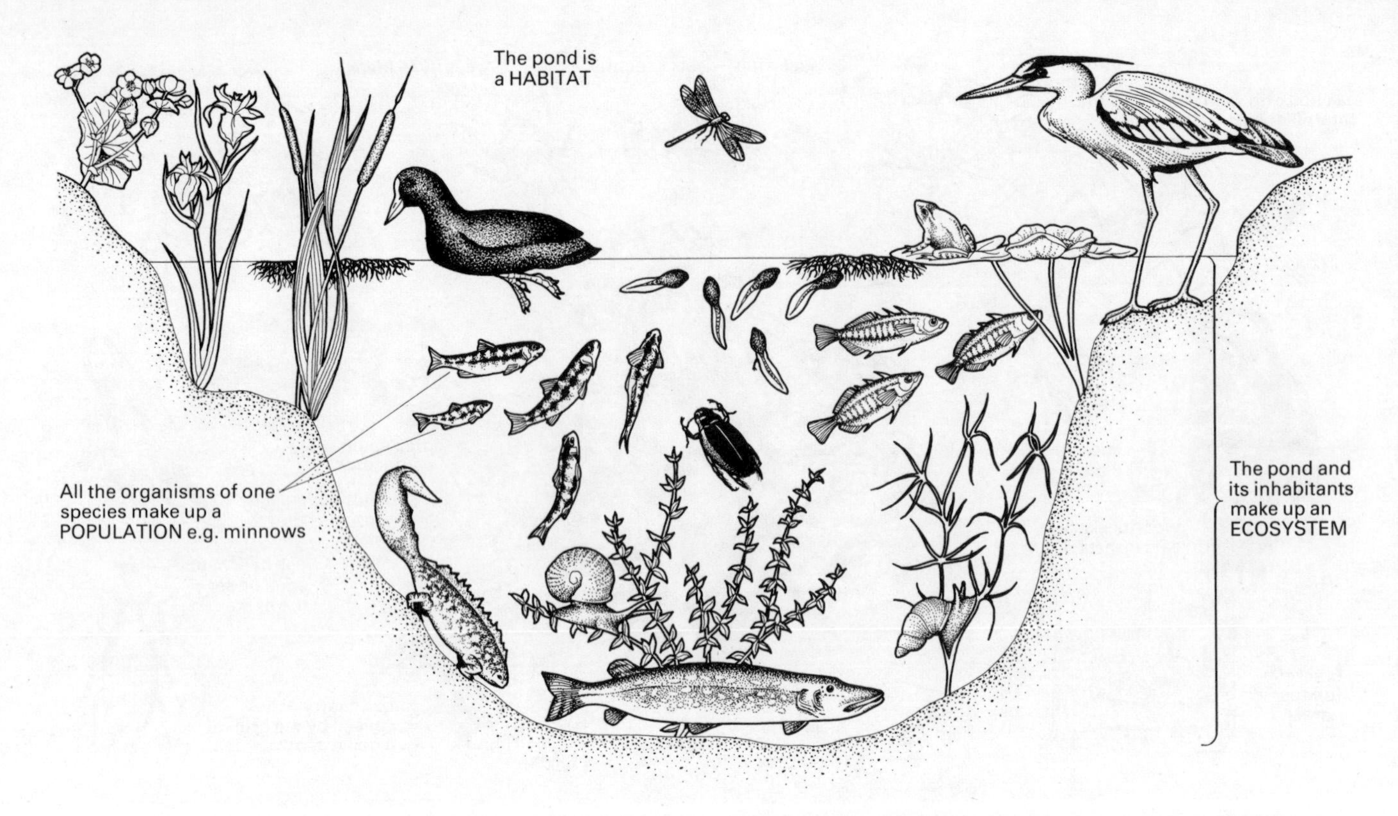

Fig. 8.40 An example of an ecosystem

Interactions in ecosystems

Some of the main interactions between organisms within an ecosystem are connected with feeding relationships.

Green plants are almost the only living things which can make their own food. They do this by using sunlight energy for a chemical process called *photosynthesis*: this will be described in Unit 9 (page 203). As well as producing food for the plants, photosynthesis also produces oxygen as a waste product. Green plants are termed PRODUCERS because they can make their own food.

Nearly all other living things depend either directly or indirectly upon plant food and they are termed CONSUMERS. Animals that eat only plants are called HERBIVORES and because they depend entirely on plant food they are termed *primary consumers*. Animals that eat other animals are called CARNIVORES. They eat herbivores and each other and so depend indirectly on plants for food. Carnivores are termed *secondary consumers*. OMNIVORES eat both plants and animals.

Table 8.4 Examples of producers and consumers

Producers	Consumers	
	Primary – herbivores	Secondary – carnivores
grass	zebra	lion
wheat	mouse	owl
blackberry	blackbird	cat

Food chains

In an ecosystem we will find many FOOD CHAINS such as the one shown in Fig. 8.41. Grass to rabbit to fox is a food chain which occurs when rabbits eat grass and foxes eat the rabbits.

Fig. 8.41 A food chain

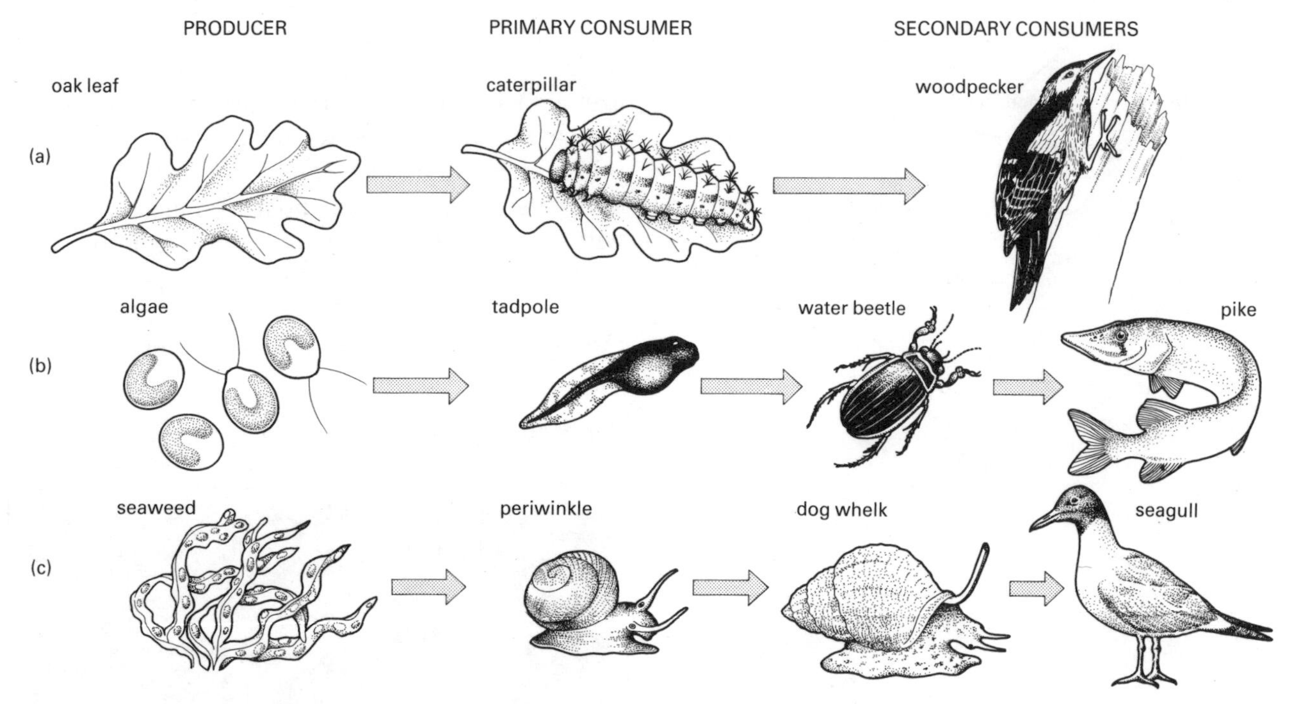

Fig. 8.42 Three food chains

All food chains begin with green plants because these are producers and the next 'links' in the chain are all consumers. Figure 8.42 shows some examples of food chains.

Pyramid of numbers

As you go along a food chain, the number of organisms at each level will be less. This is called the *pyramid of numbers*. Two examples are shown in Fig. 8.43.

The number of organisms decreases from the bottom to the top of the pyramid. You may also notice that the size of the organisms usually increases from the bottom to the top of the pyramid. Usually, carnivores are bigger than herbivores and herbivores are bigger than the plants they eat.

Food webs

Within an ecosystem, food chains like the ones in Figs. 8.41 and 8.42 are not usually as simple as this. This is because consumers usually eat more than one food, and may themselves be eaten by several different consumers. Foxes eat many types of food, not just rabbits. Rabbits also eat other plants, not just grass. Rabbits are also eaten by other animals (such as badgers), not just by foxes. Several food chains may become interlinked to form a FOOD WEB. An example is shown in Fig. 8.44.

If you did Investigation 8.7 you will probably have found some of the organisms shown in the food web of the oak tree. (You may have found many more!) If a habitat contains a large number of species the food web may be very compli-

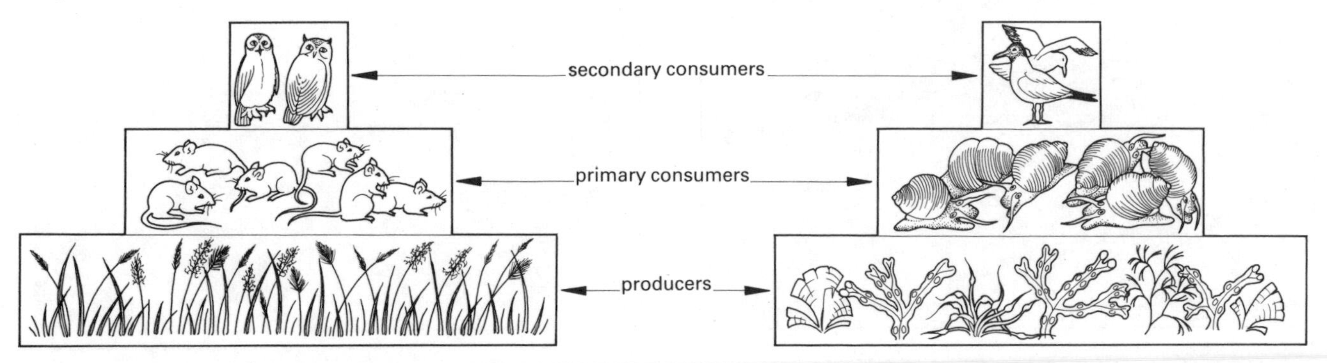

Fig. 8.43 Two pyramids of numbers

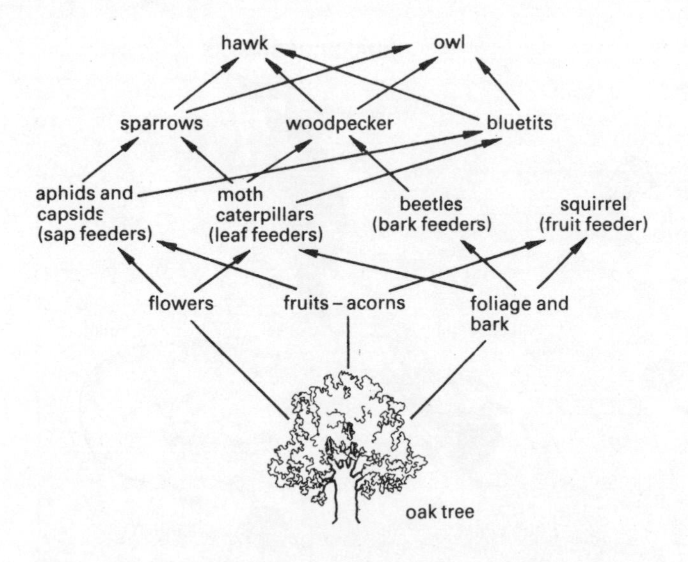

In any habitat there will be many consumers. Many will be herbivores or carnivores but there are also other types. PARASITES may be found. These are organisms which live in (or on) another living organism called the *host*. The parasite gets food from its host. It harms the host without actually killing it. Some examples are shown in Fig. 8.45.

Fig. 8.44 (left) Part of the food web of an oak tree

Fig. 8.45 (below) Some parasites

(a) Greenfly (aphids) live outside *their host plants. They suck sap from the leaves*

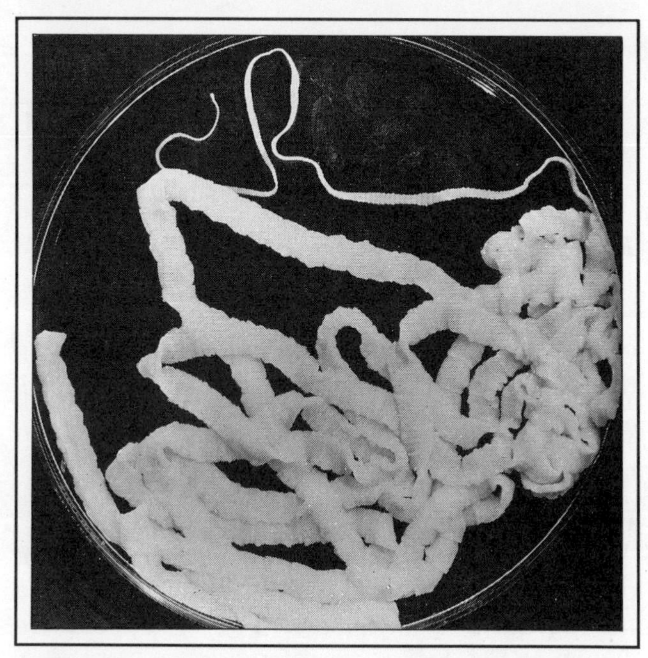

(b) Tapeworms live inside *their host – in the intestines*

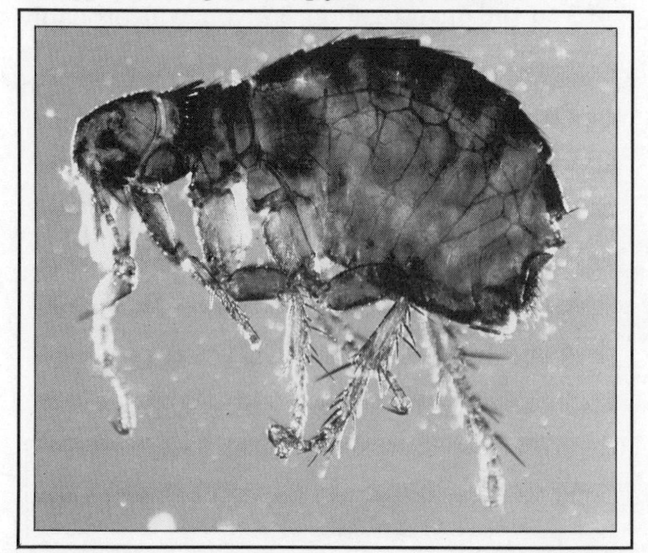

(c) Fleas live outside *the body. They suck blood from the host*

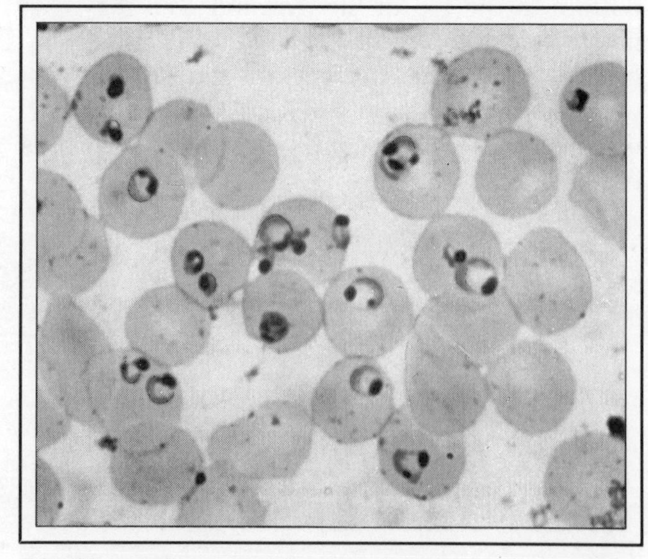

(d) Plasmodium, a microscopic parasite that causes malaria, lives in the blood. It destroys red blood cells

Assignment – Food web

Study the food web in the diagram.

1 Make lists of the producers, consumers and parasites from the food web.

2 State three things that would happen if all the tadpoles were removed from the pond.

3 If some weedkiller accidentally seeped into the pond, would it affect the animals in any way?

4 Write out two food chains that are found in the pond.

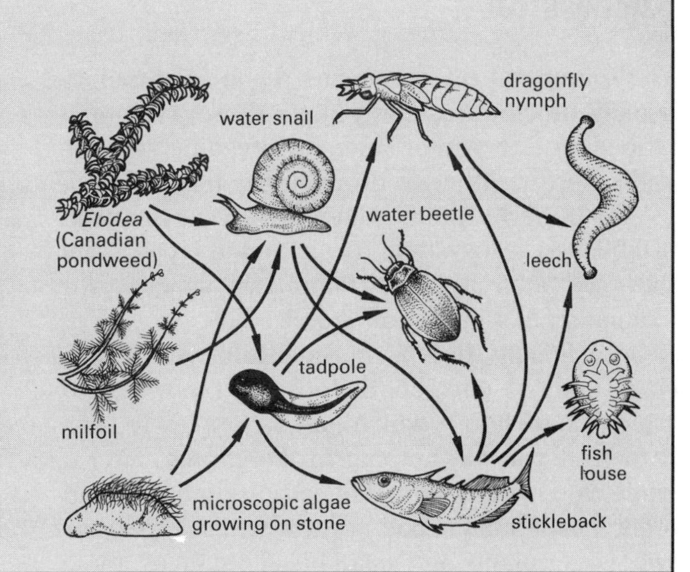

SCAVENGERS are animals such as vultures and crows which eat dead animals.

As well as producers and the various types of consumers, there will be another group of living organisms: DECOMPOSERS. Decomposers are fungi and bacteria which decay both dead plants and animals to release mineral salts into the soil. These mineral salts are needed by plants to keep them healthy and for their proper growth. Decomposers are therefore very important in any habitat even though they are not easily seen. Did you find any when you investigated an oak tree? Even if you did not there were probably some at work in various parts of the tree and around it. One important job that decomposers do is to decay all the leaves that fall from deciduous trees, such as oaks, every autumn. Decomposers may also be termed SAPROPHYTES because of their method of nutrition. A saprophyte is a living organism which feeds on dead or decaying tissue.

CONSUMERS (carnivores, herbivores, scavengers and parasites)

PRODUCERS (green plants)

dead animals and plants

DECOMPOSERS (fungi and bacteria)

nutrients released into the soil

Fig. 8.46 Decomposers decay the bodies of producers and consumers, releasing valuable nutrients into the soil, which can then be used again

Feeding relationships

Within an ecosystem there are many relationships between different species. Many of these are feeding relationships, some of which have already been mentioned.

Predation A predator is usually larger than its prey. The prey is an organism which the predator kills for food. For example, a fox kills a rabbit, a lion kills a zebra. Both predator and prey are animals.

Parasitism A parasite is an organism living in or on another living organism which is termed the host. A parasite gets food from its host without actually killing the host. Examples include a flea on a dog, a tapeworm in a human. See Fig. 8.45.

Symbiosis This is when two organisms living together mutually benefit each other. Nitrogen-fixing bacteria live in swellings on the roots of some plants. These bacteria make nitrates for the plants and in return obtain their food from the plant. This is explained more fully on page 215.

Competition Two or more organisms may compete for the same thing. Carnivores may compete with each other for the same food animals. Herbivores may compete with each other for the same food plants. Plants may compete with each other for light, water and space. Birds may compete for territories or nesting sites.

Succession

The numbers of organisms in a community usually fluctuate slightly like those of populations, due to winter/summer, breeding seasons and severe weather or drought, but they generally remain fairly stable from year to year in a well established ecosystem. In a stable ecosystem there is often a main (dominant) species, for example, oaks in an oak wood.

Occasionally there is an opportunity for a new community to develop. Suppose you moved into a brand new house which had a completely bare garden. If you did nothing to the garden, after a while you would find that various grasses and other small plants had grown. After a year or two, larger plants and small bushes such as gorse and brambles may have appeared. Shrubs such as hawthorn will develop later and after several years small trees will also probably be present. This process, where a series of changes in the organisms found within a habitat occurs, is called SUCCESSION. Some plants follow others which have prepared the way, and the animals found depend on the plants present.

A good of example of succession occurring is on sand dunes. Here marram grass is the first plant to grow. As marram grass and other early colonisers (colonisers are the first organisms, the founders of a community) such as sea couch grass die, they decay and nutrients from their breakdown are added to the sand to make it more fertile. Because the sand becomes more fertile, other plants such as sea holly and wild thyme appear and will be followed by others as the sand gradually changes to a sandy type of soil rather than just pure sand.

Old rubbish tips may also be colonised, and also land at the side of newly made motorways. The speed of growth of a community will depend more upon the type of soil than upon the location.

While changes occur in the vegetation, changes will also occur in the animals present. To begin with there will probably be only decomposers and a few small invertebrates. As more plants appear, more insects will be found in the community and, later, birds and small mammals.

As soon as large trees have grown it is unlikely that there will be more natural changes in a community. The community is then said to have reached its climax. A CLIMAX COMMUNITY is the final group of organisms found in a particular area when conditions have become more or less stable. The type of climax community depends on the habitat and on the type of soil.

Two thousand years ago most of northern Europe was covered in trees. The natural climax community was forest. Britain looks very different today, as trees have been cut down for use in building or for fuel. Land has been cleared for planting, crops and grazing animals have also been introduced. More recently, land has been cleared to build towns, roads and other necessities of an industrialised society.

Another example of succession is what may

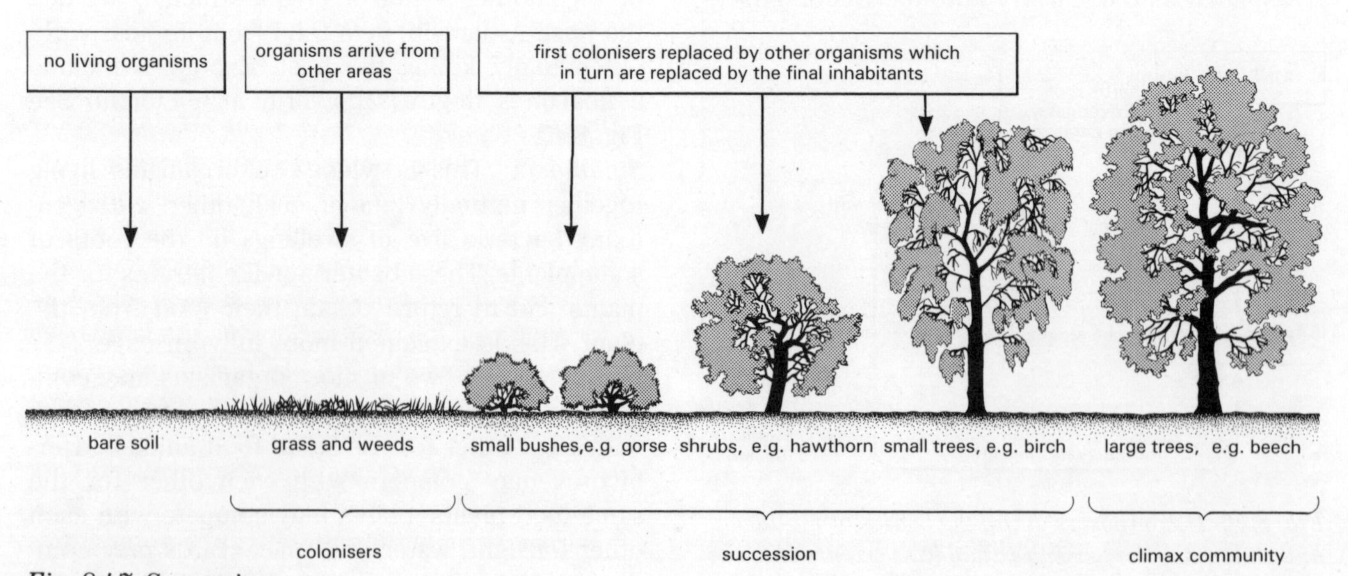

Fig. 8.47 Succession

happen on newly exposed bare rock. The first organisms that could inhabit it would be lichens. These can live on the rock surface and can gradually break up the surface so that dust particles and humus collect in any cracks and crevices. Mosses and other small plants can then grow in the 'soil', and as they do the humus builds up. More soil is formed which gradually covers the surface of the rock and more different types of plants will grow. As the layer of soil covering the rock becomes thicker, plants with larger roots are able to grow, then shrubs and eventually trees.

If you look at a newly cut rock face such as where a motorway is being built, you may see rock with only a few inches of soil on top. You will notice that, because there is only a shallow layer of soil, only shallow-rooted plants such as grasses can grow there.

Assignment – A new garden

A new housing estate has been built on a piece of land that was previously an oak wood. A few oak trees have been left growing in some of the gardens. One new householder notices some oak seedlings growing in her garden. She has no trees in her garden so she leaves the seedlings, hoping that they may develop into trees. She finds that after reaching a height of about 20 cm all the seedlings on one side of her garden die, but that those on the other side do not.

1 Explain how oak seedlings may start growing in a new garden.
2 Suggest two reasons why all the oak seedlings on one side of the garden die.

Human effects on ecosystems

Changes in communities and ecosystems can easily be caused by human beings; for example, when a grazing animal is introduced or removed. Rabbits, which were brought to Britain by the Romans for their meat, eat tree and shrub seedlings as well as grass and other small plants. Because they do this, places like downs remain mainly grassland. A similar effect is seen where sheep graze. If an area in the middle of a field normally grazed by sheep is fenced off, the vegetation in that area will soon change. It will flourish and grow higher.

Other examples of the effect of humans may be related to modern methods of agriculture. Vast areas of land such as marshes, heath, woods have been cleared to grow crops. Chemicals are used to kill both unwanted weed plants and animal pests (see page 128).

The other main effects caused by human beings are those of various forms of pollution. Food webs may be disrupted if certain organisms are killed. Sooty smoke from factories kills lichens. Chemicals discharged from factories may kill animals and plants in rivers. This topic will be dealt with more fully in Unit 10.

In the last decade or two, the effects of human activities on ecosystems have been more fully realised and more efforts are now being made to conserve certain natural ecosystems in various parts of the country.

Populations and communities

WHAT YOU SHOULD KNOW

1 A population is the number of one species of plant or animal found in a particular area.
2 The growth of a population may be exponential, when it doubles in size at regular time intervals.
3 Various limiting factors may affect the growth of a population, such as: food, space, oxygen, waste products, predators and disease.
4 The change in size of a population depends on the rates of birth, death, immigration and emigration.
5 The size of the human population can be controlled if people use various methods of birth control. Populations of plants and animals are most commonly controlled by using chemicals such as insecticides and herbicides.
5 A community is a group of organisms, animals and plants, that live together in an area. Within a community there will be populations of many types of plants and animals.

6 A habitat is the area in which a plant or animal lives, e.g. pond, seashore, garden.
7 The type of community will depend on the habitat where it is found.
8 An ecosystem is a unit formed by the interactions of the plants and animals in a community together with the non-living elements of the habitat. The non-living elements include the soil and climate.
9 Feeding relationships in an ecosystem may be represented by food chains and food webs.
10 Green plants are the starting point for food chains and food webs. Green plants are termed producers because they make their own food.
11 Animals are consumers. They may be herbivores, carnivores, omnivores, parasites or scavengers, depending on what they eat.
12 Bacteria and fungi are decomposers.
13 A climax community is one that is stable.
14 Succession is the gradual sequence of natural changes in a community from the first colonisation of an area to the development of a stable climax community.

QUESTIONS

1 Explain what is meant by the following terms (a) population, (b) community, (c) habitat (d) ecosystem. Give two examples of each.
2 If a small island became overrun with rabbits, give three methods by which the people living there could reduce the number of rabbits. For each method give the advantages and disadvantages.
3 Calculate the total population of pond snails in a small pond from the following information.

20 pond snails were found. They were marked with a spot of waterproof paint and were returned to the pond. The next day 20 snails were again collected. Of these, four had the paint spots.

4 Use Fig. 8.34 to answer this question.
(a) What was the total world population in (i) 1800, (ii) 1900?
(b) Give three reasons why the population has risen so quickly since 1800.
(c) The dotted line shows what some people think will be the population growth up to AD 2000. Give two reasons why the population growth may be slower than this and two reasons why it may be faster.

5 Imagine you are a politician who is concerned about the rapid population growth of your country. Write an article for a daily newspaper explaining
(a) *why* families should not have more than two children;
(b) *how* couples can make sure they do not have more than one or two children;
(c) the benefits to everybody in the country if the growth of population is stopped.

6 The graph shows the variation in the numbers of the rabbit population of Great Britain over a ten year period.

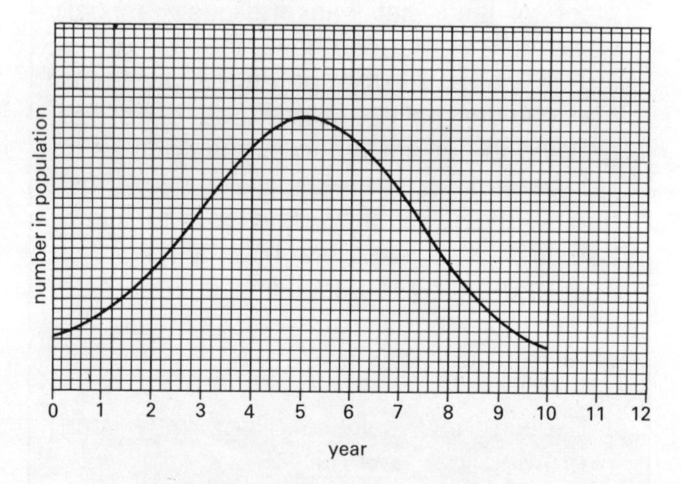

(a) State three factors which could cause a decrease in the rabbit population.
(b) Trace the graph into your book and add to it a line to show how the population of foxes might vary during the same period.
(c) What would you expect to happen in years 11 and 12? Give reasons for your answer.

7 Write out each of the following lists in the form of a food chain.
(a) seeds, owl, mouse
(b) fox, grass, hen
(c) human, wheat, pigs
(d) caterpillar, cabbage, hawk, blue-tit

8 Look at the food web on page 140 (Fig. 8.44). Copy the table below and fill it in using information from Fig. 8.44.

Oak tree community

Producers	**Consumers**	
	Herbivores	Carnivores

9 Sort out the following list into these groups – producers, consumers, parasites, scavengers, decomposers.

earthworm, mushroom, crow, deer, lion, human, grass, bees, vultures, bacteria, algae, tapeworms, mosquitos, daisy, fox, blackbird, rosebush, mould, beech tree

10 Write out a food chain that might be found in (a) a town garden, (b) the sea, (c) a pond.

11 Explain the difference between the following pairs of terms:
(a) population and community;
(b) exponential growth and stability;
(c) food chain and food web;
(d) producer and consumer;
(e) herbivore and carnivore;
(f) decomposer and parasite;
(g) colonisation and succession.

12 (a) Match the pyramids of numbers below with the descriptions.

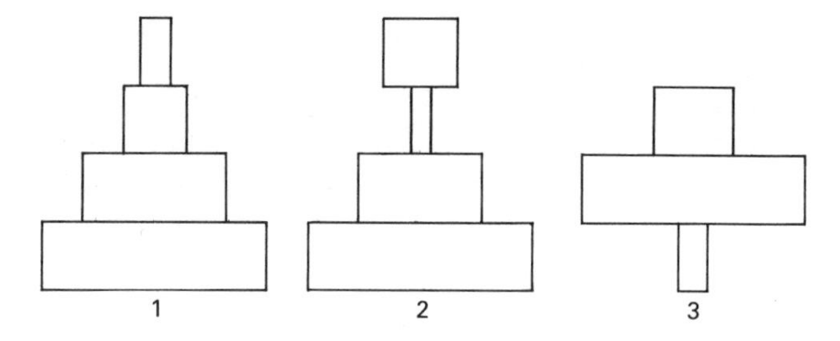

(i) A tomato plant has greenfly on it and a few ladybirds which are eating the greenfly.
(ii) Some wheat is eaten by mice which are eaten by one cat who has fleas.
(iii) Grass is eaten by caterpillars. Birds eat the caterpillars. Cats eat the birds.

(b) Copy outline 1 and fill it in using four of the organisms from the list below:

water fleas, kingfisher, whales, frogs, tadpoles, small fish, algae, grass, cats.

Reproduction, Growth and Development

HEREDITY AND GENETICS

Do you look the same as your parents? Obviously you are not the same as both parents but you will have inherited some characteristics from each of them. You have probably heard people say things like 'Jessica has got eyes like her mother', 'Rajan's nose is like his father's; 'Wesley has inherited his father's sense of humour'. The inheritance of characteristics is called HEREDITY and the branch of science that deals with how this works is called GENETICS.

The science of genetics explains how similarities and differences are inherited. It tries to forecast what characteristics will be passed on to offspring when plants or animals reproduce sexually. (See page 94.) Although in this unit many of the examples will refer to humans, characteristics are inherited in the same way in all other animals and plants that reproduce sexually.

Fig. 8.48 The Barnes family, photographed in Dorset about 1907

CHROMOSOMES, MITOSIS AND MEIOSIS

To begin to understand how genetics works, you need to know what happens in sexual reproduction (see page 94) and where and how genetic information is carried in a cell.

A new organism, whether plant or animal, starts life as a single cell which is called a zygote. In humans this is the fertilised egg. This one cell contains all the genetic information needed to decide how the body will develop. The type of cell division that occurs in the zygote and in the embryo is called MITOSIS. Each time a cell divides, two identical cells are produced each containing the same genetic information. The single cell divides into two cells, then it divides again to make four cells, then eight, and so on. Gradually the cells in the embryo change into many different types. In a human, bone cells, muscle cells, skin cells and blood cells are some of the cells that will be formed to make up the body.

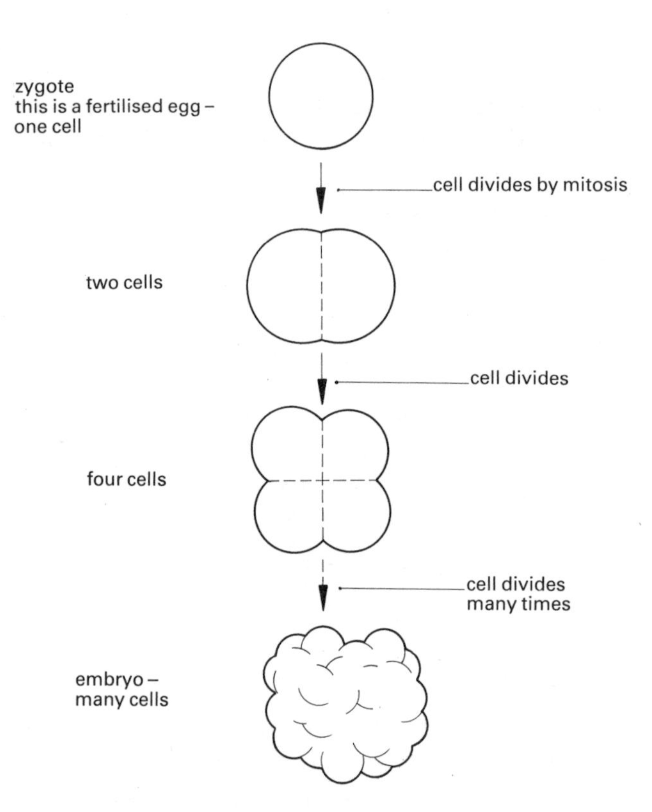

Fig. 8.49 The formation of an embryo from a zygote

Mitosis occurs when

(a) an organism is growing;
(b) cells are being repaired or replaced;
(c) asexual reproduction is taking place (see page 95).

To understand what happens in mitosis you need to know a bit more about cells. You learned about the basic structure of cells in Unit 2.

Most cells have a nucleus. The nucleus contains long thin strands called CHROMOSOMES. Most of the time the chromosomes are too thin to see except with an electron microscope. But when a cell is dividing the chromosomes get shorter and fatter and can often be seen with a light microscope (Fig. 8.50).

It is the chromosomes that carry all the genetic information. There are pairs of chromosomes in most cells. Look at Fig. 8.51(b), where the human chromosomes in Fig. 8.51(a) have been rearranged. In Fig. 8.51(a) you can see that there are 46 chromosomes. In Fig. 8.51(b) you can see that there are 23 pairs.

All the cells of an organism will contain the same number of chromosomes. In humans this is 46, that is, 23 pairs. When a cell divides for growth it is important that the two new cells should have the same genetic information as the parent cell. Mitosis ensures this.

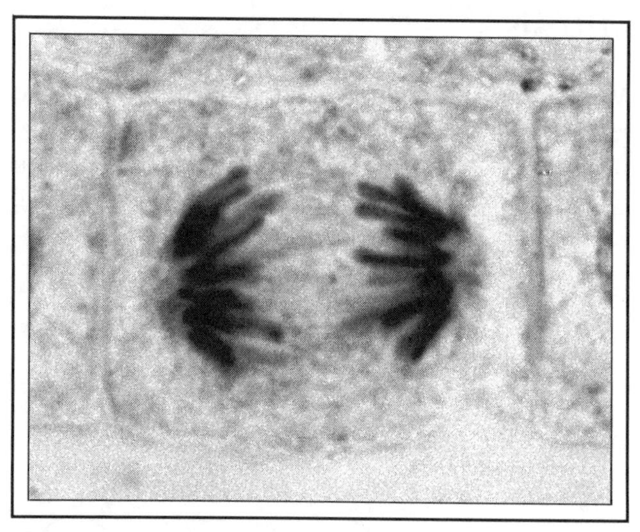

Fig. 8.50 Chromosomes in a plant cell that is about to divide

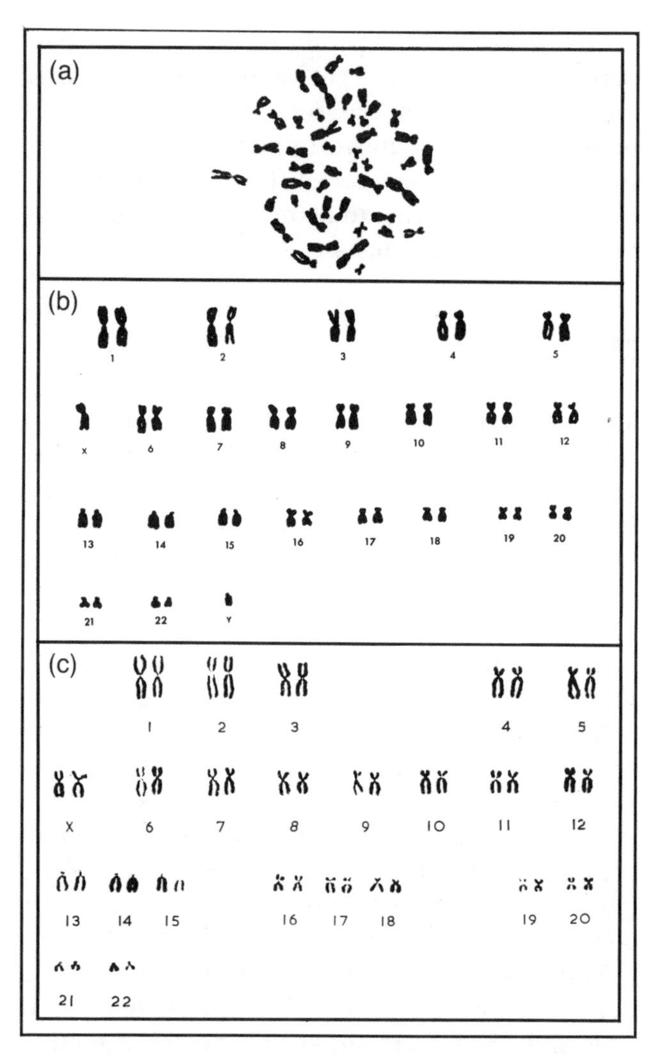

Fig. 8.51 (a) Human male chromosomes, (b) The chromosomes in (a) arranged in pairs. (c) Human female chromosomes

Mitosis

Mitosis may take place in all cells of an organism *except* when gametes are being formed in the reproductive organs. In humans it may happen in all cells *except* in the ovaries and testes when egg and sperm cells are being made.

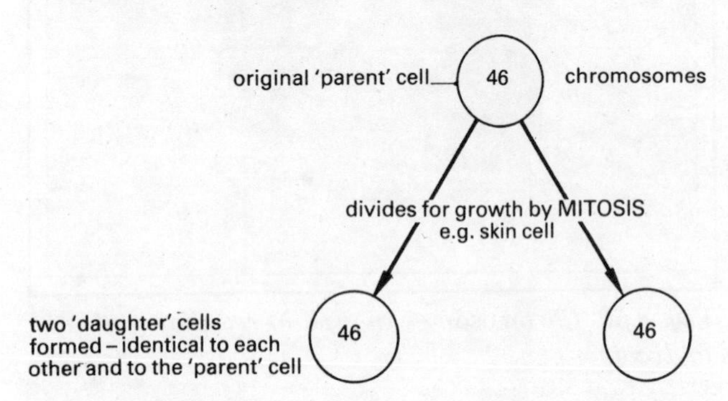

Fig. 8.52 Mitosis

So mitosis is the type of cell division that happens for growth. But what happens in the ovaries and testes and in the reproductive organs of plants? Here the gametes or sex cells are formed. In humans these are called the eggs and sperms. For eggs and sperms to be made, a different type of cell division happens.

Meiosis

Meiosis happens *only* in parts of the ovary and testis. It cannot happen in any other body cells.

All human body cells have 46 chromosomes. Some of the cells in the ovaries and testes divide by the process of MEIOSIS to make eggs and sperms. Eggs and sperms both have 23 chromosomes.

In an adult woman an egg cell ripens each month and is released (see page 111). This egg cell contains only half the number of chromosomes of a normal body cell. In an adult man millions of sperms are made in the testes. Each sperm contains only half the number of chromosomes of a normal body cell. Meiosis halves the chromosome number and for this reason is sometimes called 'reduction division'.

You should be able to see that meiosis is necessary by imagining what would happen if eggs and sperms did not each have half the number of chromosomes ... each generation would have twice as many chromosomes as the previous one.

Fertilisation occurs when an egg cell fuses with a sperm cell (see page 110). The resulting cell is called a zygote and is the beginning of a new person. As there are 23 chromosomes in the egg cell and 23 chromosomes in the sperm cell the zygote will have 46. (Fig. 8.53).

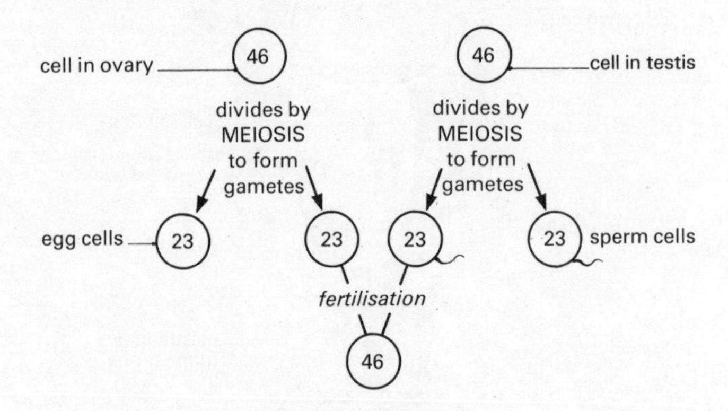

Fig. 8.53 Meiosis and fertilisation

The zygote will have 23 pairs of chromosomes, one of each pair from the mother and one of each pair from the father. After fertilisation the zygote starts to grow by dividing into 2, 4, 8, 16, etc. All these divisions will be by mitosis for growth.

Both mitosis and meiosis are complicated processes with many stages, but the important things for you to remember are where they happen, why they occur and what happens to the chromosome number.

Assignment – Chromosomes

1 Look at the diagram of the cell below.

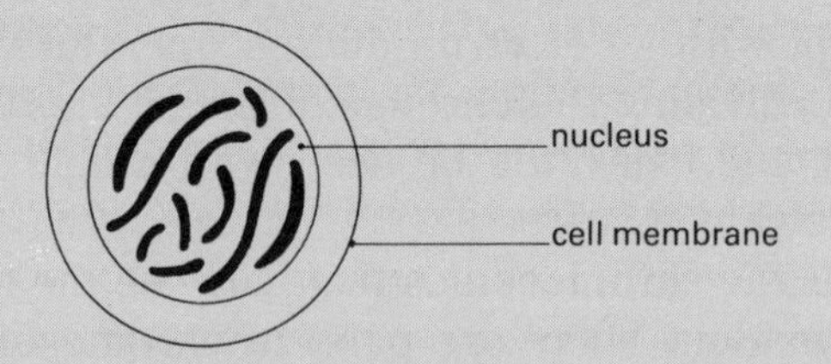

(a) How many chromosomes can you see?
(b) How many pairs of chromosomes are there?
(c) How many daughter cells would this cell have if it divided by (i) mitosis, (ii) meiosis?

2 Study Fig. 8.54 and make a list of the differences between mitosis and meiosis. You should be able to find at least three.

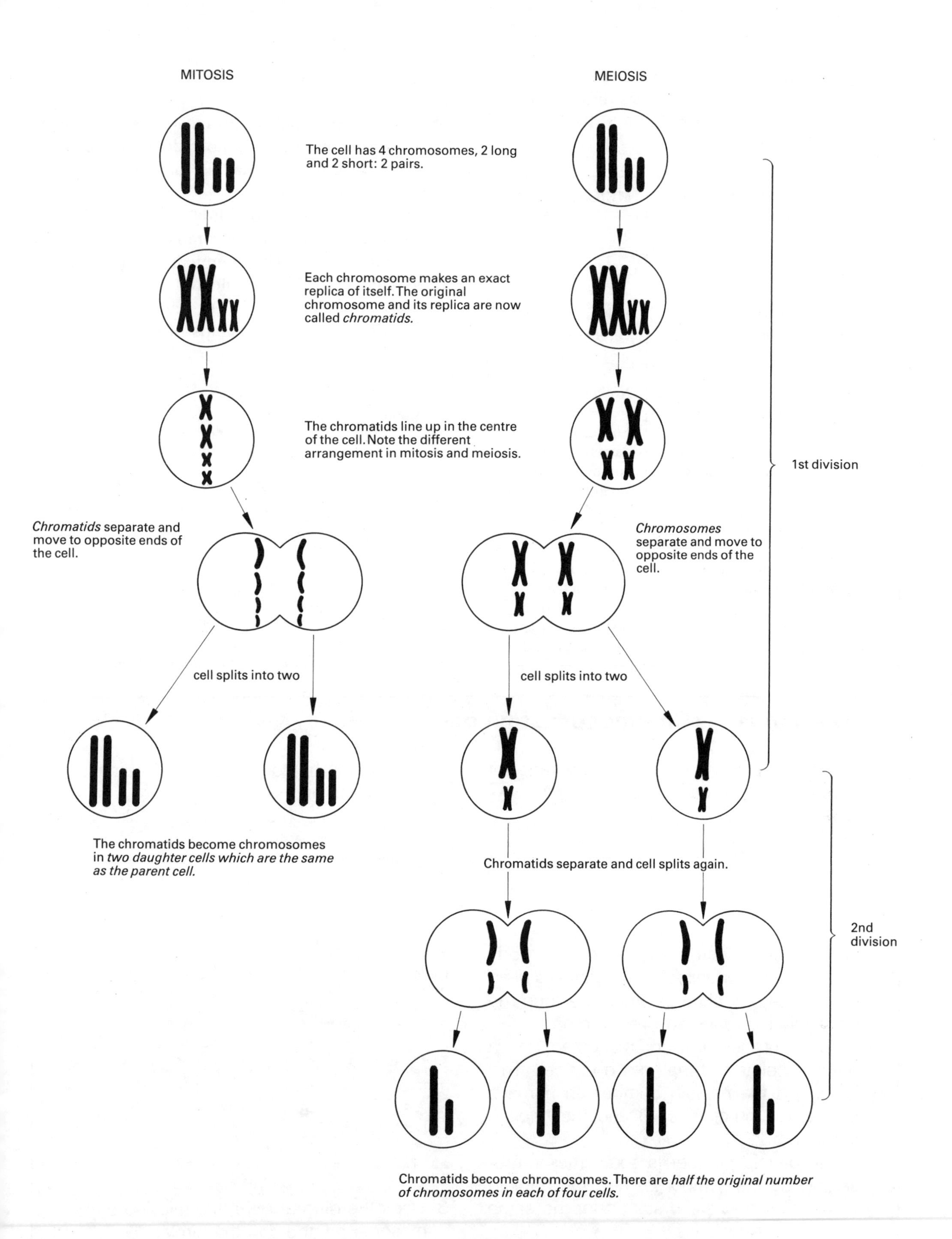

Fig. 8.54 The stages of mitosis and meiosis in a cell with only two pairs of chromosomes

As you can see in Fig. 8.54 there are several points where mitosis and meiosis differ. The two processes are compared in Table 8.4.

Table 8.4

	MITOSIS	MEIOSIS
Number of cell divisions	1	2
Resulting cells	Have same number of chromosomes All identical	Half number of chromosomes Not identical
Purpose	Growth of new cells; repair and replacement of cells; *asexual* reproduction	Formation of gametes (egg and sperm cells) for *sexual* reproduction
Occurs	in any part of an organism	only in the reproductive organs, e.g. ovary, testis (anther and ovary in plants)

Chromosomes

We have seen how chromosomes behave in mitosis and meiosis but how do they carry the genetic information that is passed from one generation to the next?

When a cell is not dividing, the chromosomes are long and thin. Each chromosome has smaller parts called GENES. The chemical which makes up these genes is very complicated. It is called DNA (deoxyribonucleic acid). On each chromosome there will be many thousands of different genes.

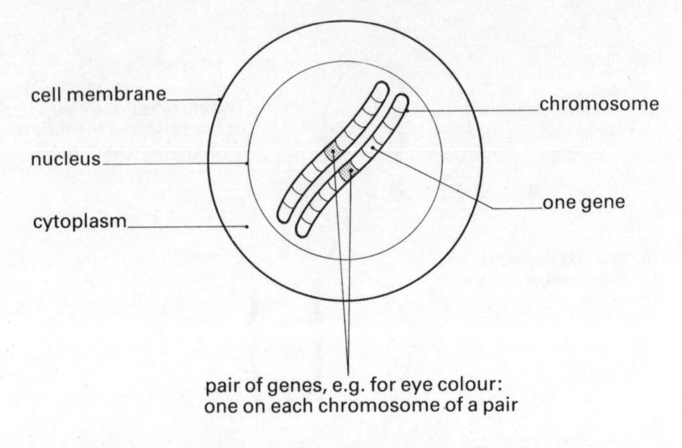

Fig. 8.55 A cell showing one pair of chromosomes

Assignment – Chromosome numbers

In every species there is a fixed number of chromosomes in each cell.

humans	46
mice	40
hens	36
peas	14
fruit flies	8

The number of chromosomes in each body cell of a plant or animal is called the *diploid* number. As all chromosomes are found in pairs this is always an even number.

The number of chromosomes in a gamete cell is half that in a body cell. This is termed the *haploid* number. In humans the *diploid* number is 46 and the *haploid* number is 23.

The diploid number is sometimes represented by the symbol $2n$ and the haploid number by n where n stands for the same number in each case.

1 What is the haploid number for humans, mice, peas and hens?

2 How many chromosomes would be present in a hen's egg and in a mouse's sperm?

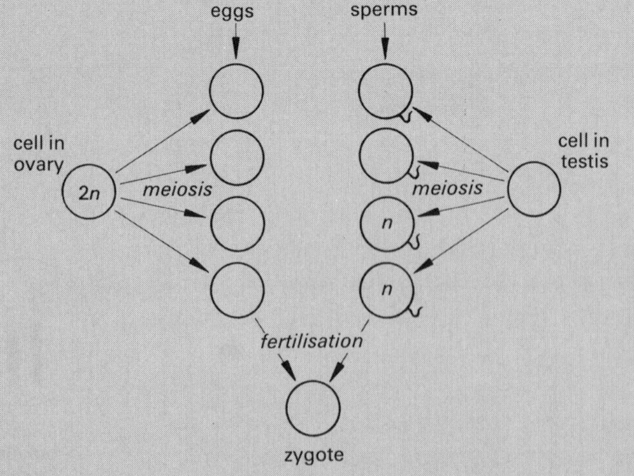

3 Copy the diagram. Put the missing chromosome numbers in the circles.

Chromosomes are found in pairs and each characteristic is controlled by a pair of genes, which both occur at the same position on their respective chromosomes. See Fig. 8.55. Each gene pair controls something that happens in the cell. It has all the 'instructions', or genetic information, on it. Genes control all an individuals's characteristics, such as eye colour, hair colour, shape of nose and height, as well as some things which are not seen, such as the ability of certain cells to produce enzymes and hormones.

Each gene has two different forms called ALLELES. For example, the gene for eye colour has two alleles, blue and brown. When there are two different forms of the same gene there are three possible combinations of alleles on a pair of chromosomes. This is shown in Fig. 8.56.

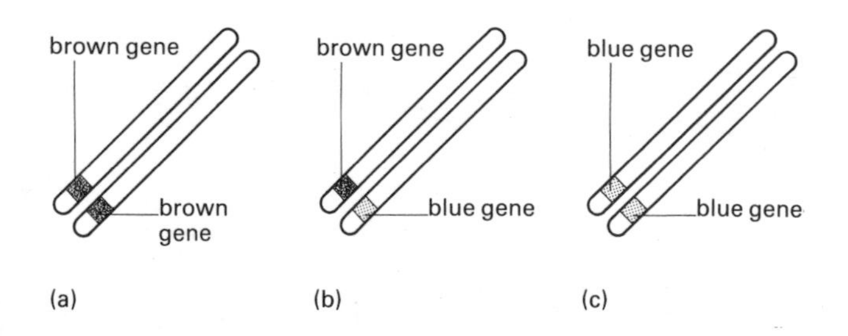

Fig. 8.56 Possible combinations of alleles for eye colour

Gamete formation and fertilisation

Earlier in this unit (page 148) we saw what happened in meiosis when gametes are formed in the reproductive organs. A sperm will contain single chromosomes, each with a set of genes from the father. An egg will also contain single chromosomes, each with a set of genes from the mother. At fertilisation these two sets of genes are brought together and the genes inherited from both parents will then control the development of the baby. The baby will show some characteristics of the mother and some of the father. Fig. 8.57 shows how this works for one pair of genes, for eye colour.

Because characteristics are inherited from both parents on thousands of different genes, millions of combinations of genes are possible in the resulting baby. This is why there is so much VARIATION between offspring. Do you know any people who both look alike and behave in the same way?

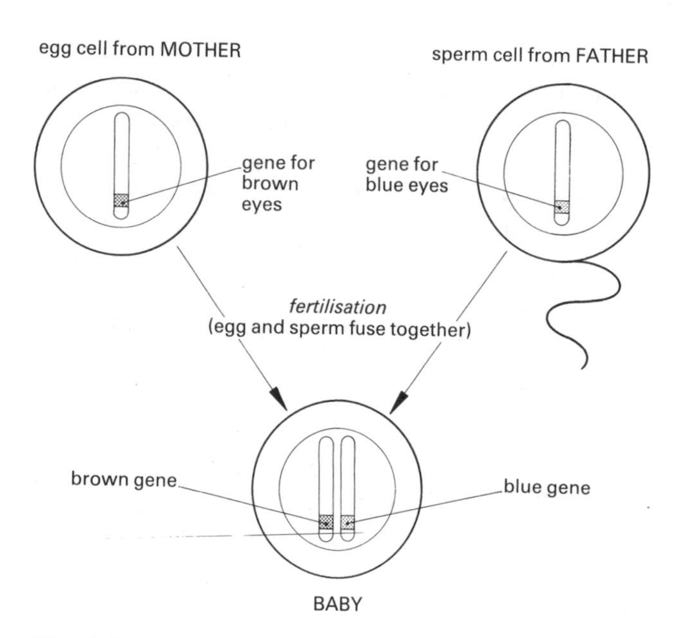

Fig. 8.57 Diagram to show the inheritance of one gene, in this case for eye colour. Will the baby have brown eyes or blue eyes?

VARIATION

Fig. 8.58 The Grebbell family of St. Budeaux, Plymouth, taken in 1905

Why do people and other animals and plants vary? The main reason is that each individual has a different combination of genes. Your combination of genes is unique – it is *very* unlikely that there is anyone else in the world who has *exactly* the same combination . . . unless you are one of a pair of identical twins (see page 115).

The combination of your genes depends firstly on how your parents' chromosomes behaved during meiosis when the eggs and sperms were being formed (see page 148). During meiosis half the chromosomes, with all their genes, go into one gamete and half into another. It is chance as to which chromosome of a pair goes into each gamete and as there are 23 pairs of chromosomes, each with its own set of genes, there are many possibilities.

Secondly, your combination of genes depends on what happens at fertilisation. When fertilisation of an egg by a sperm occurs it is again chance as to which of the many sperms, which have different combinations of genes, fertilises the egg.

This should help to explain why, if you have brothers and sisters, they are probably different

Investigation 8.8 Variation in fingerprints

There are four main types of fingerprint. No one has exactly the same combination of fingerprints. This is why the police can use them to help trace criminals.

Collect

plain paper
office ink pad
hand-lens

What to do

1 Wash and dry your hands.
2 Place the tip of your index finger on the ink pad to 'ink' it.
3 Put your inked fingertip on a piece of white paper. Roll it carefully while pressing quite hard before removing it from the paper. Take care not to move your finger, as you will smudge your 'print'.
4 Repeat steps 2 and 3 with your other fingers and thumb, and with your other hand.
5 By now you should be able to produce good clear fingerprints. Repeat steps 2 and 3 again with all your fingers but this time make the prints in your exercise book. Arrange your prints neatly as in the diagram.

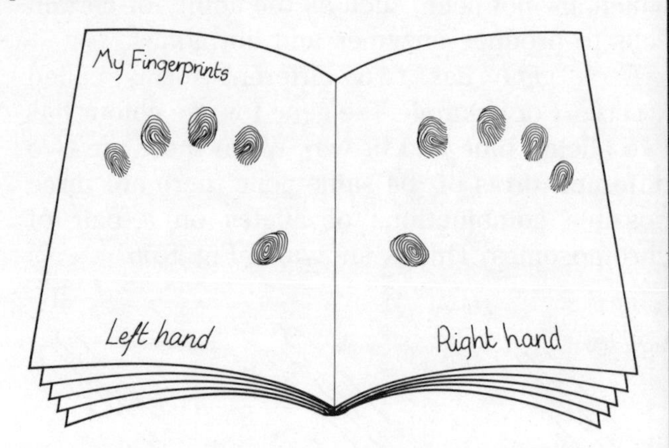

6 Look at each print with a hand-lens. Compare your prints with the photographs. Try to decide whether each is an arch, a loop, a whorl or a double whorl and label it accordingly.
7 Compare your prints with other members of the class.

Are any of your fingerprints the same? Are any of yours the same as other members of the class? If you have compared them properly you will have found that no two are alike even though they may be of the same type. You may also notice breaks in the lines if you have small scars on any of your fingers.

(a) Arch

(b) Loop

(c) Whorl

(d) Double whorl

in many ways even if you have the same parents. Look again at Fig. 8.58. Are all the children the same?

Variation is mainly due to
(a) what happens during meiosis when gametes are being formed;
(b) the random fertilisation of an egg by a sperm.
(There is more information about variation later in this unit, on page 175)

HOW CHARACTERISTICS ARE INHERITED

A brown eyed man and a blue eyed woman have a baby with brown eyes. Why are they brown and not blue?

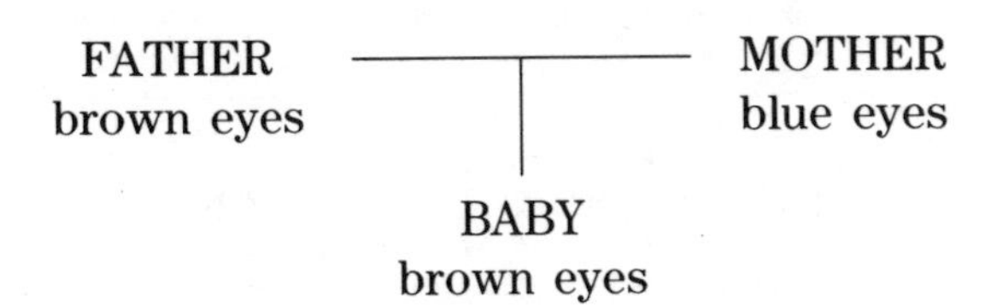

This can be explained by thinking about
- the father's genes;
- the mother's genes;
- what happens when the eggs and sperms are being made;
- what happens at fertilisation.

Each parent has a pair of genes which control eye colour (See Figs. 8.56 and 8.57). There are two different alleles, one for brown eye colour and one for blue eye colour.

In genetics, letters are used for different alleles to help us explain how characteristics are inherited. In this example we will use the letter **B** to stand for brown and **b** to stand for blue. The father has two genes which make his eyes brown, **BB**. The mother has two genes which make her eyes blue, **bb**. The egg and sperm each contain only one of these genes. Each sperm contains one **B** gene, each egg contains one **b** gene. When fertilisation takes place the gametes fuse and the **B** gene is brought together with the **b** gene. The baby will have one **B** gene and one **b** gene. It will be **Bb**.

But the baby has brown eyes even though it has a blue (**b**) gene. This can be explained because the **B** (brown) gene somehow suppresses

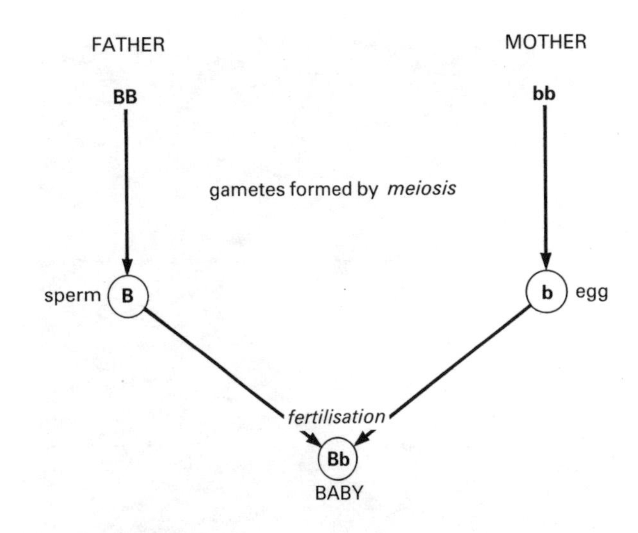

Fig. 8.59 How the baby gets its genes for eye colour

the **b** (blue) gene and only the effect of the **B** (brown) gene shows in the baby. The **B** gene is DOMINANT over the **b** gene. The **b** gene, which does not show up in the baby, is RECESSIVE.

In genetics, a capital letter is always used to represent a dominant allele and a small letter is used to represent a recessive allele.

So for eye colour,

B = brown DOMINANT
b = blue RECESSIVE

For eye colour, brown is dominant to blue, so that if one allele is **B** (brown) the baby will have brown eyes. If the baby has blue eyes then both alleles must be **b** (blue).

We have been thinking about eye colour and how it is inherited. So we will carry on using this example to explain the rest of the terms used.

The GENOTYPE of an individual is the genetic make-up of it and we use letters to represent this.

The PHENOTYPE of an individual means the physical characteristics it shows, what it looks like.

So in this case,
B = brown **b** = blue

GENOTYPE	PHENOTYPE
BB	brown
Bb	brown
bb	blue

If both genes are the same (e.g. **BB** or **bb**) the person is said to be HOMOZYGOUS for the characteristic. If the genes are different (e.g. **Bb**) the person is said to be HETEROZYGOUS for the characteristic.

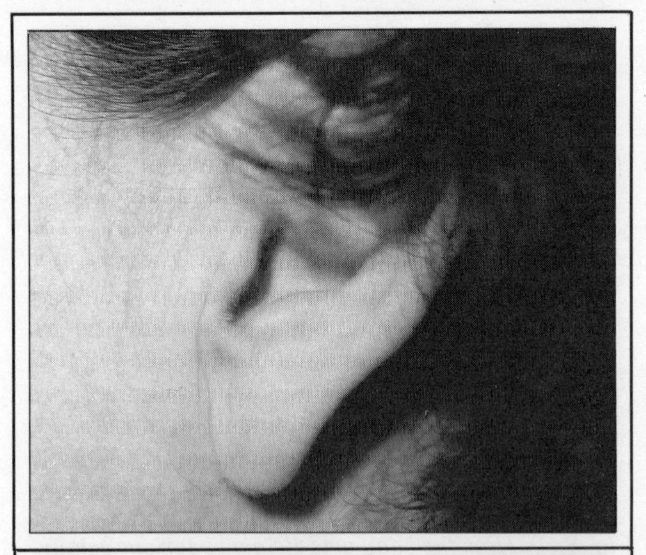

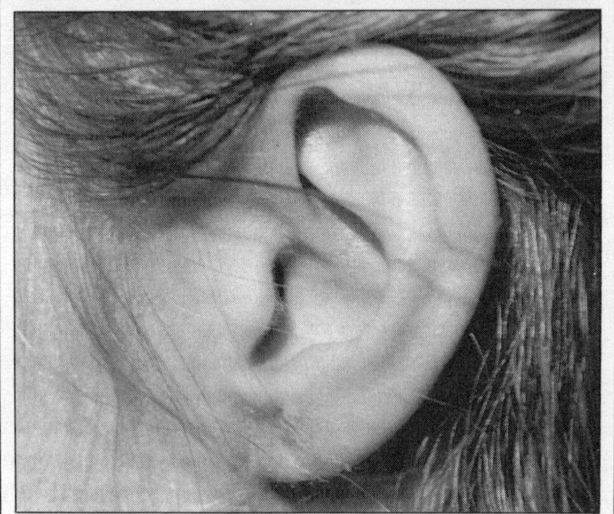

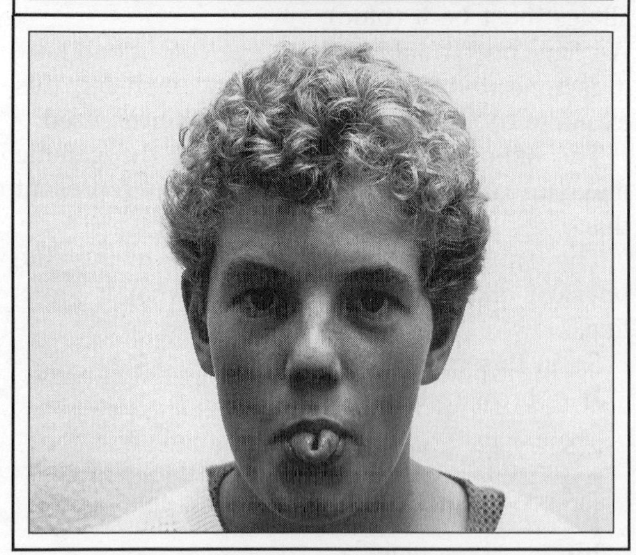

Fig. 8.60 Characteristics which have dominant and recessive forms: (a) ear lobes can be free (top) or attached (centre); (b) some people can roll their tongues like this, others cannot

Some other characteristics which have dominant and recessive forms are shown in Fig. 8.60.

Fig. 8.61 explains diagrammatically how eye colour will be inherited in children of a woman who has blue eyes and a man who has homo-

Assignment – Chromosomes and genes

The diagram below shows a pair of chromosomes with two pairs of genes. The pairs of genes control the inheritance of eye colour and type of hair in humans.

B = brown eyes **C** = curly hair
b = blue eyes **c** = straight hair

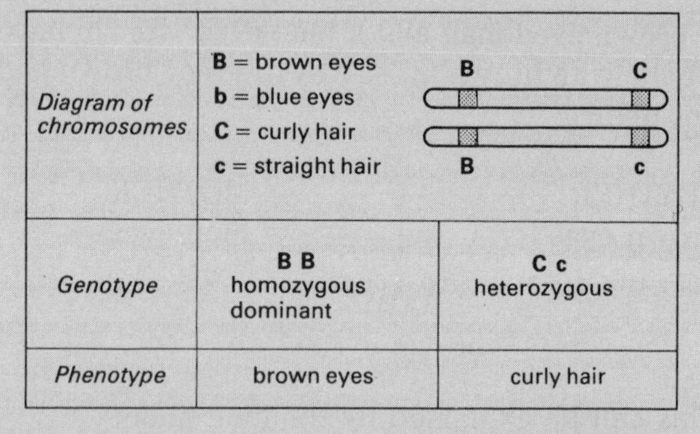

Diagram of chromosomes	**B** = brown eyes **b** = blue eyes **C** = curly hair **c** = straight hair	B C (upper chromosome) B c (lower chromosome)
Genotype	**B B** homozygous dominant	**C c** heterozygous
Phenotype	brown eyes	curly hair

The table below the diagram shows how you describe the genotypes and phenotypes for this example. Study the table and the diagram, then look at the example below. This refers to two pairs of genes which control certain characteristics of pea plants.

T = tall **R** = round peas
t = dwarf **r** = wrinkled peas
G = green pods
g = yellow pods

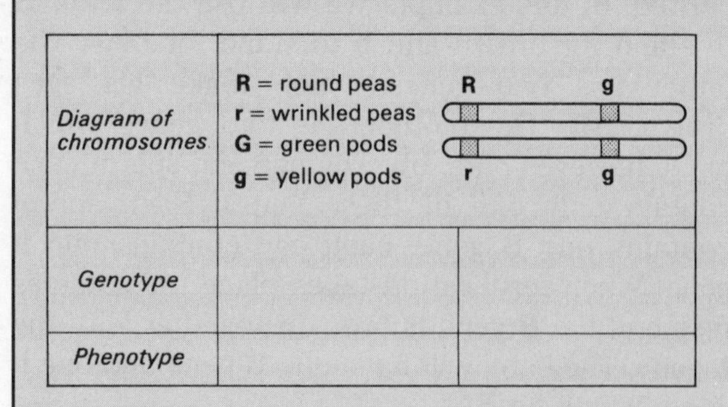

Diagram of chromosomes	**R** = round peas **r** = wrinkled peas **G** = green pods **g** = yellow pods	R g (upper chromosome) r g (lower chromosome)
Genotype		
Phenotype		

Copy the diagram and the table. Fill in the table.

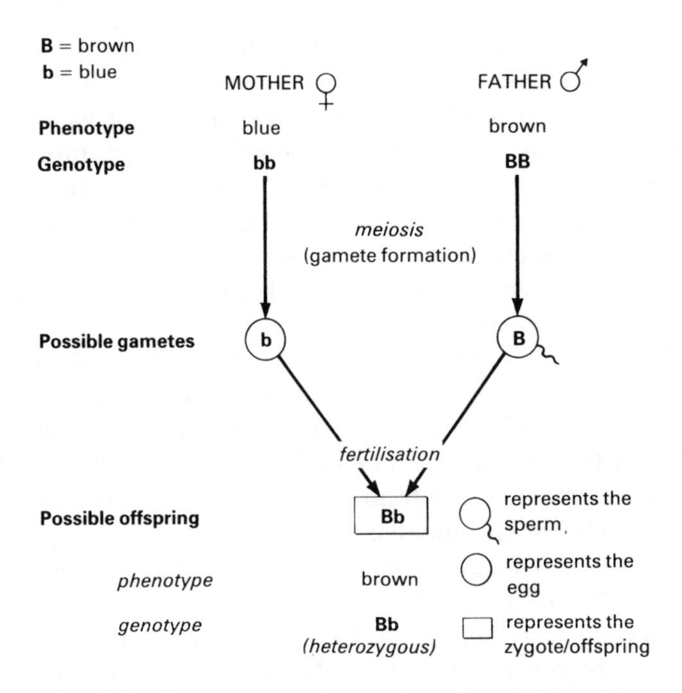

Fig. 8.61 Inheritance of eye colour from blue-eyed mother and homozygous brown-eyed father

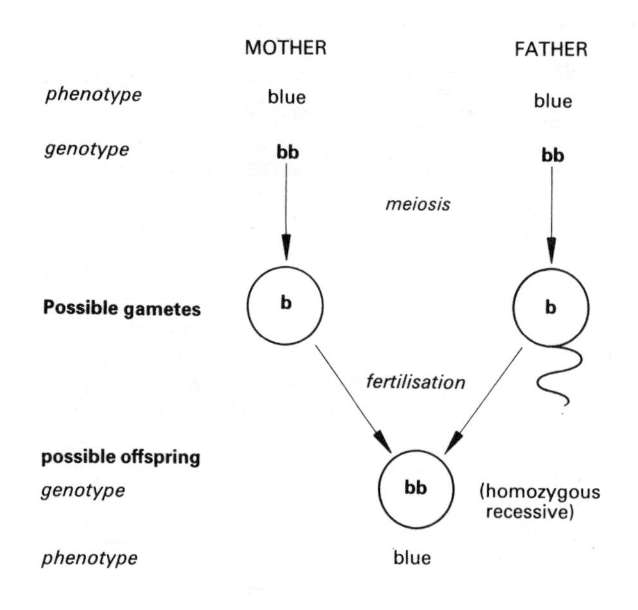

Fig. 8.62 Inheritance of eye colour when both parents have blue eyes

zygous brown eyes. This is a good way to set out your answers when doing genetics problems.

If both parents had had blue eyes all the children would have blue eyes. See Fig. 8.62.

If both parents had had brown eyes they could each have had the genotypes **BB** or **Bb**. This could have given any of the alternatives shown in Fig. 8.63.

In (a) (**BB** × **BB**), all offspring would have brown eyes and all would have the same genotype.

In (b) (**BB** × **Bb**), all offspring would have brown eyes but the genotype could be either **BB** or **Bb**.

In (c) (**Bb** × **Bb**), we can see the following.
(i) There are two different genotypes for brown eyes.
(ii) It is possible for two brown eyed parents to have a blue eyed child if both are heterozygous (**Bb**), that is, if both have a recessive gene.
(iii) The ratio of brown:blue is 3:1. This means it is more likely that they will have a brown eyed baby. The chance is 3:1. It does *not* mean that if a couple have four children three will have brown eyes and one will have blue. This shows what could happen each time random fertilisation happens (see page 152). So it would be possible for them to have four blue eyed children if every time an egg containing a **b** gene was fertilised by a sperm containing a **b** gene.

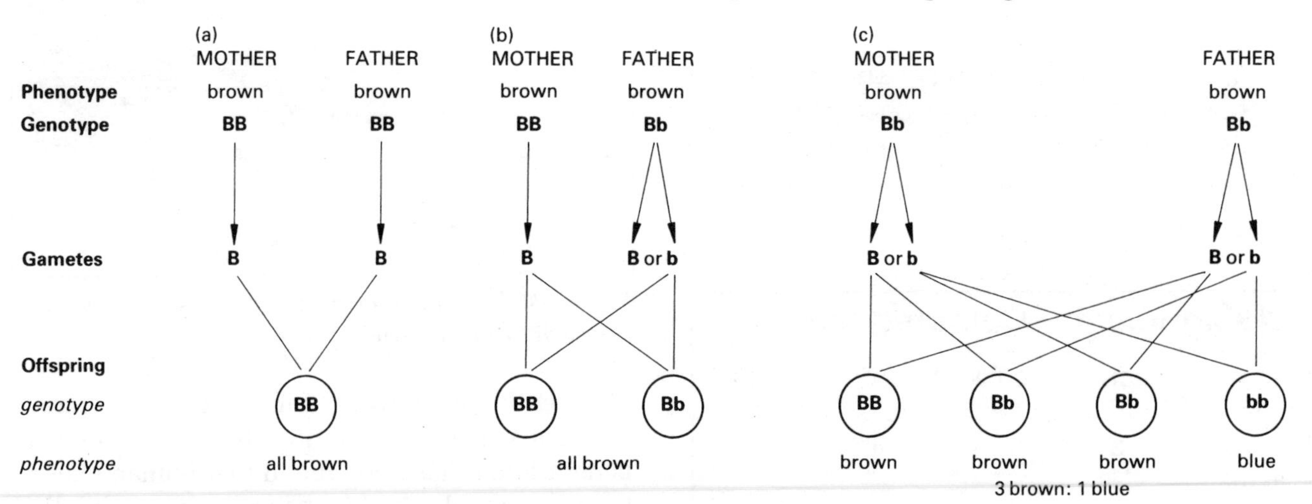

Fig. 8.63 Inheritance of eye colour when both parents have brown eyes

Assignment – Phenotypes and genotypes

Look back at Fig. 8.56. What are the genotypes for (a), (b) and (c)? Put your answers in a table like the one below.

	Genotype	Phenotype	Homozygous or heterozygous
(a)			
(b)			
(c)			

Assignment – Hair colour

Dark hair **D** is dominant to blond hair **d**. Using these symbols work out the following.

1 What are the possible genotypes for (a) a person with dark hair and (b) a person with blond hair?
2 What are the possible hair colours of children from the following?
  (a) A homozygous dark haired man and a blond woman.
  (b) A heterozygous dark man and a heterozygous dark woman.
  (c) A heterozygous dark man and a blond woman.
3 If a child has blond hair, what are the possible genotypes and phenotypes of its parents?

Assignment – Coloured mice

In mice black coat colour **B** is dominant to brown **b**. What are the possible outcomes of mating a brown male with a black female?

How could you find out if the genotype of a brown eyed person was homozygous **BB** or heterozygous **Bb**? With humans this is not easy. If a couple of brown eyed people have any blue eyed babies they both must have a recessive gene and be **Bb**, as in (c) in Fig. 8.63. But if all the children have brown eyes the genotype of either parent could be either **BB** or **Bb** as in (a) and (b) in Fig. 8.63.

Most genetics research has been done with plants and small animals which breed quickly, which have large numbers of offspring and which can be kept easily in laboratories or under strict experimental conditions. Obviously this is not possible with humans!

An Austrian monk called Gregor Mendel (1822–84) first discovered the way that characteristics are inherited. He did experiments with pea plants, studying the inheritance of various of their characteristics such as height, flower colour, seed shape, etc. Since Mendel's time the rules that he deduced from his work on the pea plants have been found to apply to all other animals and plants. A small fruit fly called *Drosophila* has also been used for a lot of research, as the insect has some very clear characteristics of eye colour, size and shape of wings and it breeds very quickly. You may have some *Drosophila* in your school laboratory.

Fig. 8.64 The fruit fly Drosophila *is used in genetics research (14 times life size)*

Recently, by studying family trees and characteristics which are inherited from generation to generation, it has been found that human characteristics are inherited in the same way as in other living organisms.

True breeding, back cross

As we have seen above, one phenotype may represent more than one genotype. For example, with pea plants, tall **T** is dominant to short **t**. A tall plant may have the genotype **TT** (homozygous) or **Tt** (heterozygous). With this plant the easiest way to find out whether the tall plant is homozygous or heterozygous is to self-pollinate it, that is, transfer pollen from the anther to the stigma using a paintbrush. (See pollination on page 101.) When the resulting seeds (peas) are grown, if the plants from them are all tall we can conclude that the parent was **TT**. If they are a mixture of tall and short then the parent must have been **Tt**. This is shown in Fig. 8.65.

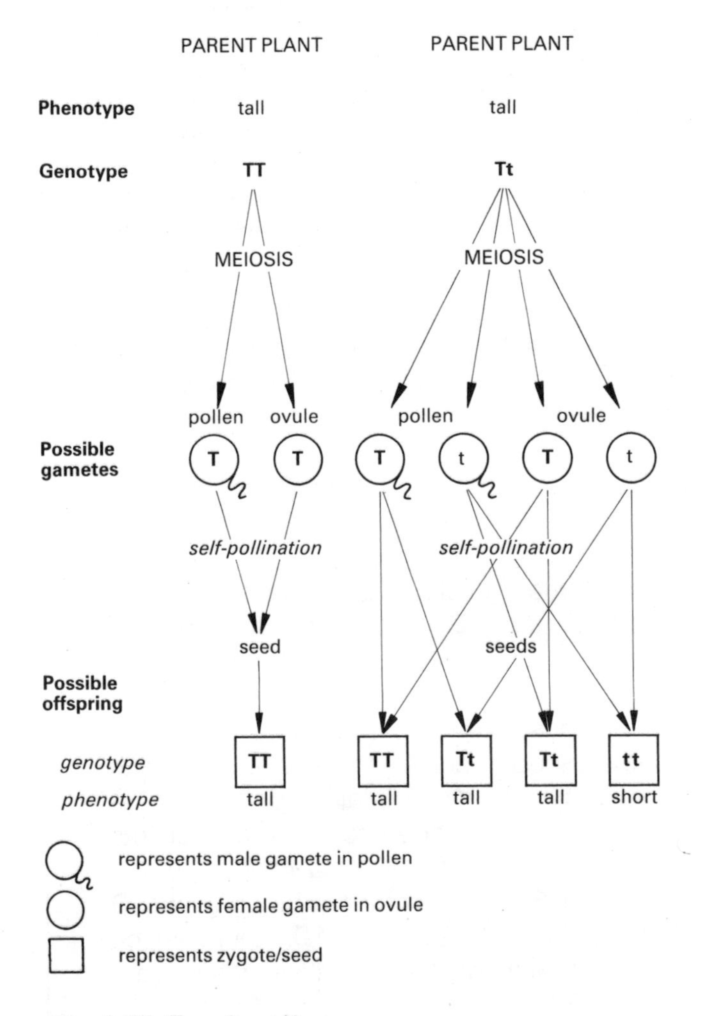

Fig. 8.65 True breeding

If a homozygous plant is self-fertilised it will produce offspring which are the same as the parent. The same happens if any homozygote is crossed with another homozygote. This is known as *breeding true.*

Another way that you could find out if the tall plant is **TT** or **Tt** is to cross it with a short plant, **tt** homozygous recessive. This is called a *back cross*. If all the resulting seeds grow into tall plants the original tall plant was **TT** but if some of them grow into tall plants and some into short plants it was **Tt**. This is explained in Fig. 8.66. Back crosses always have to be used with animals as it is not possible to self-fertilise them!

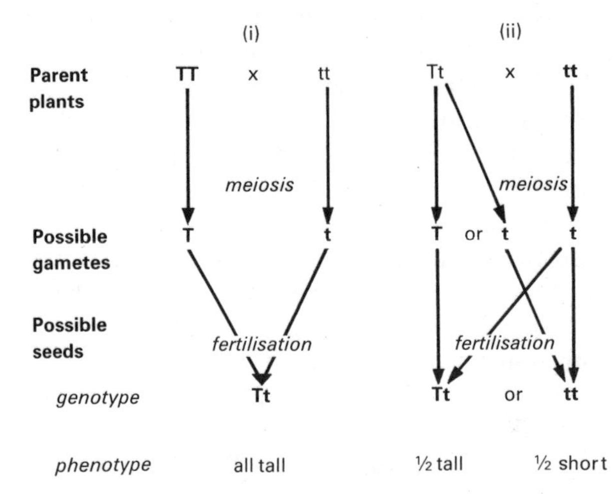

Fig. 8.66 Back cross

Assignment – Back crosses

1 In pea plants, green pods **G** are dominant to yellow pods **g**. If a plant has green pods describe two ways that you could find out if it is a true-bred plant.

2 In mice, albinoism is recessive to normal characteristics.
(a) What are the possible genotypes of a normal mouse?
(b) How could you find out if a normal mouse is homozygous or heterozygous?

3 In *Drosophila*, long wings are dominant to vestigial wings. Three pairs of flies, all with long wings, are mated. What can you deduce about the genotypes of each of these pairs if their offspring have wings as follows:

	Offspring
Pair 1	101 long
Pair 2	45 long, 49 vestigial
Pair 3	76 long, 24 vestigial

3:1 ratio and other 'results'

In Fig. 8.65, when the heterozygote plant was self-fertilised, we can see that the resulting offspring were **TT**, **Tt**, **Tt**, **tt**. There are three tall plants for every short one. That is a 3:1 ratio of tall:short. So if you see some results that have a 3:1 ratio this usually means that the parents were heterozygotes, in this case **Tt**.

If all the offspring are the same, and are recessive, then both parents must be homozygous recessive. If all the offspring were **tt** then both the parents must also have been short **tt**.

If the offspring are $\frac{1}{2}$ dominant and $\frac{1}{2}$ recessive, for example $\frac{1}{2}$ tall and $\frac{1}{2}$ short, one parent must be homozygous recessive **tt** and one must be heterozygous **Tt**, as in Fig. 8.66(ii).

If all the offspring are the same and show the dominant characteristic, their parents could have been either of the following combinations:

both homozygous dominant **TT**

one homozygous dominant **TT** and one heterozygous **Tt**

one homozygous dominant **TT** and one homozygous recessive **tt**.

All the ratios of 'results' are summarised in Table 8.5.

Table 8.5

Results of offspring (phenotypes)	**Parents must be (genotypes)**
3:1 tall:short	**Tt & Tt** both heterozygous
1:1 tall:short	**Tt & tt** one heterozygous one homozygous recessive
all tall	**TT & TT** both homozygous dominant **TT & Tt** one homozygous dominant one heterozygous **Tt & Tt** both heterozygous
all short	**tt & tt** both homozygous recessive

Assignment – Human blood groups

In some cases there may be more than two alleles of a gene, although one individual can have only two of the possible alleles. An example of this is found in human blood groups. These are controlled by three alleles **A**, **B** and **O**.

There are four blood groups **A**, **B**, **AB** and **O**.
A and **B** genes are equally dominant.
O gene is recessive to both **A** and **B**.

Blood group (phenotype)	Genotype
A	**AA** or **AO**
B	**BB** or **BO**
AB	**AB**
O	**OO**

1 A woman is blood group **A** and a man blood group **B**. If they have a child what blood group could it have? Write down all the possibilities.

2 Is it possible for parents who both have the blood group **A** to have a child with blood group **O**? Explain your answer.

3 A man being sued for paternity claims not to be the father of a child of blood group **B**. If the mother is blood group **O** and the man is group **AB**, could he be the father of the child? Explain your answer.

Investigation 8.9 The inheritance of tongue rolling

Some people can roll their tongues, others cannot. Look at the photograph on p. 154.

Collect

a mirror

What to do

1 Put out your tongue and try to roll it as in the photograph.
2 Look in the mirror and decide whether you are a roller or a non-roller.
3 Look at other members of the class. Count how many can roll their tongues and how many cannot. Write down the two totals.
4 Work out what proportion of your class is able to roll their tongue.

$$\frac{\text{total no. rollers}}{\text{total no. in class}}$$

5 Which do you think is dominant, rollers or non-rollers?
6 How confident are you in your conclusions? How could you be more certain?

Assignment – More about tongue rolling

1 Find out if other members of your family can roll their tongues. Then construct a family tree of the phenotypes (rollers and non-rollers) as in the example below.

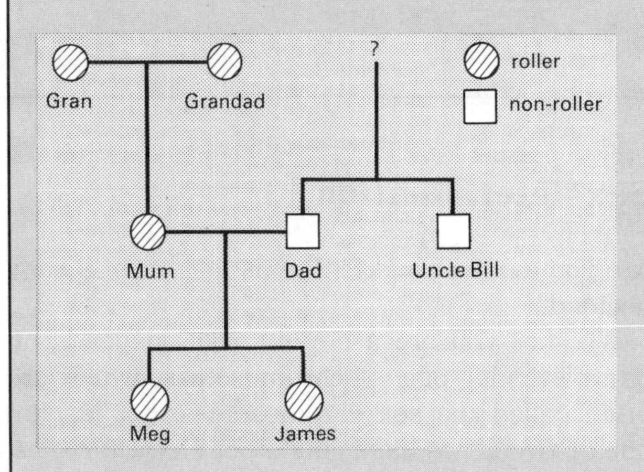

2 Can you work out any of the genotypes? Use the symbols

R = roller
r = non-roller

3 In this example, do you think Dad's parents would be rollers or not?

Assignment – Ear lobes

Look at the family tree and at the photographs on p. 154.

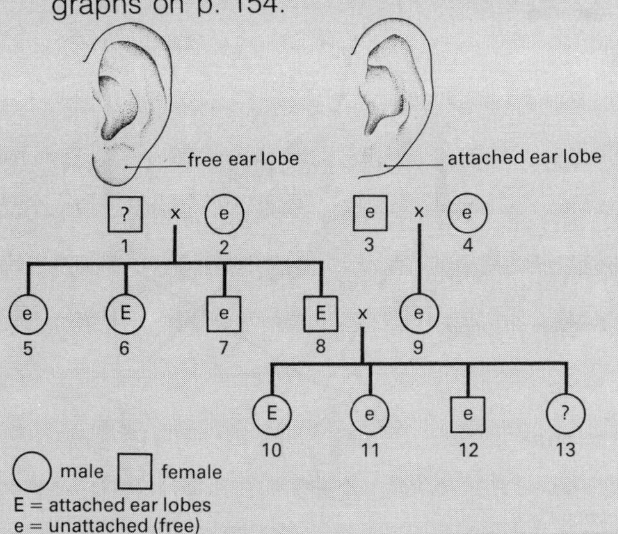

1 (a) Which type of ear lobe is dominant?
(b) Which men have attached ear lobes?
(b) Which women have free ear lobes?
(c) What do you think the phenotype and genotype of **1** would be? Explain your answer.
(d) What are the possible phenotypes and genotypes of **13**?
(e) If **6** married a homozygous dominant male (attached ear lobes), what would you expect the ears of their children to be like?

2 Make your own family tree for ear lobes. Write the names of your family under the symbols instead of numbers which are used in the example given.

Assignment – *Drosophila* eye colour

In the fruit fly *Drosophila*, red eye colour is dominant to brown eye colour. Two flies were crossed and the following offspring were obtained:

red eyed flies	32
brown eyed flies	10

What do you think was the genotype of the parents? Explain your answer using suitable symbols and diagrams.

2 If a brown eyed fly was crossed with a *pure breeding* red eyed fly, what would you expect the genotypes and phenotypes of their offspring to be? Explain your answer using symbols and diagrams.

Sex determination

Are you a boy or a girl? How is the sex of a baby decided?

Whether you are a boy or a girl depends on one particular pair of chromosomes, which are often called the sex chromosomes, not just on one or two genes like many other characteristics. Humans have 23 pairs of chromosomes. Twenty-two of these pairs are the same but the twenty-third pair, the sex chromosomes, may not be the same. Look back at Fig. 8.51 (page 147). Can you see a difference in the twenty-third pair of chromosomes between a man and a woman?

There are two types of sex chromosomes: a long one called the **X** chromosome and a shorter one called the **Y** chromosome. In women there are two **X** chromosomes. In men there is one **X** chromosome and one **Y** chromosome. The female genotype is **XX** and the male genotype is **XY**.

But how and when is the sex of a baby decided? When meiosis takes place in the female's ovaries all the eggs produced will contain an **X** chromosome. But when sperms are formed by meiosis in the testes, half the sperms will have an **X** chromosome and half will have a **Y** chromosome.

When fertilisation occurs, if an **X** sperm fertilises the egg the baby will be **XX** and be a girl. If a **Y** sperm fertilises the egg the baby will be **XY** and be a boy. This is shown in Fig. 8.67.

Because half of a man's sperms contain an **X** chromosome and half contain a **Y** chromosome, there is an equal chance of the egg being fertilised by an **X** or a **Y** sperm. So each time a couple have a baby there is a 50:50 chance that it will be a boy, or a girl. The population as a whole is made up roughly of equal numbers of men and women because of this.

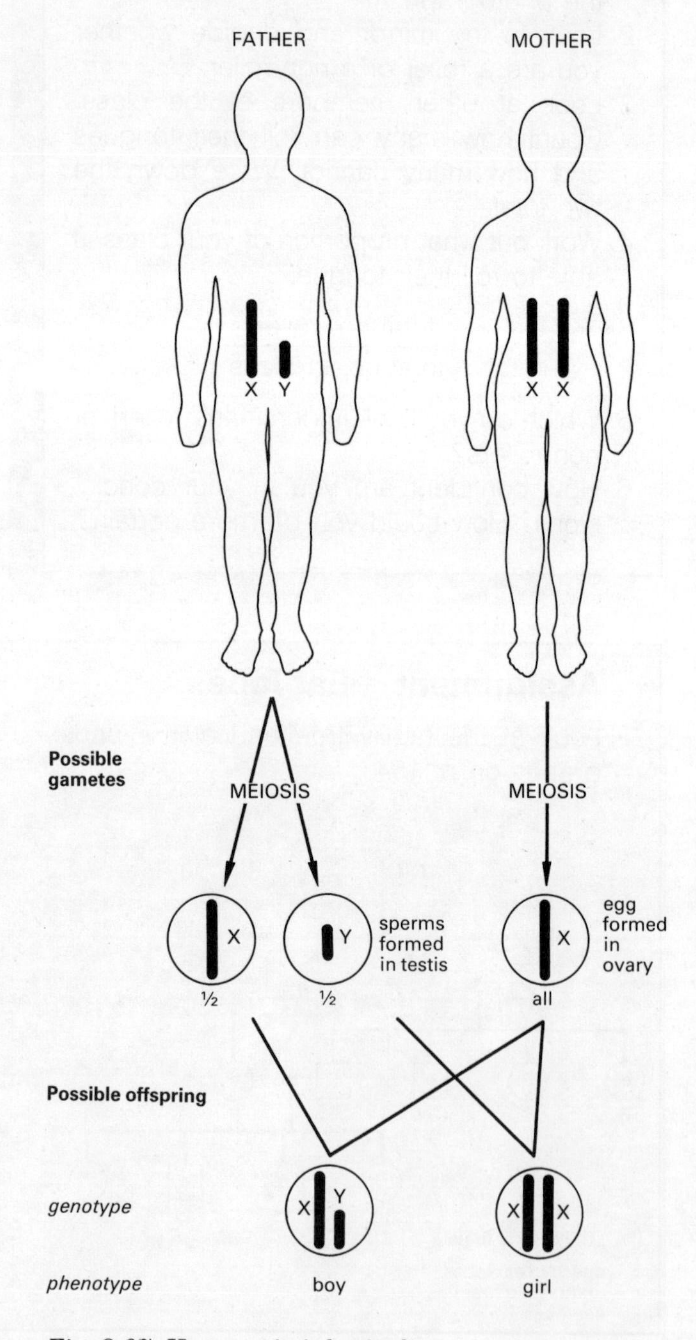

Fig. 8.67 How sex is inherited

Assignment – Do you want a boy or a girl?

'We already have two daughters and would like to have a son. Is there any way that this can be arranged?'

50% of couples with two children have two of the same sex. Many of them will have asked this question. There are now several ways that the sex of a child may be influenced at fertilisation. Two of these are described below.

Diet before conception

Two French doctors, J. Stolkowski and J. Choukron, think that the diet of the mother before conception (fertilisation) can be used to produce a boy or a girl. For a boy, the mother should eat foods rich in potassium and sodium and foods rich in calcium and magnesium should be avoided. For a girl the diet is reversed. Anyone trying this method should be careful that, overall, their diet remains balanced, or there may be a danger of birth abnormalities.

The French doctors say that, between 1979 and 1980, they obtained 22 boys and 17 girls as requested. There were 7 failures but they say that in these cases the women did not stick to the diet.

Separating the sperm

A Californian biologist, R. Ericsson, claims to have found a method for getting boys. His method is to take the father's semen and separate the 'male' sperms from the 'female' sperms. The 'male' sperms are then used to inseminate the mother.

Out of 84 births using this method, 65 were boys and 19 were girls, i.e. 77% were boys, which is more than the average of 52.5% boys at birth in the USA and the UK. The same method cannot be used for getting girls.

There are also seasonal and other influences on the number of boy and girl babies born. In late autumn and winter there is an increase in the number of boys born. There is also an increase in the number of boys born during wartime. It has been suggested that the reason is that when intercourse is more frequent – during spring and summer, and when soldiers return for short leaves during wartime – the chances of having a boy are greater. Younger parents and newlyweds are also more likely to have boys.

1 For the 'diet' method:
   (a) suggest a list of foods a woman should eat, and a list of those she should *not* eat if she wants a girl;
   (b) say why it is important that a woman should have a balanced diet which includes the 'special' foods.
2 Which of the methods described above do you think is most reliable? Give reasons for your answer.
3 Should parents be able to choose the sex of their baby? What do you think? Give advantages and disadvantages that may arise if all parents were able to choose the sex of their child.
4 The article mentions 'male' sperm and 'female' sperm. Explain what this really means.

Assignment – Acquired and inherited characteristics

There are two types of characteristics. Some are inherited from your parents via genes, such as brown eyes. Others are acquired characteristics which develop during a person's life (for example, sun tan). Acquired characteristics cannot be passed on to offspring. Many characteristics are a mixture of inherited and acquired, for example, weight.

1 Divide the following characteristics into two groups, inherited and acquired:

   hair colour, shape of nose, ability to read, rickets, height, well developed muscles, type of ear lobe, cancer

2 Can you add to either list?
3 Are any of these characteristics a combination of inherited and acquired?

Assignment – Sex linkage

Certain characteristics are linked with the sex chromosomes. Many of these sex-linked characteristics are seen only in men. This is due to the different structure of the **X** and **Y** chromosomes. Look at the diagram below which shows the positions of two genes.

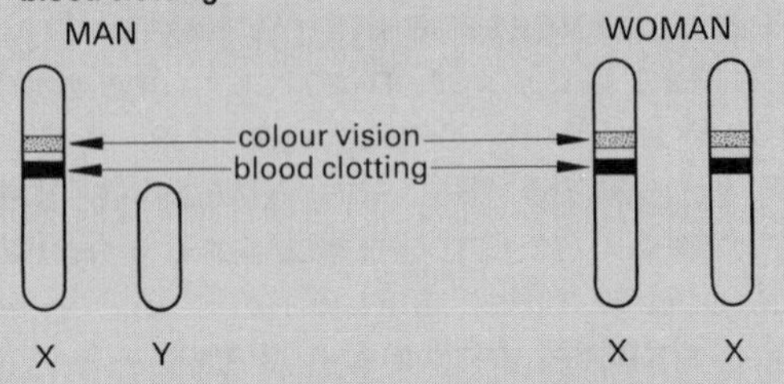

1 How many alleles (genes) are present for blood clotting in (a) a woman, (b) a man?

The dominant gene **H** allows the blood to clot normally. The recessive gene **h** causes haemophilia, a disease where even a bruise or a small scratch will go on bleeding for a very long time.

A woman has three possible genotypes but a man has only two because the **Y** chromosome does not have a blood-clotting gene of any kind.

	Genotype	Phenotype
WOMAN	X^HX^H X^HX^h X^hX^h	normal carrier haemophiliac
MAN	X^HY^- X^hY^-	normal haemophiliac

2 If a 'carrier' woman and a normal man have children, is it likely that the children will be normal? Explain your answer.

3 The gene for colour blindness **c** is recessive to the gene for normal vision **C**.
(a) What would be the genotype of a colour-blind man?
(b) A man and his wife both have normal vision. They have three sons, two of whom are colour blind. Explain how this may happen.

APPLICATIONS OF GENETICS

Selective breeding

Gradual change in species is brought about by evolution, which is described later in this unit (page 169), but people have also changed species by selective breeding. Animals have been bred from parents which have favourable characteristics. For example, with cows, the ability to produce large quantities of milk is obviously desirable. From each successive generation individuals have been selected which had the most useful phenotypic characteristics and, over the years, all the domestic animals – pigs, cows, sheep, etc. – have developed so that they are now very unlike their ancestors. Think of the differences between a pig and a wild boar.

It is also possible to change plant species by selective breeding. In plants such as the potato, the storage organs are important to help the plant survive the winter. Gradually, over many generations, the largest tubers have been selected for breeding and not eaten. So the modern potato plant has been developed from a wild South American ancestor. The modern potato has much larger tubers than would be needed by the wild plant to last through the winter.

Many other characteristics have been selectively bred by farmers in the past and now by research scientists. Some of these are listed below.

(a) Higher yields: of milk from cows; of wool; of various cereal crops and of fruit.
(b) Resistance to disease, for example, in wheat, types have been developed which are resistant to a very serious fungal disease called rust.
(c) Early maturity in both plants and animals. In the case of plants this means that two or more crops can be grown in one season.
(d) Increased length of productive season. This means that certain crops such as strawberries have a much longer season and therefore usually give a higher yield per plant as a result.
(e) Adaptation to unfavourable conditions; for example, crops have been developed that will survive in very dry places, very cold places or in certain types of poor soil.
(f) Improved taste and eating quality; for example, seedless oranges.

Fig. 8.68 Selective breeding has changed the pig into an animal very different from his wild ancestor

Selective breeding is sometimes called artificial selection, as people are artificially selecting animals and plants to breed together instead of allowing them to breed naturally. Natural selection is dealt with on page 175.

Artificial insemination (AI) of cattle has helped with selective breeding in cattle in this country, especially with dairy cows. A farmer does not need to keep a bull to service (mate with) his cows. There are numerous AI centres around the country where semen is collected from bulls which are known to have favourable characteristics. The semen is frozen and stored and can then be used to inseminate cows. Insemination means putting the semen into the cow.

Assignment – Horned and hornless cattle

A farmer would like all his cows to be hornless. Most of his herd have no horns but occasionally a few horned calves are born. The farmer knows that the gene for horns is recessive **n** to the gene for no horns **N**. This means that some of his hornless cattle are probably heterozygous **Nn**.

The farmer does not have a bull of his own so he uses AI for all his cows.

What should the farmer do to get a completely hornless herd?

A dairy farmer can choose which bull is best for each of the cows in the herd. The farmer may decide that, for some cows, a bull who has good characteristics for high milk yields will be better than a bull with good characteristics for putting on weight quickly (desirable for beef cattle). This depends on what the farmer hopes to do with the calves when they are born.

Genetic engineering

Genetic engineering is a branch of genetics which is involved in trying to change the actual genes of an organism. The idea is to produce either a better characteristic in an organism or to produce a new species.

Some examples are given below.

(a) Genes have been added to bacteria to make them produce antibiotic drugs and other useful chemicals.

(b) Other bacteria have been changed to make them 'eat' oil which has been spilled on the sea.

(c) Growth hormones, and other hormones such as insulin, have been made artificially by genetic engineers.

Protein engineering is a branch of genetic engineering. Protein engineers are attempting to construct new genes that will make new proteins. These new proteins could be used instead of naturally occurring ones. For example, enzymes carry out many chemical reactions that could be useful in industrial processes, but they are too easily denatured. (This means that their structure is changed in a way that cannot be reversed. Boiling an egg changes the protein in it by denaturing it – just think of trying to change a hard-boiled egg back into a fresh one!) Protein engineers will try to make new proteins which carry out the same reactions as enzymes but which are not denatured at high temperatures.

As well as making new genes, genetic engineers also hope to remove some bad genes such as those that cause hereditary diseases like haemophilia and cystic fibrosis.

Assignment – Cloning

As well as tinkering with genes to produce new and different species of plants and animals, it is now possible with some species to produce many identical animals by a technique called cloning.

This has always been possible with plants. Many plants can be produced from the same parent by taking cuttings (see asexual reproduction, page 97). All are the same as the parent.

Identical animals can be produced from one parent by using cells taken from the parent. This has already been done with frogs.

The nucleus of a frog's egg is killed using ultraviolet rays. A cell is taken from the intestine of the frog and the nucleus from this is put into the frog's egg. The frog egg then grows into a tadpole and eventually into an adult in the usual manner. However, the adult frog is identical to its parent. All its genes will be identical to its parent's genes, as they are those of the nucleus of a cell from the intestine and not a combination resulting from fertilisation.

This technique of cloning could be profitable in animal breeding to make exact copies of high-yielding milk cows, prize bulls, good wool-producing sheep, race-winning horses, etc. It may also be possible to produce human clones by similar techniques but although the clones would have identical bodies their personalities may vary and environmental influences would also produce variations, as with identical twins.

1 Would it be a good idea to have human clones or would it cause problems? List some advantages and disadvantages.
2 Who would have the right to create clones?
3 Who would be used as parents for human clones?

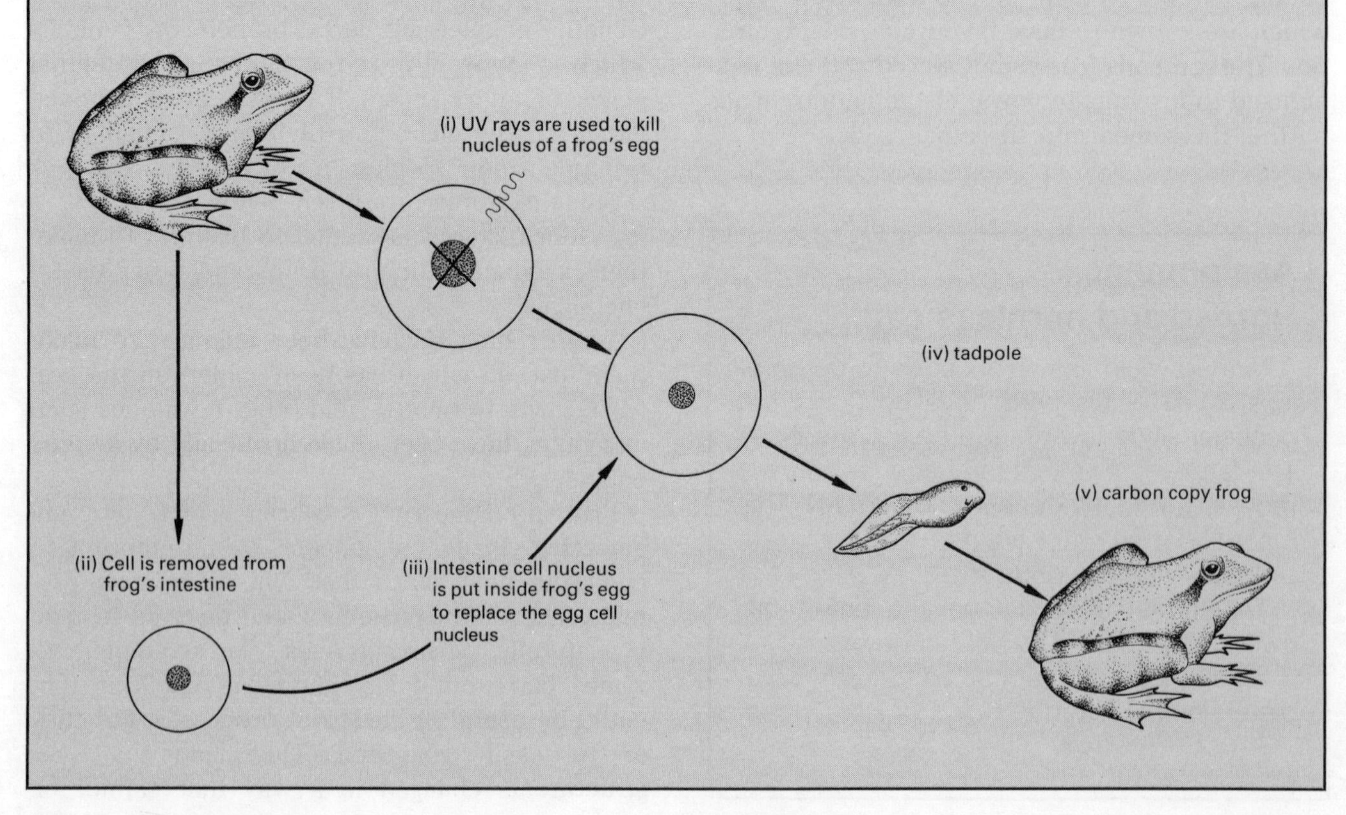

Assignment – The genius baby

The genius baby begins its climb

DORON BLAKE, the first deliberately designed boy "genius", celebrates his first birthday on Wednesday. Last year The Sunday Times revealed the baby was one of a group about to be born as a result of an extraordinary genetic experiment designed to enhance the number of highly intelligent people in future generations.

The Repository of Germinal Choice, a California-based charity founded by an eccentric millionaire, Robert Graham, is a sperm bank containing contributions from Nobel Prize winners and outstanding younger scientists and mathematicians.

A woman who wants a superior baby can apply to the bank and if accepted she chooses a donor. She is then sent a canister of liquid nitrogen containing a month's supply of his frozen sperm with which she can inseminate herself.

Afton Blake, a 41-year-old unmarried psychologist with a lucrative Los Angeles practice, chose Doron's father from a portfolio of donors which listed not only his attributes – looks, intelligence, genetic potential – but also such minor defects as a tendency to haemorrhoids and impacted wisdom teeth.

Donor number 28, as he was listed, is a brilliant computer scientist at a European university, as well as an accomplished musician and athlete.

Doron (curiously, an anagram of donor) has inherited his father's Nordic good looks but what about his intelligence? At four months, psychologists from the university of California's child development centre estimated he had an IQ of 200. ('They were jumping up and down with excitement he was scoring so high on their tests,' says Afton Blake.)

She relies on her own observations: 'He has an athletic figure and is very agile. Last week I found him four rungs up a ladder. He says things like "doggie", "oh boy", and "oh dear". What he really needs now is a computer.'

Sunday Times, 21 August 1983

Marjorie Wallace

1 (a) What is the Repository of Germinal Choice?
  (b) What happens there?
2 How can a woman try to conceive a 'genius' baby?
3 (a) Are the characteristics listed in the portfolio of donors phenotypes or genotypes? Explain your answer.
  (b) What type of men are chosen to be donors? Do you think selection of donors should be done like this, in some other way or not at all? Give reasons for your answer.
4 Do you think this type of practice should be allowed? If you were the president of the USA would you allow it to continue? Give reasons for your answer.
5 If all women could choose the father of their children as described here, would they still need to get married? Discuss your answer.
6 List some of the advantages and disadvantages to society of being able to produce 'genius' babies.

Assignment – Genetic counselling

If a couple know of a possible inherited disorder in the family, a genetic counsellor can estimate the risk of it occurring in their children. The estimate is based on family history and special tests. Genetic counsellors offer advice and support but do not tell parents what they should do. The final decision about having children or possibly having an abortion is left to the parents.

Diseases caused by a chromosome disorder are investigated with the aid of a microscope. In a normal human being there are 46 chromosomes in 23 pairs in each cell, so it is easy for odd numbers to be spotted. For example, people with Down's Syndrome (mongolism) have 47 chromosomes.

Diseases caused by abnormal individual genes are recognised by the characteristic signs of the condition, by family histories and by special medical tests. Phenylketonuria is a metabolic disease which causes mental retardation. It is a 'recessive' condition which may occur if both parents 'carry' the abnormal recessive gene. This cannot be detected prenatally (before birth), but if detected at birth it can be treated successfully.

The most common method of detecting abnormality in pregnancy is by amniocentesis. A small quantity of amniotic fluid is collected by inserting a needle through the mother's abdomen into the uterus. Laboratory tests can then be done on the baby's cells, which will be floating in the fluid. Although amniocentesis is quite a safe procedure, it does involve a slight risk of miscarriage. The most common reason for offering this procedure is the increased chance of Down's Syndrome in 'older' mothers. Other reasons may be a previous baby with Down's Syndrome or chromosome abnormality: the presence of a 'carrier' situation in one parent. Chromosome studies may also be done to find the sex of the foetus in sex-linked conditions where a male baby would be at high risk, for example, haemophilia.

Ultrasound scans are now given routinely in many hospitals. They are used to check the position of the foetus and placenta and to estimate the age of the foetus so that the expected delivery date can be worked out more accurately. With continuing improvements in equipment some physical malformations may now be seen. Another new method of detecting such malformations is by fetoscopy. This is the introduction of a small viewing tube into the uterus through which the foetus may be directly examined. This method has a high risk of damaging or aborting the foetus.

Not all congenital malformations are genetic in origin. Environmental factors may also be involved. The drug thalidomide caused severe limb abnormalities. Heart, ear and eye defects and mental retardation are the result of a mother getting rubella (German measles) in early pregnancy.

1 List all the disorders mentioned and the causes of each.
2 List the methods used for detecting abnormalities before birth.
3 What do you understand by the following:
   (a) 'older' mothers;
   (b) a 'carrier' situation;
   (c) congenital malformations?
4 (a) Why is it so important that all girls should have rubella vaccinations?
   (b) Why is 11 to 13 the recommended age for rubella vaccinations?
5 How can a genetic counsellor help a couple who have a family history of inherited disease?

WHAT YOU SHOULD KNOW

1 In the nuclei of all cells are thread-like structures called chromosomes. The chromosomes are in pairs.

2 A chromosome has smaller parts called genes. Genes are made of DNA (deoxyribonucleic acid). Genes control all of an individual's inherited characteristics (for example, eye colour).

3 Genes are found in pairs, one on each chromosome of a pair. Each gene has two or more different forms called alleles (for example, brown and blue for eye colour).

4 Mitosis is a type of cell division which occurs when
(a) an organism is growing;
(b) cells are being made or replaced;
(c) asexual reproduction is taking place.
The resulting cells are identical to the original 'parent' cell.

5 Meiosis is a type of cell division which occurs only when gametes, eggs and sperms are being formed in the reproductive organs. The resulting cells have only half the number of chromosomes.

6 In humans all body cells have 46 chromosomes, that is, 23 pairs.
In the gametes, eggs and sperms, there are 23 single chromosomes.

7 At fertilisation when gametes fuse the full number of chromosomes is restored in the zygote.

$$\text{egg}(23) + \text{sperm}(23) = \text{zygote}(46)$$

8 Offspring inherit characteristics from both parents. These characteristics are carried in the genes of the chromosomes in the gametes. The study of how genes pass on inherited information is called genetics.

9 The genotype of an organism is its genetic make-up. This is inherited from the individual's parents. The genotype is usually expressed by using letters (for example, with eye colour **B** = Brown and **b** = blue).

10 The phenotype of an organism is its visible physical characteristics. This is due to its genotype and environmental influences.

11 If the genes in a pair of genes (alleles) are identical they are termed homozygous (for example, **BB** or **bb**). If the genes in a pair are not identical they are termed heterozygous (for example, **Bb**).

12 One of a pair of heterozygous genes may be dominant; its characteristic will show up in the appearance (phenotype) of the individual. The characteristic of the other gene of the pair will not show in the phenotype; this gene is termed recessive. (For example, an individual with genotype **Bb** will have brown eyes. **B** (brown) is dominant and **b** (blue) is recessive).

13 Sex in mammals is determined by the **X** and **Y** chromosomes. Males are **XY** and females are **XX**.

14 Artificial selection of characteristics can be used to breed new types of organisms. These may be useful in various branches of agriculture and horticulture.

15 In genetic engineering the genes of an organism are changed to produce more favourable characteristics or a new and different organism.

QUESTIONS

1 In a human, how many chromosomes are found in (a) a sperm cell in the testis; (b) a skin cell; (c) a fertilised egg in the uterus (womb); (d) a heart cell; (e) a hair cell?

2 (a) In which parts of the human body would you expect to find mitosis occurring in (i) a two-year-old; (ii) a fifty-year-old?
(b) In which parts of an old oak tree would you expect to find mitosis occurring?

3 Explain the difference between mitosis and meiosis.

4 Explain the difference between the following terms:
(a) dominant and recessive;
(b) heterozygous and homozygous;
(c) genotype and phenotype;
(d) hybrid and pure-bred.

5 Explain what is meant by the following:
(a) artificial selection;
(b) genetic engineering;
(c) cloning.

6 Copy the table below. From the list of terms, select those that best fit the descriptions in the table and enter in the space provided.

Description	Term
(a) An individual bearing identical genes for a given character	
(b) The genetic make-up of an individual	
(c) A permanent change in the chemical composition of a gene	
(d) Genes controlling the same characteristics but having different effects	
(e) The number of chromosomes found in a gamete	
(f) The characteristic that will show up in a heterozygous individual	

alleles, heterozygous
diploid, homozygous
dominant, mutation
genotype, phenotype
haploid, recessive

7 Mr and Mrs Singh have three girls. They are sure that their next child will be a boy. Do you agree? Explain your answer.

8 In *Drosophila* straight wing **S** is dominant over curved wing **s**.
(a) If you crossed a homozygous straight winged fly with a curved wing fly what would you expect in the offspring?
(b) If you obtained 300 straight winged flies and 99 curved-wing flies, what would the parents have been? Explain your answer.
(c) If you obtained 56 straight winged flies and 60 curved wing flies, what would the parents have been? Explain your answer.

9 In guinea pigs, short hair is dominant to long hair. If you had a short haired female and you wanted it to have some long haired babies, what type of male should you buy? Explain your answer.

10 Defective vision in dim light (night blindness) in humans is caused by a rare dominant gene. A man suffering from night blindness married a normal woman and half their children were normal and half were 'night blind'.
Using the symbol **N** for the dominant gene and **n** for the recessive gene, state the genotypes of
(a) the father;
(b) the mother;
(c) the normal children;
(d) the 'night blind' children.

Reproduction, Growth and Development

EVOLUTION

The world is full of millions of different kinds of plants and animals. They are all adapted to the habitat in which they live. Have you ever thought about how they got there, or how they have changed to suit the place where they live?

Until the nineteenth century, most people believed that all living things were created, at the same time, by God. Most people also thought that the different species had not changed since they were created.

During the nineteenth century scientists such as Charles Darwin and Alfred Wallace made other suggestions about the origins of life on earth. Darwin put forward a theory of EVOLUTION. He suggested that all the different species of living things found in the world today have descended from simpler forms over millions of years. He thought that the simple ancestors gradually changed or evolved into the living things we see today. But where did these simple ancestors come from? Scientists now think that the first forms of life on earth were single-celled plants and animals which gradually changed and developed into more complicated forms. Look at Fig. 8.69.

Assignment – Creation science

In Little Rock, Arkansas, USA, science teachers have developed a simple way of giving seven-year-olds a concept of the earth's geological age. They stretch a piece of string from end to end of a large classroom and they put a marker roughly in the middle to indicate the point in time when the first forms of life appeared. A few centimetres from the end which represents the present they add another marker to show when the human species evolved.

If a new Arkansas law is passed making it compulsory to teach 'creation science' along with the scientific theory of evolution, the teachers will have to do another demonstration. This time the string will be less than a centimetre long to demonstrate that the earth is only 10 000 years old rather than 4500 million. Man's origins will be marked close to the beginning as described in the book of Genesis in the Bible.

The children will be left to decide for themselves which version is true.

1 What are the main differences between the two theories?
2 If you were a seven-year-old, which theory do you think you would believe? Give a reason for your answer.

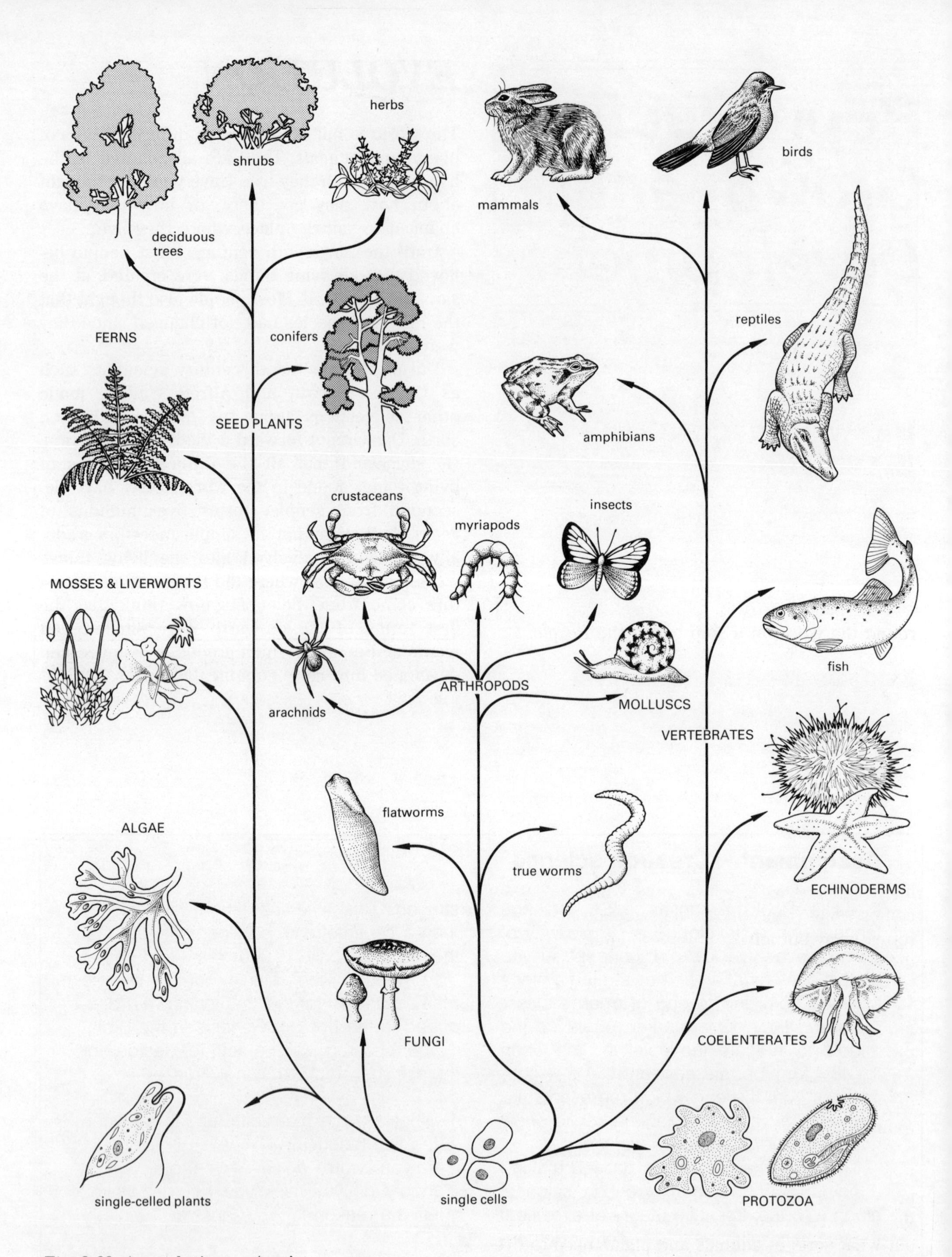

Fig. 8.69 An evolutionary 'tree'

Assignment – Theories of evolution

1 Find out about Lamarck's theory of evolution and Biblical views of evolution. (Read Genesis in the Bible.)
2 How does Lamarck's theory differ from Darwin's?
3 What evidence is there to disprove the idea of creation which is outlined in the Bible?

DARWIN'S THEORY OF EVOLUTION

In 1859 Darwin published a book called the *Origin of Species* which gave evidence for his theory of evolution and which explained his ideas. He wrote the book after studying plants and animals in many countries while sailing round the world in a ship called the *Beagle*.

Evidence for evolution

Nowadays most people believe that evolution has taken place. Darwin gave the following evidence for it in his book.

1 Fossils
Usually when a plant or animal dies it decays. Sometimes a FOSSIL is formed when plant or animal remains are buried by mud which later becomes sedimentary rock. The animals may have rotted away but left an impression in the rock, or the hard parts of animals (that is, the bones) may have gradually been replaced by rock.

Look at the photographs of fossils in Fig. 8.70. Which of them look like plants and animals that are alive today?

Geologists can usually find out how old a particular rock layer is. So the age of any fossils found there will also be known. (See Unit 1.)

Many different types of fossils have been found all over the world which are not like present-day plants and animals. From studying these fossils scientists have been able to work out what sorts of animals and plants used to live on the earth and when they lived.

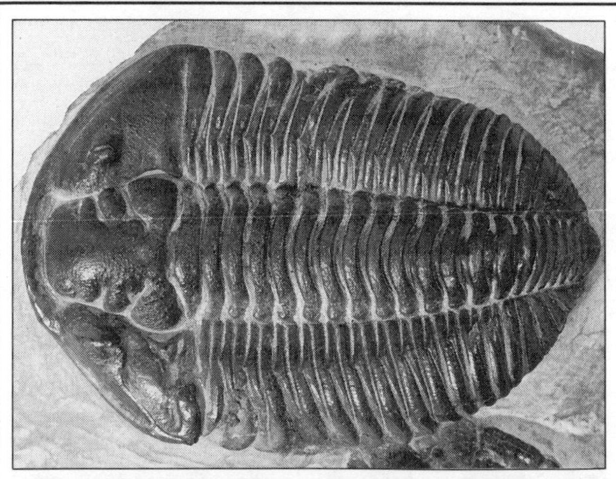

Fig. 8.70 Fossils

The fossil record shows that animals have changed from one form to another very gradually over millions of years.

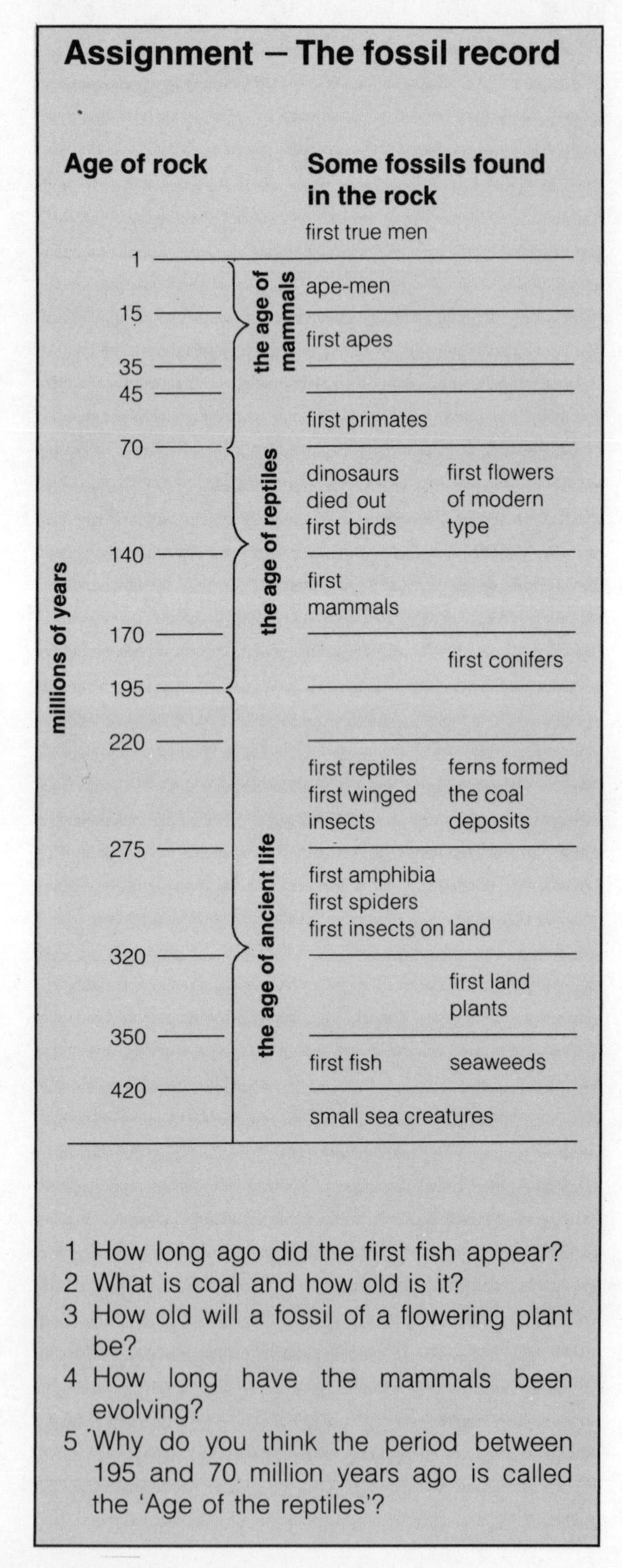

Assignment – The fossil record

Age of rock (millions of years)		Some fossils found in the rock	
		first true men	
1–15	the age of mammals	ape-men	
15–35	the age of mammals	first apes	
35–45	the age of mammals		
45–70	the age of mammals	first primates	
70–140	the age of reptiles	dinosaurs died out first birds	first flowers of modern type
140–170	the age of reptiles	first mammals	
170–195	the age of reptiles		first conifers
195–220	the age of ancient life		
220–275	the age of ancient life	first reptiles first winged insects	ferns formed the coal deposits
275–320	the age of ancient life	first amphibia first spiders first insects on land	
320–350	the age of ancient life		first land plants
350–420	the age of ancient life	first fish	seaweeds
420 +	the age of ancient life	small sea creatures	

1 How long ago did the first fish appear?
2 What is coal and how old is it?
3 How old will a fossil of a flowering plant be?
4 How long have the mammals been evolving?
5 Why do you think the period between 195 and 70 million years ago is called the 'Age of the reptiles'?

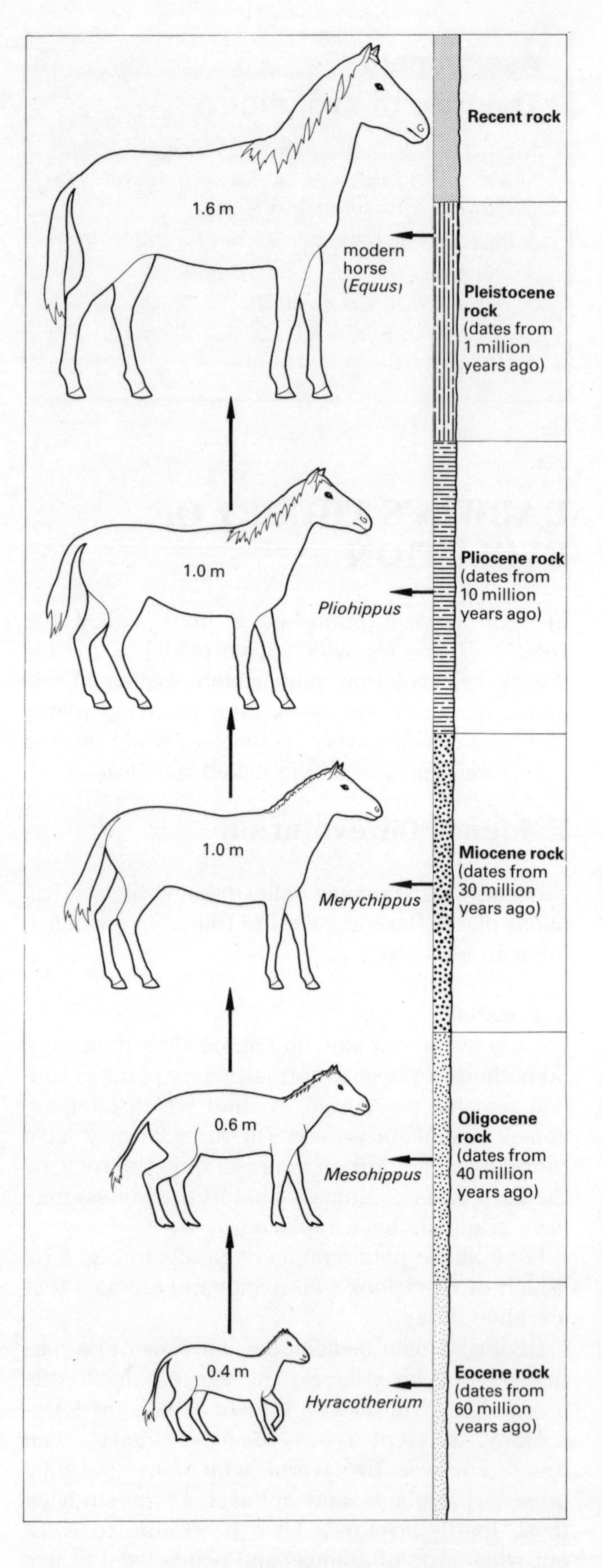

Fig. 8.71 The evolution of the horse

Assignment – Human skulls

The first human fossils have been found in rocks which are about 1 million years old – *Homo erectus* is one. Study the diagrams of the skulls of *Homo sapiens* (modern man) and *Homo erectus*.

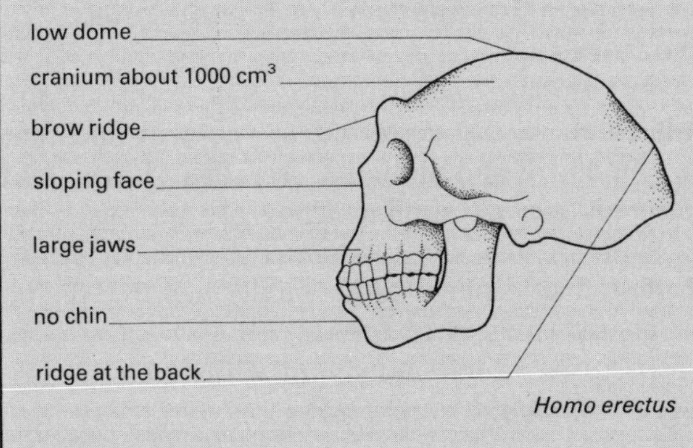

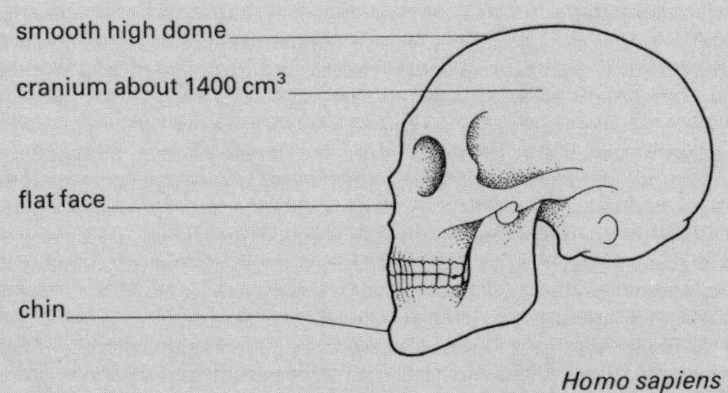

1 Draw a table like the one below. Look for differences between the skulls and fill in the table. The first difference has been done for you and you should be able to find four more differences.

Feature	Skulls of	
	Homo erectus	*Homo sapiens*
1 face shape	sloping face	flat face
2		
3		
4		
5		

2 How has the skull changed? Why do you think it has changed?

2 Geographical distribution

If you think about where certain animals are found in the world, this may give you some clues about how these species have evolved. Where do you find kangaroos? Where do you find lions? Most pouched mammals (which are called marsupials), such as kangaroos, are found only in Australia. Giraffes and lions are found only in Africa. Llamas and jaguars are found only in South America.

Look at the maps of the world in Fig. 8.72. Perhaps common ancestors of these mammals originated in North America or Asia and then went southwards. When the southern continents were cut off by the sea, did the mammals develop differently in their respective continents?

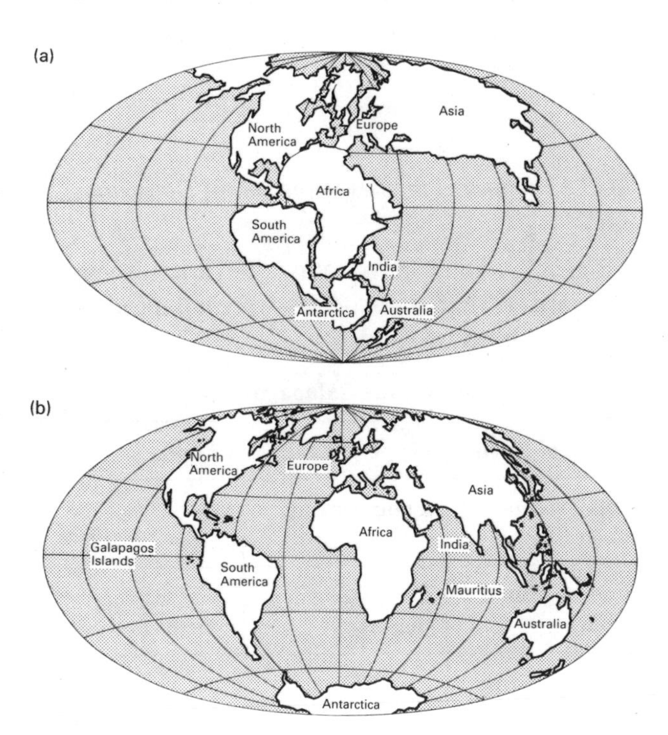

Fig. 8.72 (a) It is thought that the continents once formed one large land mass. (b) About 200 million years ago the continents began to drift apart, until they lie in the position they are in today

Barriers such as the sea or mountains have led to some species being found only in a small area. When Darwin visited the Galapagos islands (see Fig. 8.72), he noticed that there were different species of finches on different islands. The main difference between the species was the shape of their beaks. Finches found on the mainland of South America all have short, straight beaks for crushing seeds. On the Galapagos islands there are six main types: look at their different beaks

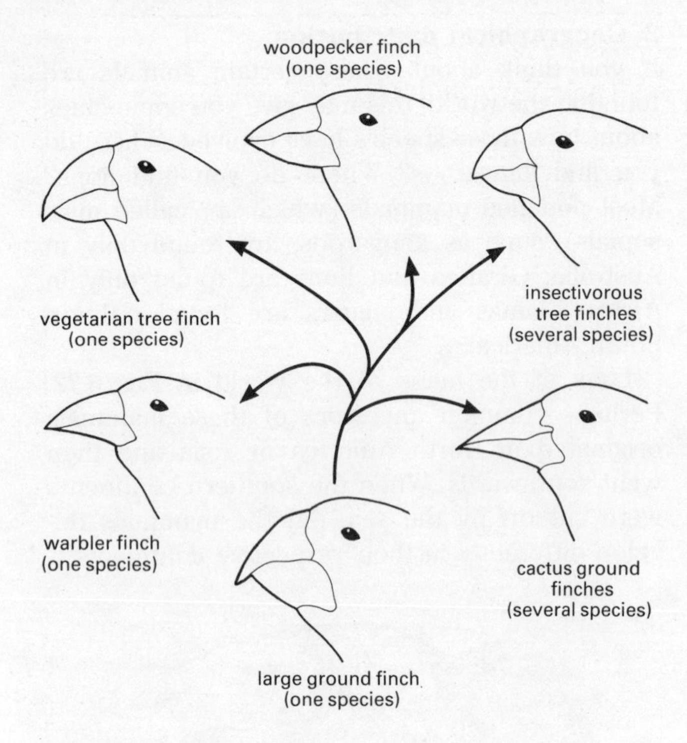

Fig. 8.73 Types of finches found on the Galapagos Islands

in Fig. 8.73. Darwin watched the finches eating and realised that their beaks were adapted for a particular kind of food.

Table 8.6 Finches on the Galapagos islands

Name of finch	Type of beak	Food
Large ground finch	short, straight	seeds
Cactus ground finch	long, straight; has a split	nectar from cactus flowers
Vegetarian tree finch	curved, parrot-like	buds and fruits
Insectivorous tree finch	small	beetles and other small insects
Warbler finch	slender	small insects which it catches while flying
Woodpecker finch	like a woodpecker, pointed	insects: it makes a hole in a tree with its beak, picks up a stick with its beak and pokes the insects out

Assignment – Finches key

Using the information in Table 8.6, construct a key which would enable someone to distinguish between the different types of finch. (See Unit 6, page 4.)

3 Comparative anatomy or homologous structures

If you compare the structure of a group of animals, such as the vertebrates, you will find that the limbs of amphibians, reptiles, birds and mammals are all similar. They are all based on the same design, the *pentadactyl limb*. It is called pentadactyl because it usually has five digits, that is, five fingers or toes.

Structures which show similarities like this are called *homologous structures*.

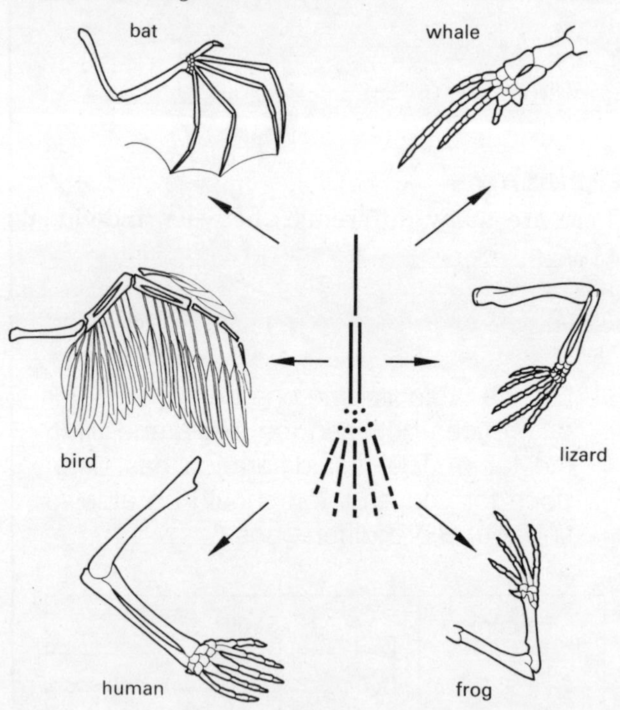

Fig. 8.74 Fore-limbs of different animals all have the same basic structure

Look at Fig. 8.74. What do these animals use their limbs for? Do they all run? They are used for different activities (jumping, flying, swimming, grasping) according to where the animal lives or what it eats. Even though they are now used for different functions, the limbs are similar enough to suggest that all of the animals shown had a common ancestor. Look at the evolutionary tree in Fig. 8.69.

Another example of the way in which homologous structures give evidence for evolution is shown by organisms which possess structures that do not seem to be used for anything. These structures are called *vestigial*. For example, some snakes such as pythons have small limb bones. One way of explaining the presence of these bones is to suggest that snakes have evolved from other reptiles which used their limbs for walking.

MECHANISM OF EVOLUTION – THE THEORY OF NATURAL SELECTION

How has evolution taken place? Darwin gave evidence for his theory of evolution, some of which has been described above. But as well as giving evidence to show that evolution has occurred, Darwin also suggested how the process of evolution may have taken place.

Variations

There are many differences between individuals of a species, even brothers and sisters. Look back at Fig. 8.58 on page 151. Features such as height, hair and eye colour, shape of face, ability to run fast and intelligence vary from person to person. Differences between these and other characteristics are examples of VARIATION.

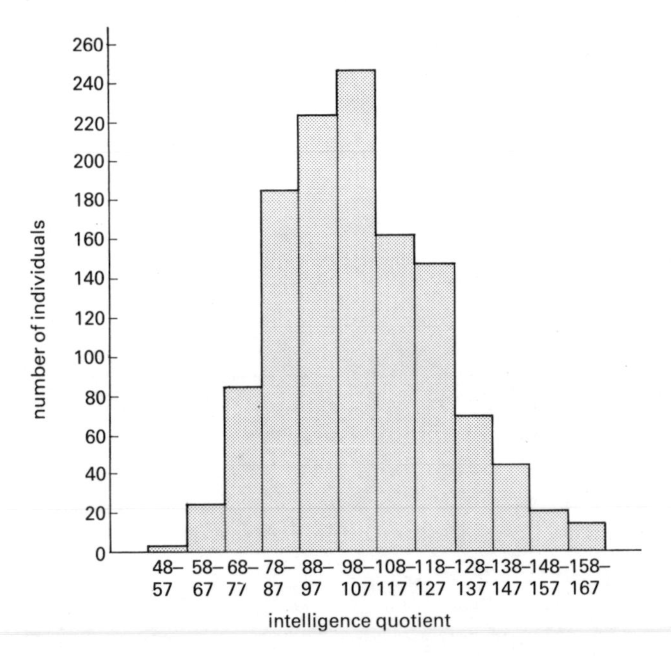

Fig. 8.75 IQ shows continuous variation

Investigation 8.10
Variation in height

If you measure the heights of a group of people of the same age, you will probably find many different heights between the shortest and the tallest.

Collect

tape measure
graph paper

What to do

1 Fix the tape measure vertically onto the wall, with 0 cm touching the floor.
2 Measure the height of each pupil in your class. Record all results on the blackboard.
3 Divide the heights into 5 cm groups. For example, if the shortest pupil is 120 cm and the tallest 140 cm you could put the results in a table like the one below.

Height/cm	Number of pupils
120–125	
126–130	
131–135	
136–140	

4 Construct a bar chart on graph paper. Put height on the horizontal axis and numbers of pupils on the vertical axis. Your chart will show how height varies in your class.
5 Look at your chart and answer the following:
   (a) To which group of heights do (i) most pupils belong, (ii) fewest pupils belong?
   (b) Which group contains the largest number of pupils?
6 Work out the average height of your class. (Add all the heights together and then divide by the number of pupils.)

If you measure the heights of any large group of people of the same age, as in Investigation 8.10, you will probably find that there are many different heights between the shortest and the tallest. They can be arranged to show a steady gradation from very short to very tall. This kind of variation is called CONTINUOUS VARIATION. Other examples of continuous variation are intelligence (see Fig. 8.75), hair colour and weight.

Some characteristics have no intermediate forms. For example, people are either male or female, they can either roll their tongue or can not (see page 154). These are examples of characteristics known as DISCONTINUOUS VARIATION.

Variation in a population happens as a result of:

(a) what happens when gametes (eggs and sperms) are formed by meiosis – the way the chromosomes end up in the gametes (see page 148);

(b) what happens at fertilisation – that is, the random combination of eggs and sperms (see page 151);

(c) mutation – new different genes may arise by accident (see page 177);

(d) environmental influences (e.g. the diet of a person will affect their weight).

Assignment – Variation in snails

A type of snail, *Cepaea nemoralis*, which is found in many different habitats, shows much variation in its colour and the number of bands on its shell. Some of the different types are shown in the photograph.

Study the table, which shows an analysis of results of some snails collected from different habitats.

Habitat	Shell colour					
	yellowy green		pinkish		brown	
	banded	unbanded	banded	unbanded	banded	unbanded
Short grass	10	60	1	25	5	50
Hedge with many dead leaves and bare earth	–	9	–	15	–	48
Deciduous wood with some light brown dead leaves	15	4	10	2	–	4

1 Plot a bar graph of these results.
2 What patterns can you see from these results? You should be able to see at least three.
3 How does the habitat affect what snails will live there?

Survival of the fittest

Darwin noticed that in any type of habitat most plants and animals have more offspring than the habitat can support. There is always a struggle for existence as animals and plants compete for food, space and shelter (see limiting factors, page 127). Within a species there is variation between individuals (for example, some may be taller, some may run faster). Those with favourable variations will stand a better chance of survival. Tall plants will be able to get more light; deer that can run very fast can escape from their predators. This is often referred to as 'survival of the fittest'. These survivors will then probably reproduce and have offspring which also have the same favourable characteristics. Individuals with unfavourable variations may die before they are able to reproduce and so will not pass on their characteristics. This whole process is called NATURAL SELECTION.

For example, in a population of early humans, any man with particularly good eyesight was more able to find lots of food. This man would live longer and father more babies to whom the characteristic of good sight was passed. He was 'selected by nature as the fittest to survive'. Darwin believed that this is how evolutionary change occurs.

With plants, the 'fittest' are those that are best able to use the environment in which they live. They compete for light, water and nutrients from the soil. For example, in a dense forest the fastest growing trees get more light and are more successful.

Another example is the evolution of the giraffe which is shown in Fig. 8.76.

Ancestral giraffes probably had necks that varied in length. The variations were hereditary.(Darwin could not explain the origin of variations.)

Competition and natural selection led to survival of longer-necked offspring at the expense of shorter-necked ones.

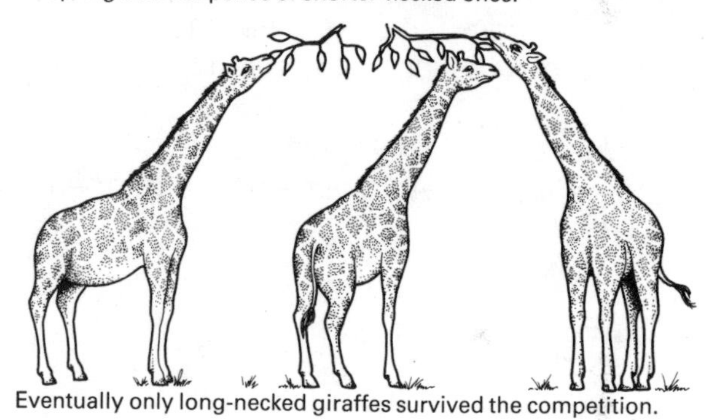

Eventually only long-necked giraffes survived the competition.

Fig. 8.76 The giraffe's long neck could have evolved by natural selection

Mutations

A MUTATION is a sudden change in the genetic make-up (genotype) of an organism. For example, a baby may be born with four toes. A mutation occurs during meiosis when eggs and sperms are being formed, either in the chromosome or in individual genes.

In a chromosome mutation, a piece of the chromosome may turn around the wrong way, a piece may break off or an extra bit may be added. For example, Down's Syndrome (mongolism) is caused by a baby having an extra chromosome. In a gene mutation an individual gene may be changed.

Mutations usually happen by chance but the chance of occurrence is increased by radioactivity and by certain chemicals. Mutations are usually harmful. In humans they usually cause defects which make a child mentally or physically handicapped. Occasionally mutations may be beneficial. A particular individual may be produced which is better adapted to the environment. It will survive, reproduce and gradually spread the new gene throughout the population. One example of this happening has been seen in the peppered moth. This shows how natural selection may take place as a result of a mutation.

During the first half of the nineteenth century all the peppered moths caught had greyish-white wings with a few darker markings. Towards the end of the nineteenth century, a few dark forms, which looked nearly black all over, were found in the North of England and in the Midlands. The black form is thought to have been the result of a mutation in the 1840s.

The peppered moth rests on tree trunks and is fed on by thrushes. If the tree trunks are covered in greyish lichen, the light moths are well camouflaged but the dark forms would show up very clearly. So the thrushes would eat mostly the dark forms. But in industrial areas, where trunks are blackened with soot, the reverse is true. The white form shows up best and the thrushes eat mostly white moths. Gradually the numbers of black moths increased in industrial areas, as the black colouring was a favourable variation.

Other examples of natural selection are when new types of bacteria have been formed as a result of a mutation which makes them resistant to penicillin or some other drug, or when head lice become resistant to different chemicals. The new forms are at an advantage and spread quickly, as they are resistant to the chemicals.

If the organisms are disease causing, it can produce a lot of problems. The research departments of many chemical companies and universities are constantly trying to find new drugs which will kill mutant forms of disease-causing organisms.

Assignment – Peppered moths

1 How has the distribution of peppered moths changed from 1890 to 1970?

The Clean Air Act of 1956, which regulates pollution from factories, has had an effect on the distribution of the moths.

2 What do you think the long term effect of the Clean Air Act will be on the peppered moth?

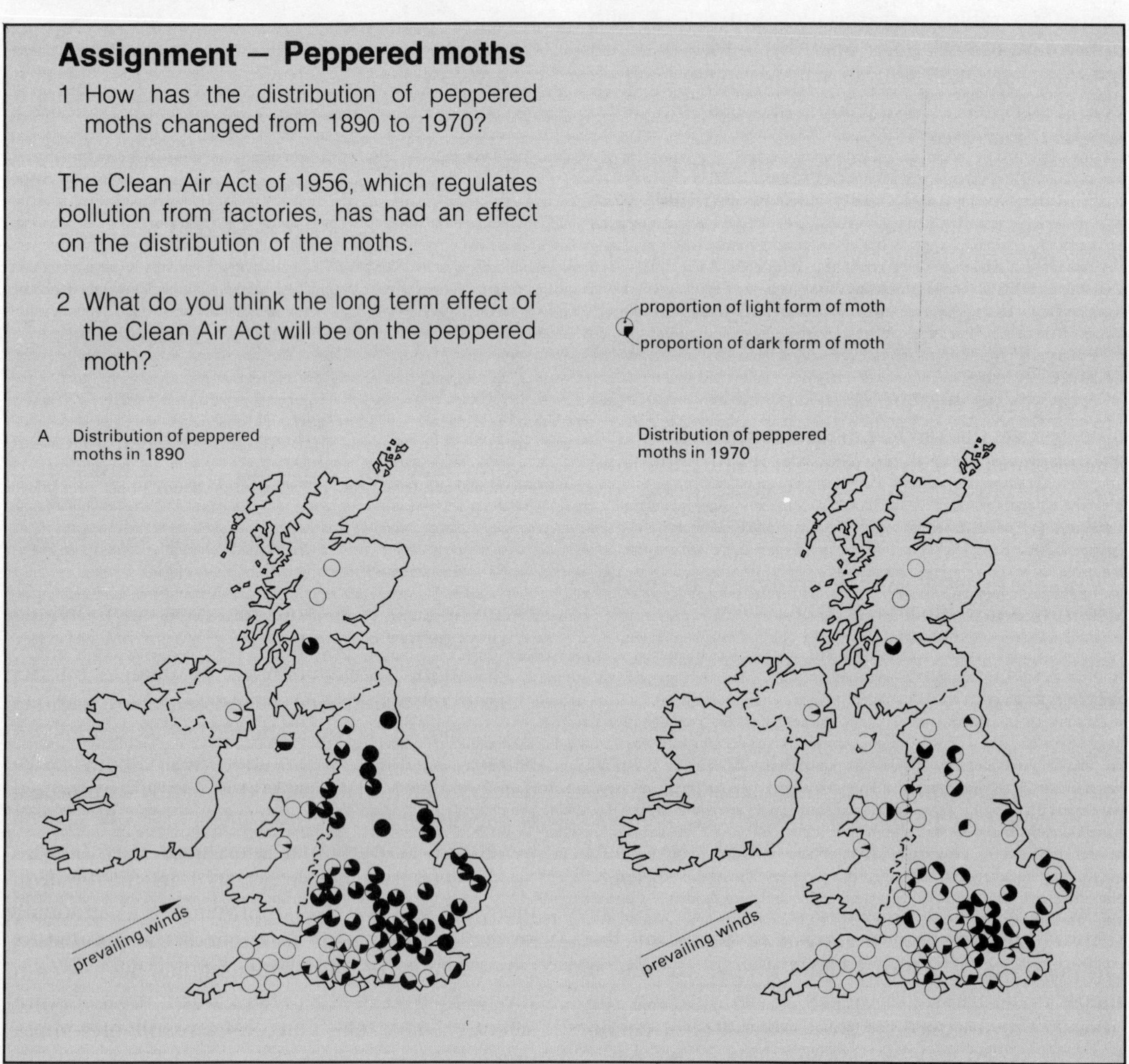

Assignment – Antibiotics

Bacteria are normally killed by antibiotics like penicillin. Mutations in some bacteria make them either completely or partly resistant to certain antibiotics. Because bacteria reproduce very rapidly and are often present in large numbers, the chance of a mutation occurring is greater than for many other organisms. If a group of bacteria is exposed to an antibiotic, any resistant forms may survive and breed while the non-resistant forms will be killed.

1 Design an experiment which would show whether a sample of bacteria included any resistant forms.
2 Maria, Manjit and Michael are all ill. They are all given the same dose of penicillin tablets which they should take for a week.
   Maria takes all her tablets and gets better.
   Manjit takes all her tablets but is still quite ill at the end of the week.
   Michael stops taking the tablets after two days when he feels better. However, several days later he becomes ill again.
   Suggest explanations for each of these results.

Evolution

WHAT YOU SHOULD KNOW

1 The theory of evolution tries to explain how present-day animals and plants came into existence. It suggests that the first forms of life were single-celled organisms which have gradually changed over millions of years.
2 Evidence for evolution can be seen from
   (a) fossils;
   (b) geographical distribution of similar species;
   (c) comparative anatomy.
3 There is great variation in the characteristics of any species. Variation can be continuous, e.g. height, or discontinuous, e.g. the ability to roll the tongue or not.
4 The theory of natural selection suggests that evolution could have happened because:
   (a) plants and animals produce more offspring than can possibly survive;
   (b) some of the offspring have small variations which help them to survive better – 'survival of the fittest';
   (c) some of the useful variations may be inherited by future generations.
   Over millions of years the variations build up until an organism is very different from its ancestors.
5 The type of variation that can be inherited is caused by mutations. A mutation is a sudden change in the genetic make-up of an organism.

QUESTIONS

1 Make a table to show arguments for evolution and arguments against it.
2 In what way do fossils suggest that evolution has happened?
3 The dodo was a large, slow moving bird which could not fly. It evolved on the island of Mauritius which is in the Indian Ocean far away from the mainland of Africa. When people came to the island and brought mainland animals it soon became extinct.
   (a) How do you think the dodo evolved in the first place?
   (b) Why did the dodo become extinct so quickly?
4 Humans have an appendix but it does not seem to have any use. If it becomes infected it may have to be removed.
   Suggest why we have an appendix and why it does not seem to have any use now.
5 Explain the difference between genetic variation and environmental variation.
6 In any type of habitat the organisms that are best adapted to that habitat are the most successful. Explain the following.
   (a) A population of organisms which can reproduce sexually usually becomes adapted to its environment more quickly than do organisms which can only reproduce asexually.
   (b) Evolution does not stop even when organisms are well adapted to their environment.
7 Rats are normally killed by a poison called Warfarin. However, some individual rats are not killed by the poison, and these appear to be on the increase. Suggest an explanation for this.

9. Energy 2

REACTIVITY

METALS

It is possible to group the elements into 'metals' and 'non-metals' on the basis of their chemical properties *and* their physical properties. The *chemical* properties depend on the electronic arrangement of the individual atoms. This arrangement determines how the element reacts with other elements. The *physical* properties depend on how the atoms are arranged within the structure. The arrangement gives rise to such things as appearance, density and strength. In this section we will look only at the physical properties. Remember, however, that the chemical properties are just as important as the physical properties.

Fig. 9.1 Some uses of metals

Table 9.1 Physical properties of metals and non-metals

Metals	Non-metals
1 Good conductors of electricity	Poor conductors of electricity
2 Good conductors of heat	Poor conductors of heat
3 High densities	Low densities
4 High melting point	Low melting point
5 Shiny surface (metallic lustre)	Non-shiny surface
6 Very strong	Weak and brittle (break easily)
7 Malleable (can be altered in shape by hammering)	Non-malleable
8 Ductile (can be drawn into wires)	Non-ductile

After studying Table 9.1, you may find that you know some metals that do not have all the above properties. For example, the metal mercury is liquid at room temperature. The metal sodium has a density less than that of water and so will float on water. However, these have the chemical properties that distinguish metals from non-metals. If most of the physical and chemical properties of a substance are those of a metal, then the substance is described as one.

Fig. 9.2 Close-up of galvanized steel

Assignment – Metals and non-metals

You are given two blocks of equal volume and painted the same colour. One is a metal and the other is a non-metal. Describe, with brief experimental details, how you would use the physical properties of metals and non-metals to distinguish between them.

Why are metals so useful?

In Fig. 9.1, there are a few examples of the uses of metals in everyday life. You can probably think of many more.

Over the years, many materials have been developed to replace naturally occurring substances. These are called synthetic materials. Perhaps plastics are the most important of these. So far, it has not been possible to produce 'synthetic metals'. We have become so dependent on metals that it is impossible to imagine what life would be like without them. You might like to suggest ways in which our lifestyle would change if there were no metals around. What is it about metals that makes them so useful?

The simple answer is that metals have a number of very desirable properties. You have already learned (in Unit 6, page 19) that the Periodic Table is a way of classifying the elements according to their chemical properties. Over 70 of the 92 naturally occurring elements can be called metals. About half of these are used at present, but many only in very small quantities.

The structure of metals

It is not apparent, just by looking at a metal, that it is made up of crystals. Normally, the crystals are so tiny that it is not possible to see them. However, with galvanised iron, the crystals are very large and easily seen. Galvanised iron is iron that has been dipped into molten zinc. As the zinc solidifies, it produces large crystals.

But it is the outermost electrons of the atoms that make metals so different. These are free to move through the structure, leaving the remainder of the atom (a positive ion) fixed in position.

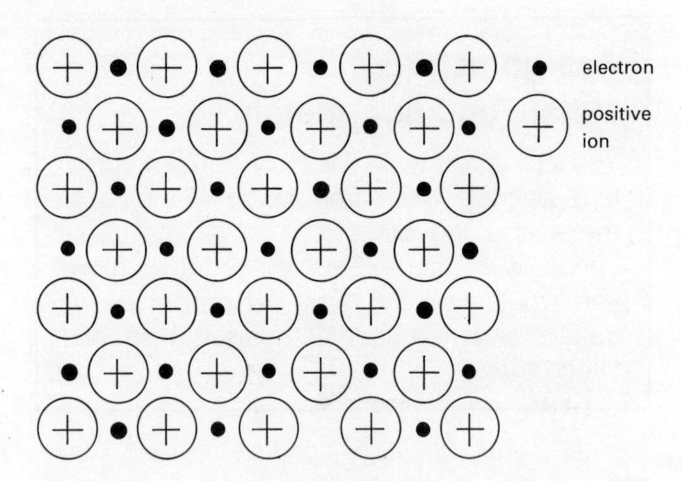

Fig. 9.3 The structure of a metal

These mobile electrons are responsible for the fact that metals are good conductors of electricity and heat.

Uses of metals

Why is one particular metal selected for one particular task? It might be because of one of its physical properties (for example, its high melting point or its low density). It might be because of one of its chemical properties (for example, its lack of reactivity with the oxygen in the air). But there is another important factor – the cost of the metal. This will depend on the cost of extraction of the metal ore from the earth and the cost of extraction of the metal from its ore. The four metals most widely used at the moment are iron, aluminium, copper and zinc.

Iron

Iron is the most widely used metal today. However, the cast iron (or 'pig iron') obtained from the blast furnace is very brittle and not suitable for use as it is (you read about this in Unit 1). It is usually made into steel. This involves removing various impurities from the cast iron and adding other metals. Steel is the general name for a wide range of alloys that contain iron as their main component with a smaller amount of carbon (up to 1.5%). Most alloys are mixtures of two (or more) *metals* but iron forms alloys with non-metallic carbon. A small amount of carbon produces a mild steel that is soft and easily shaped. Increasing the amount of carbon produces a steel that is harder but more brittle. You can probably think of many uses of steel.

There are two main problems with using iron in the form of steel. One is the high density of iron (see Table 9.2 on page 184) and the other is that iron corrodes easily. Corrosion is the general name for the reaction between a metal and the chemical substances around it. This will be discussed on page 192. It is possible to make 'stainless steel', which does not corrode, but it is very expensive.

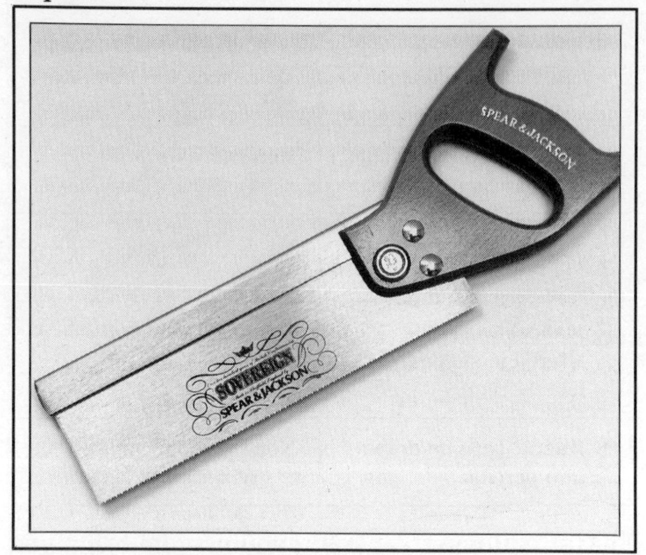

Fig. 9.4 Steel is used for tools

Aluminium

Although it is the most abundant metal in the earth's crust, it is expensive to extract aluminium metal from its ore. But the uses for it are increasing all the time. Two of its properties make it especially useful. It has a low density compared with iron. On its own, aluminium is not very strong. But mixing it with other metals produces an alloy that can be as strong as steel. Its most important use has been for building the bodies and wings of aircraft. Aluminium, or its alloys,

Fig. 9.5 Aluminium is used in vehicle bodies

are being more and more used in the manufacture of motor vehicle bodies and the superstructure of ships.

Also, aluminium is very resistant to corrosion (see page 192 for the reasons). The non-poisonous nature of aluminium leads to a number of uses associated with food. Foil is used for the protection of food in packets or for cooking purposes. Aluminium cans are used for drinks and many saucepans are made from this metal (it is, also, a good conductor of heat).

Although aluminium is not as good a conductor of electricity as copper, the National Grid system which carries electricity around the country uses aluminium conductors rather than copper ones. The smaller density of aluminium makes it cheaper to install cables of this metal. However, the aluminium cables contain a central core of steel. The steel provides enough strength to support the weight of the cable between the pylons.

Copper

Copper's most important use is based on the fact that it is the second best conductor of electricity (silver is the best) and that it is easy to make it into wires (ductile). About half of the world's production of copper is used for electrical purposes. Every wire that you see or use (except for the conductors in the National Grid) will have copper as the conductor.

Copper is a very good heat conductor and has many applications in this area. Copper is used for saucepans, the radiators of cars and for tubes in boilers. Copper piping has many uses as it is easy to bend and is very resistant to corrosion. On its own, copper is not very strong but alloying it with other metals can greatly increase its strength. Alloys of copper are very resistant to corrosion.

Zinc

Perhaps the most important use of zinc is as the negative electrode (cathode) of the dry cell. It is very resistant to corrosion and can be used as a roofing material. The covering of iron (or steel) with a thin layer of zinc, called galvanising, has already been mentioned. Zinc does not corrode in the same way as iron or steel and forms a protective layer over it. Zinc is used in a number of alloys, such as brass.

Table 9.2 compares the four metals. In the column headed 'Estimated life of reserves', the figures assume our present rate of consumption. In practice, this rate of consumption will probably increase and the estimate will become less.

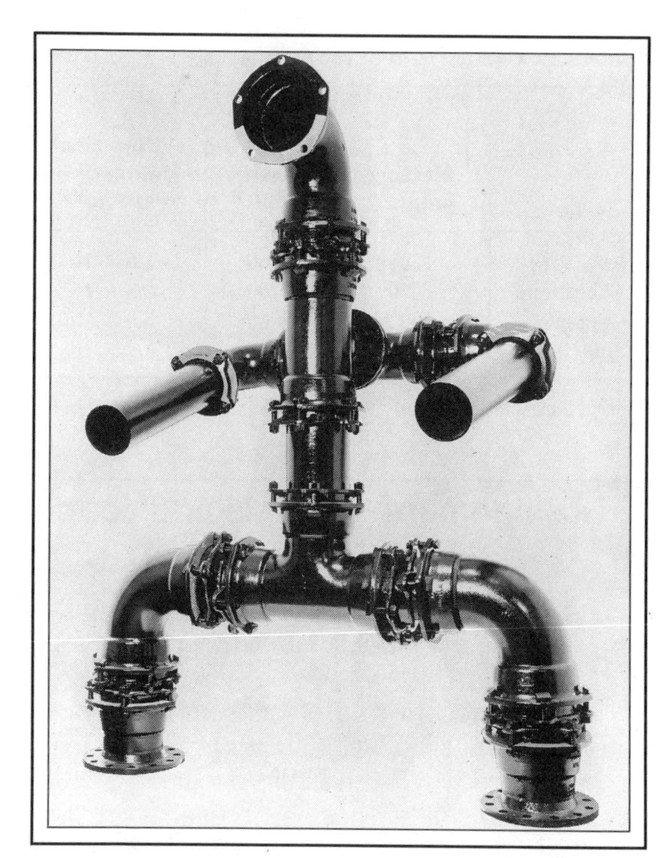

Fig. 9.6 Copper pipes

Fig. 9.7 An important use of zinc

Table 9.2 Comparison of the four metals

	Melting point/°C	Density/ kg m^{-3}	Electrical conductivity/ $\Omega^{-1} m^{-1}$	Cost per tonne (1 tonne = 10^3 kg) (1985)	World consumption/ tonnes per year (1982)	Estimated life of reserves/years
Iron	1535	7860	1.0×10^7	£70	784×10^6	250
Aluminium	660	2700	3.8×10^7	£700	10.7×10^6	100
Copper	1083	8920	5.9×10^7	£1000	7.0×10^6	20
Zinc	420	7140	1.7×10^7	£400	4.1×10^6	20

(With electrical conductivity, the higher the value the better the metal is as a conductor.)

Assignment – Metals and economics

1 If the price of a particular metal increases, what do you think will happen to the rate of consumption? Explain your answer.

2 The present rate of consumption could decrease as well as increase. Give one reason why each could occur.

3 Which metal has at present (a) the largest (b) the smallest reserves?

4 The column headed 'estimated life of reserves' is based on the reserves known now. Give two reasons why this might be an under-estimate.

5 'There may be no metal coins by the year 2000.' Explain why this statement could be true. Use Table 9.2 to help you answer the question.

6 Because of the increasing demands of technology, metals are now being used that a few years ago were considered of little importance. Examples are: beryllium, niobium, silicon, tantalum, titanium, uranium and zirconium.

Select one of these metals (or another that you have read about) and write a short report on it. Your report should include: how it is extracted from its ore; cost; useful properties; present and possible future uses.

Alloys

A metal as extracted from its ore may not have the properties that are required. The metal might be too soft, too weak or perhaps corrode too easily. In fact metals are very rarely used in their pure state. It is possible to change the metal's physical properties by heating and/or hammering. Both these processes can change the crystal structure of the metal and, thus, its physical properties. With most metals, heating the metal and then cooling it rapidly will produce a metal that is harder but more brittle. Allowing the metal to cool slowly produces a softer metal that is easier to shape. Hammering the metal, when either hot or cold, can also change its physical properties. In the past, this has been the only way of changing the physical properties of a metal.

Fig. 9.8 A blacksmith's forge

Table 9.3 Some common alloys and their uses

Name	Main constituents	Uses of the alloy	Comments
Mild steel Hard steel	iron, 0.5% carbon iron, 1.5% carbon	cars, bridges, washing machines cutting tools	by adding small amounts of different metals, steels with many different properties can be produced
Stainless steel	iron, chromium, nickel	cutlery, razor blades, car exhausts	very corrosion resistant
Alnico	iron, aluminium, nickel, cobalt	permanent magnets	very powerful magnets
Duralumin	aluminium, copper, magnesium	aircraft, ship superstructures, vehicle bodies	much stronger than aluminium but higher density
Brass	copper, zinc	castings, screws, ships' propellers	harder than copper or zinc but more easily worked, corrosion resistant
Bronze	copper, tin	1p and 2p coins, castings	corrosion resistant, hard wearing
Cupro-nickel	copper, nickel	'silver' coins	corrosion resistant, hard wearing
Nichrome	nickel, chromium	heating filament in electric fires	very high melting point
Solder	tin, lead	fixing electrical components or copper pipes in plumbing	very low melting point
Wood's metal	bismuth, lead, tin, cadmium	fire detectors in alarm systems	melting point is 70 °C

An ALLOY is usually a mixture of two or more metals. With steel, one of the constituents is usually carbon (a non-metal) but normally the constituents of alloys are all metals. Iron (or steel) on its own corrodes but alloyed with other metals it is possible to produce 'stainless steel' which does not. Aluminium is not very strong as a pure metal. But alloying it with other metals to produce 'duralumin' makes a very strong material.

Usually, the physical properties of an alloy are a mixture of the properties of its components. If a strong metal is alloyed with a weak one then the strength of the alloy is usually between the two and will depend on the relative percentages of each metal present. Years ago, making alloys was a very hit-or-miss affair but it is now a very exact process. If you need an alloy with very definite properties then it can probably be made.

Table 9.3 gives a small number of examples of the many alloys used today.

Metals and the future

Although the actual quantity of metal present in the earth is virtually limitless, metals are classed as a 'non-renewable' or 'finite' resource. The reason is that metals can only be extracted from the top few kilometres of the earth's crust. The ores that have been used in the past were the ones which were easiest to get out of the ground and had a high concentration of ore. But as these are gradually used up, so ores that are more difficult to extract or of lower concentration are used. Together with the increasing demand for metals, this means that costs will rise.

With the present known reserves, some metals could run out early in the twentieth century. These metals are copper, zinc, tin, magnesium, silver and mercury. The phrase 'present known reserves' has been used because circumstances may change. New sources of ores may be discovered in the near future or improved technology may affect the amount of metal that can be got from an existing ore. For example, the amount of magnesium (and other metals) in the seas of the world is enough to satisfy demands for many years. At the moment, it is not economic to extract it. Other metals will be in short supply by the year 2000. These are lead, copper and tungsten.

The unique properties of metals ensure our dependence on them. How can we make them last longer? One answer is thought to be recycling. But the main problem here is that the metal that you want to recover is usually mixed with other metals and non-metals (as in a motor car, for example). At the moment, it is too costly to extract the metal to use it again. But with dwindling reserves and ever rising costs, recycling is going to become more and more important. (See Unit 10, p. 281)

Fig. 9.9 An open-cast copper mine in Chile

REACTIVITY SERIES

We have seen that there is a great deal of variation in the physical properties of metals. The same thing happens with the chemical properties. We will look at only a few of the many reactions of metals with other substances.

The history of the use of metals tells us a great deal about the reactivities of metals. You have learned about the extraction of a metal from its ore in Unit 1. The easier it is to extract a metal from its ore, the earlier in history it was used (see Fig. 9.10).

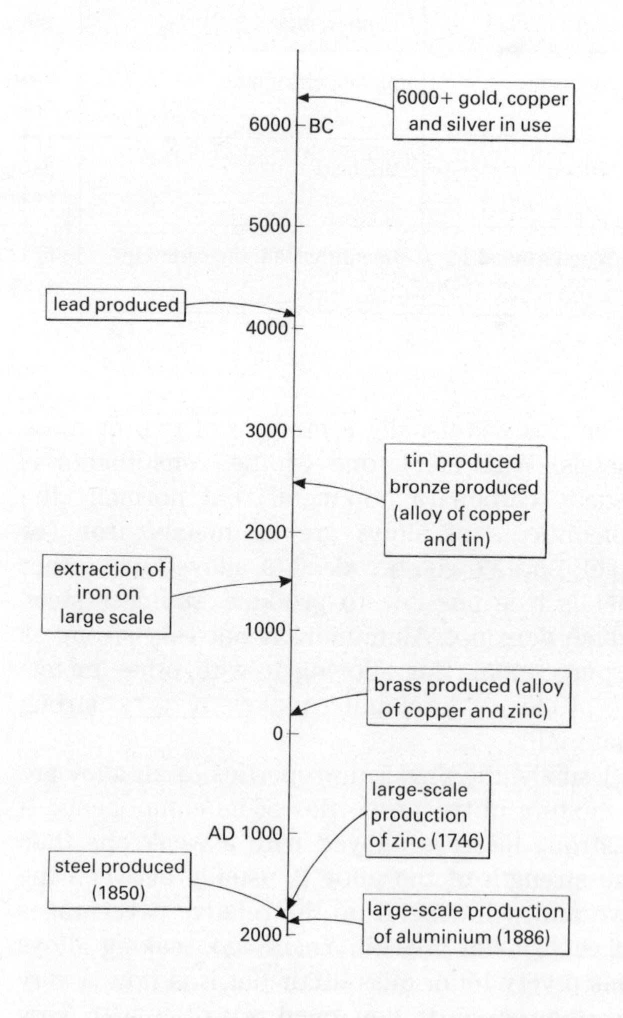

Fig. 9.10 The use of metals in history

Producing a reactivity series

The less reactive a metal is, the easier it is to extract it from its ore. This is because the chemical bonds between the metal and the other elements are weaker. The metals can be placed in a series, in order of reactivity. At one end of the series, there will be the metals that do not form any bonds at all. At the other end, the metals will be so reactive that it is very difficult to remove the metal from the elements with which it is combined.

How could such a REACTIVITY SERIES be produced? One method would be to see how the metal reacts with water. The more reactive the metal, the more vigorous the reaction. For example, the reaction of potassium with water is very violent. (Note: this reaction must only be demonstrated by your teacher.)

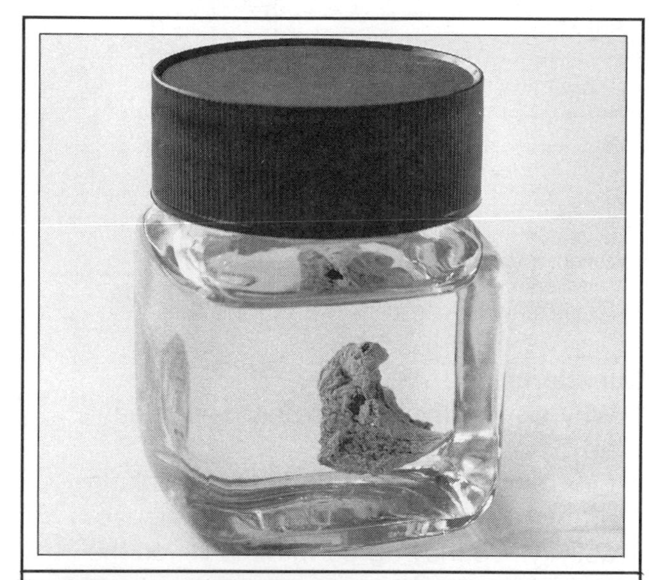

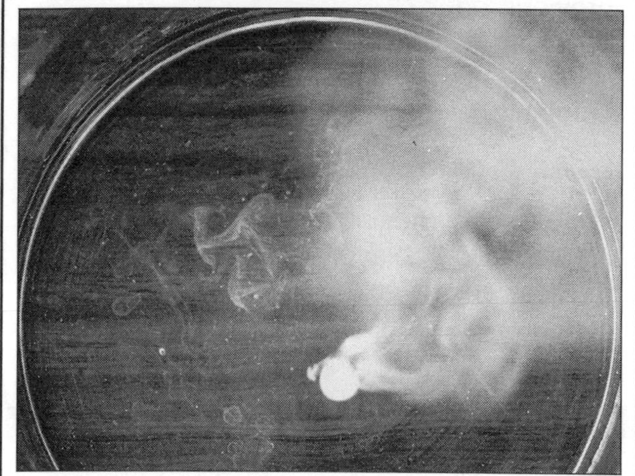

Fig. 9.11 (above) Potassium is kept under oil (below) The reaction of potassium with water

The reaction is given by the following equation

$$2K(s) + 2H_2O(l) \rightarrow 2KOH(aq) + H_2(g)$$

With less reactive metals, their action with water may be slower or may not happen at all. It is also possible to study the reaction of a metal with oxygen or acids. In all cases, a similar pattern of reactivity is produced. Table 9.4 shows the commoner metals in the order of their reactivity. Potassium is the most reactive in the list and platinum the least reactive.

Table 9.4 Reactivity series of metals

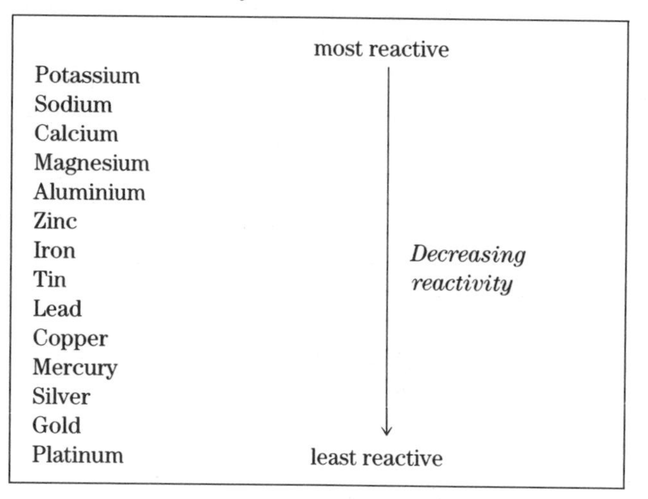

Metal	
	most reactive
Potassium	↓
Sodium	
Calcium	
Magnesium	
Aluminium	
Zinc	
Iron	*Decreasing reactivity*
Tin	
Lead	
Copper	
Mercury	
Silver	
Gold	
Platinum	least reactive

Assignment – The reactivity series

1. Why are the metals potassium and sodium always stored under oil?
2. What difference would you expect between sodium and calcium when reacting with water?
3. Magnesium ribbon reacts very slowly with water at room temperature. Mention two ways that the speed of this reaction could be increased.
4. When platinum is extracted from the earth, would you expect it to be combined with any other elements? Explain your answer.
5. Here is a list of metals: copper, magnesium, lead, sodium, zinc and tin.
   Put them in the order of their reactivity with the most reactive first.
6. In each of the following, say whether a reaction will take place or not. If a reaction will occur, then write the equation for the reaction.
   (a) Iron is added to lead nitrate solution, $Pb(NO_3)_2$.
   (b) Copper is added to iron(II) sulphate solution, $FeSO_4$.
   (c) Magnesium is added to silver nitrate solution, $AgNO_3$.

Investigation 9.1
To establish the position of various metals in the reactivity series

A metal will displace a less reactive metal (i.e. one lower in the reactivity series) from a compound. If this occurs, then the displacing metal is higher up the reactivity series than the displaced metal. Soluble metal salts will be used.

Collect

4 test tubes
metal samples (copper, zinc, magnesium, iron, lead)
solutions (nitrates or other soluble salts of the above metals)
test tube rack
emery paper
thermometer

What to do

1 Pour a small amount of *one* solution into each of the test tubes.
2 Thoroughly clean each metal sample with emery paper. Touch the metal as little as possible after cleaning it.
3 Place the metal samples in the test tubes.

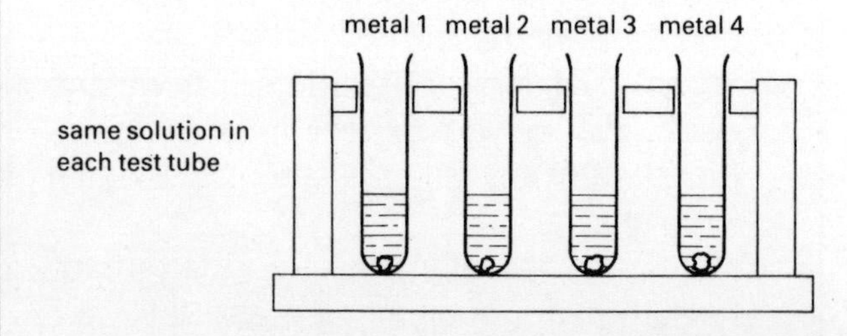

4 Note any changes, such as the solution changing colour, a powdery metal being deposited or a rise in temperature.
5 Record your results in a table like this:

Solution \ Metal	Zinc		
Copper nitrate	Y		

Reaction: Y (yes) N (no)

6 Repeat with the other solutions

Questions

1 Why do you think the metals were cleaned beforehand?
2 From your results, arrange the metals in order of increasing (or decreasing) reactivity. Do they agree with the list on the previous page?
3 When iron is added to copper(II) nitrate solution, copper is precipitated. The equation for the reaction is

$$Fe(s) + Cu(NO_3)_2(aq) \rightarrow Fe(NO_3)_2(aq) + Cu(s)$$

Write down equations for two other reactions that took place.
4 If you were given iron(II) chloride solution, as well as the solutions above, what results would you expect?

In Investigation 9.1 we chose displacement reactions where it was not necessary to supply energy. A more reactive metal was displacing a less reactive metal. Energy, in the form of heat, is released but the amount may be so small that it is not easy to detect.

Some of the most important displacement reactions are used in the production of metals from their ores. Energy must be supplied for these reactions to occur. Usually, a metal is found either as an oxide or sulphide. In Unit 1, you saw that lead could be obtained from its ore, lead(II) oxide, by heating it strongly with carbon (carbon acts just like a more reactive metal). The equation is:

$$2PbO(s) + C(s) \rightarrow 2Pb(s) + CO_2(g)$$

The higher the metal in the reactivity series, the higher the temperature that is needed. While lead can be extracted in the laboratory using a

Bunsen burner, this would be impossible if iron ore were being heated with carbon. A much higher temperature is needed (as in a blast furnace). In practice, only the metals from zinc downwards can be extracted using heating processes. With aluminium and metals above it in the reactivity series, electrical energy must be used for extracting the metal from its ore. This is done using electrolysis which you learned about in Unit 4.

ELECTRICITY AND METALS

The simple cell

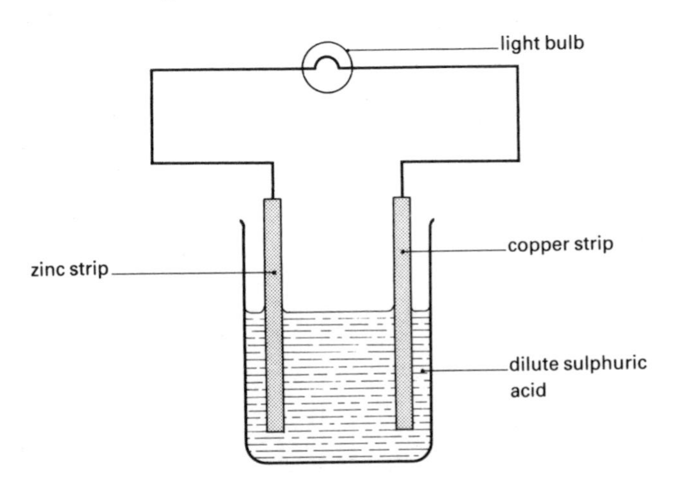

Fig. 9.12 The simple cell

If the arrangement shown in Fig. 9.12 is set up (the metal strips should be cleaned beforehand), then the bulb lights up showing that a current is flowing. A little later, some changes will have occurred:
(a) gas bubbles form on the copper strip – the gas turns out to be hydrogen;
(b) the zinc strip is being slowly eaten away;
(c) the bulb is no longer lit.

This is a similar experiment to that performed by Volta about two hundred years ago. It is called a SIMPLE CELL and was the first battery. It produces electrical energy from chemical changes. Electrolysis works the other way round: electrical energy is used to produce chemical changes. The 'simple cell' does not work very well and, over the years, it has been improved. Today, the commonest form of cell is called the DRY CELL. What differences can you see between the dry cell and the simple cell?

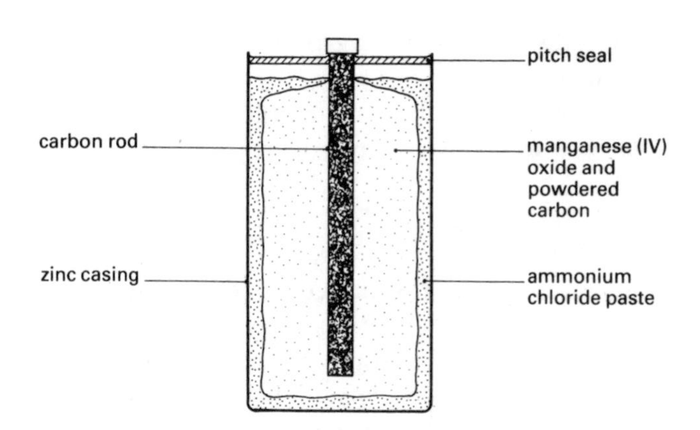

Fig. 9.13 The dry cell

Assignment – Cells

Find out answers to the following questions.
1 The dry cell is not the only cell that is used nowadays. Find out about other kinds that are used (e.g. in calculators and digital watches).
2 A cell that cannot be recharged (such as the dry cell) is called a primary cell. Rechargeable cells are called secondary cells. Find out what you can about these and in what circumstances they are used.
3 One of the newer types of cell is called a fuel cell. Why is this device becoming more and more important?
4 What are solar cells? What are the advantages and disadvantages of using them?

How the simple cell works

1. The higher up the reactivity series the metal is, the more readily it will form positive ions. Zinc is higher than copper and zinc atoms go into solution as zinc ions.

$$Zn(s) \rightarrow Zn^{2+}(aq) + 2e^-$$

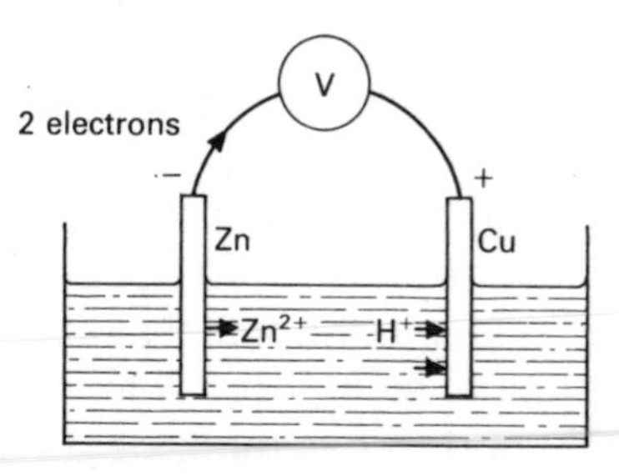

Fig. 9.14 The action of the simple cell

2. Two electrons from each zinc ion flow through the conductor connecting the plates to the copper electrode.
3. The electrons then discharge two hydrogen ions as a molecule of hydrogen gas at this electrode.

$$2H^+(aq) + 2e^- \rightarrow H_2(g)$$

You can see that the above statements explain why the zinc electrode slowly dissolves and hydrogen gas forms on the copper. For the electrons to move, there has to be a difference in voltage and this is recorded by the voltmeter in Fig. 9.14. After a time the voltage drops – this is why the light went out in the experiment shown in Fig. 9.12. The voltage drops because of the presence of the hydrogen gas on the copper, blocking access for further hydrogen ions. Removing the bubbles would restore the original voltage.

The electrochemical series

From Investigation 9.2, you can see that the voltage has to be measured between two different metals. In fact, we have used copper as the 'reference' metal. This means that all the voltages are measured with respect to copper. In practice, the 'reference' element is hydrogen. You may

Investigation 9.2
To determine the voltage differences for different pairs of metals

Using the fact that a pair of different metals can produce a voltage, it is possible to build up another series like the reactivity series.

Collect

- copper strip
- strips of other metals, e.g. zinc, magnesium, iron, lead
- dilute sulphuric acid (0.1M)
- connecting wires and clips
- high resistance voltmeter
- emery paper
- ethanol
- beaker
- tweezers
- eye protection

What to do

1 Pour the acid carefully into the beaker so that it is about half full.
2 Clean the copper strip with the emery paper. Then, handling it with tweezers, wash it in ethanol before placing it in the beaker as shown in the diagram.
3 Clean each of the other metal strips before using them. Place the metal strip in the beaker opposite the copper. Connect the voltmeter to the metal strips, the positive terminal being connected to the copper. Record the voltage.

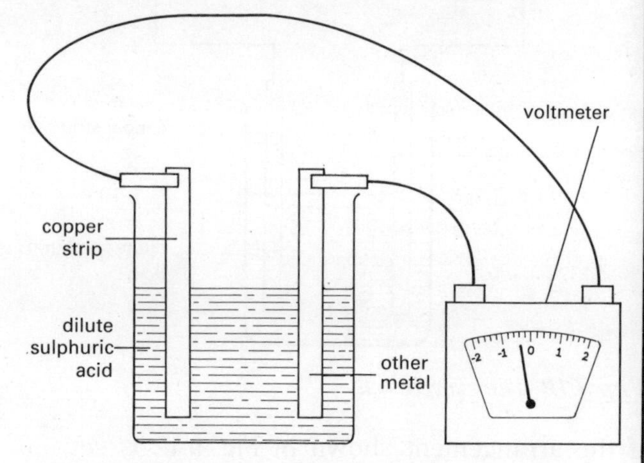

4 Repeat this procedure with the other metals you have been given. Some will react vigorously with the acid so you will have to take the readings quickly. If the copper becomes covered with bubbles of hydrogen gas, remove it and wash it in ethanol.

Questions

1 Which pair of metals produced (a) the largest, (b) the smallest voltage?
2 What voltage do you expect you would get if both metals are copper? Try this if you are not sure.
3 If you wanted to light a small bulb using two metals and an acid, which two would you use and why? Again, you might like to try this out.

think that this is strange since hydrogen is a gas. But it has been found in studying chemical reactions that it is more useful to use hydrogen as the reference.

Table 9.5 shows the voltage produced between a metal and a hydrogen electrode. This table is usually called the ELECTROCHEMICAL SERIES. To get the values shown, complicated apparatus is needed and it will not be discussed here. Can you see any similarities between the electrochemical series shown here and the reactivity series shown on page 187? Can you see any differences between the two?

Table 9.5 Electrochemical series

Element	Voltage/V
Potassium	−2.92
Calcium	−2.87
Sodium	−2.71
Magnesium	−2.37
Aluminium	−1.66
Zinc	−0.76
Iron	−0.44
Tin	−0.14
Lead	−0.13
Hydrogen	0.00
Copper	+0.34
Mercury	+0.80
Silver	+0.81
Platinum	+1.20
Gold	+1.50

Uses of the electrochemical series

The electrochemical series is very useful and here are a few of its applications.

1. Only metals above hydrogen in the series will release hydrogen gas when placed in an acid. For example,

$$Zn(s) + H_2SO_4(aq) \rightarrow ZnSO_4(aq) + H_2(g)$$

The higher the metal is in the series, the more violent the reaction. Metals below hydrogen will not release hydrogen from an acid.

2. The series tells us if one metal will displace another metal in aqueous solution. A metal higher up the series will displace one lower down. Your results from Investigation 9.2 should show that this is correct.

3. The series indicates how reactive a metal will be. You have seen the similarity between the electrochemical series and the reactivity series. But with the reactivity series, the products that are formed in the reaction will help to determine where the metal is placed. For example, calcium is placed higher than sodium in the electrochemical series. But sodium reacts far more vigorously with water than calcium does. When calcium reacts with water, the oxide layer formed protects the calcium underneath. So the reaction seems to be slower.

4. It is possible to work out the maximum voltage that could be produced using two different metals. For instance, if the two metals were zinc and aluminium then the resultant voltage would be 0.90 V (1.66−0.76).

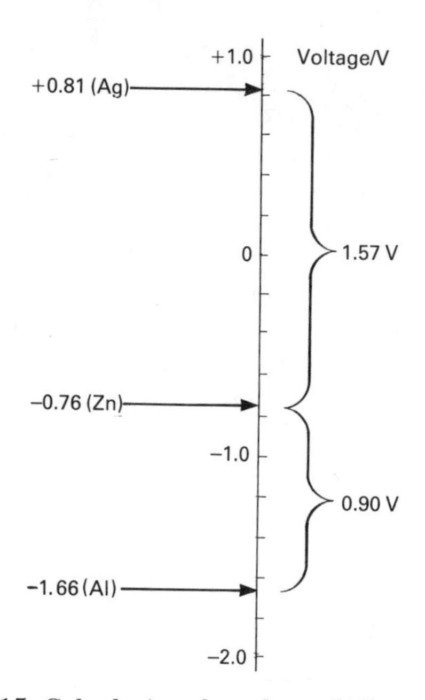

Fig. 9.15 Calculating the voltage difference

If the two metals were zinc and silver then the resultant voltage would be 1.57 V (0.76 + 0.81).

Assignment – The electrochemical series

Your experimental results used copper as the reference rather than hydrogen. Calculate the theoretical values you should have got. Do they agree with your experimental values? Can you suggest reasons for any differences?

CORROSION OF METALS

We have seen that many metals are reactive. After they have been extracted from their ore, metals will start to react with the gases and liquids around them. Apart from oxygen and water, the reactions can involve carbon dioxide and the chemical substances that pollute the atmosphere.

Fig. 9.16 An example of corrosion

To most of us, CORROSION means 'rust'. Each of us is familiar with the rust on bicycles or cars. Corrosion of iron (rust is an oxide of iron) is probably the most expensive form of corrosion there is. Yet iron is the most widely used metal. The cost of corrosion is many millions of pounds each year. These costs are for replacement of corroded objects or for preventative measures, such as painting. The two main types of corrosion are 'oxide layers' and 'electrochemical processes'.

Oxide layers

If a reactive metal is exposed to the oxygen in the air then it will react with it. This reaction can be very quick for a highly reactive metal such as sodium. If a piece of sodium is cut in two (it is very soft for a metal) then the shiny surface quickly dulls as the sodium reacts with the oxygen in the air.

$$4Na(s) + O_2(g) \rightarrow 2Na_2O(s)$$

The sodium oxide produced is very soft and easily flakes off exposing fresh sodium underneath. This then reacts with the air, and so on.

On the other hand, aluminium forms an oxide layer that remains attached to the aluminium underneath. This oxide layer is very strong and stops any further reaction occurring. In practice, this oxide layer can be further increased in thickness by an electrolytic process called 'anodising'. Other 'useful' metals such as zinc and tin produce a tough outside layer of oxide.

Unfortunately, iron behaves like sodium – but at a much slower rate. The oxide formed is very weak and it easily flakes away, exposing fresh iron underneath. But the formation of rust is much more complicated as we shall see.

Electrochemical processes

In the previous section we saw that if two different metals are placed in an acid and connected by a wire then a current flows. In the simple cell (page 189), the zinc slowly dissolves away while hydrogen gas is deposited on the cathode. This will occur if the acid is replaced with any aqueous solution of ions (i.e. an electrolyte). It is always the metal which is higher in the electrochemical series that dissolves. It is interesting to demonstrate this effect using iron as one of the electrodes. In the examples shown in Fig. 9.17 the electrodes are connected by an ammeter to show that a current is flowing.

You will see that a current is flowing in both cases – but in opposite directions. With the iron/lead combination, it is the iron that dissolves.

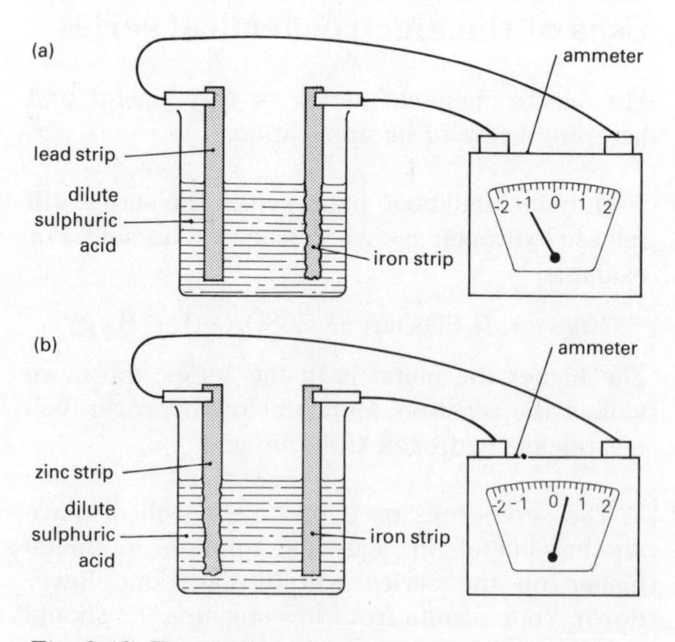

Fig. 9.17 The rusting of iron

Which metal is higher in the electrochemical series? With the iron/zinc combination, it is the zinc that dissolves. Which metal is higher in the electrochemical series?

What about iron on its own? Pure iron would corrode on its own but what accelerates the rate of corrosion is the fact that iron is always used as an alloy. One of the constituents is carbon, which acts in the same way as a metal, but other metals are usually present. If conditions are correct, then many 'simple cells' are set up within the metal that result in the corrosion of the iron. Rust is hydrated iron(III) oxide. This means it is iron(III) oxide combined with water. Investigation 9.3 will let you look at the factors that affect the rate of rusting of iron.

Investigation 9.3 To find out what affects the rate at which iron rusts

The rate at which iron rusts depends on a number of external factors. The purpose of the investigation is to see how these affect the rusting rate of iron nails.

Collect

iron nails
emery paper
metal samples
test tubes
sulphur dioxide 'atmosphere'
carbon dioxide 'atmosphere'
indicator (potassium hexacyanoferrate(II) solution)

Read step 1 before you collect your equipment

What to do

1 As the number of experiments is very large, you will set up only a few of them. You will have to decide with the rest of the class and your teacher how this is going to be done.

The iron nail will be (a) by itself; (b) greased; (c) painted; (d) attached to tin or (e) attached to magnesium.

It will be placed in one of the following environments: (a) dry air; (b) damp air; (c) carbon dioxide-rich air; (d) sulphur dioxide-rich air; (e) oxygen-free water; (f) oxygen-rich water; (g) tap water; (h) sea water

2 Before setting up your experiments, clean one of the nails with emery paper and then place it in a test tube containing some of the indicator. The first sign of rusting is the formation of Fe^{2+} ions. The indicator will turn blue when these ions are present.

3 For each arrangement that you use, check that the iron nail is carefully cleaned first. This applies also to the magnesium and tin, when they are used. Label the test tubes when they are set up and put them away in a suitable place.

4 Draw a suitable chart to record all the results of the different experiments set up by the class.

5 Over the next week, note any changes that occur and record the results on your chart.

Questions

1 How quickly does iron start to rust (as shown by the indicator)?
2 Is oxygen essential for rusting?
3 Is water essential for rusting?
4 Does iron rust faster in seawater or in tap water?
5 What disadvantages can you see in salting roads in winter to melt ice?
6 Does the presence of carbon dioxide or sulphur dioxide affect the speed of rusting?
7 What differences did you observe between the rusting of the iron/tin combination and the iron/magnesium combination?

Prevention of rusting

Investigation 9.3 shows that rust is only produced in any quantity when both oxygen and water are present. Pure water contains only a small number of ions. Increasing the number of ions increases the rate of rusting. What additional ions are present in sea water? The presence of acidic gases (such as sulphur dioxide and carbon dioxide) speeds up the rusting rate.

To prevent rusting, it is essential to stop oxygen and water reaching the surface of the iron. The following assignment will allow you to investigate some of the methods used to try to prevent this happening.

Sacrificial protection

We have seen that if two metals are in direct contact or linked by a conductor then the metal that is higher in the electrochemical series dissolves leaving the other one intact. The more reactive metal is 'sacrificed' to protect the less reactive one. The hull of the ship in Fig. 9.18 has large blocks of zinc attached to it. The zinc corrodes rather than the iron of the hull to which it is attached. At intervals the zinc blocks have to be replaced.

Underground pipes are protected in the same way. It would be very expensive to have to replace such pipes. A bag of magnesium scrap is attached to the underground pipe by a metal cable (Fig. 9.19). What do you think happens? Why is it necessary to be able to get to the magnesium scrap easily?

Fig. 9.18 Blocks of zinc attached to this ship's iron hull protect it from rusting

Assignment – Corrosion and the motor car

1 The exhaust system of a motor car is usually made of mild steel. This easily corrodes and the 'life' of such an exhaust system can be as little as two years. A stainless steel system costing about three times as much would last indefinitely. But very few drivers will fix such exhausts to their cars. Why do you think this is?

2 With today's technology, it is possible to build a 'corrosion free' motor car. Admittedly, it would be more expensive than a normal car. But can you think of other reasons why this has not been done?

Assignment – Preventing rust

Here is a list of some methods used for preventing the rusting of iron:

(a) coating the surface with grease or a sealant;
(b) painting;
(c) alloying with other metals;
(d) coating with another metal.

For each method, find out where it is used and its advantages and disadvantages.

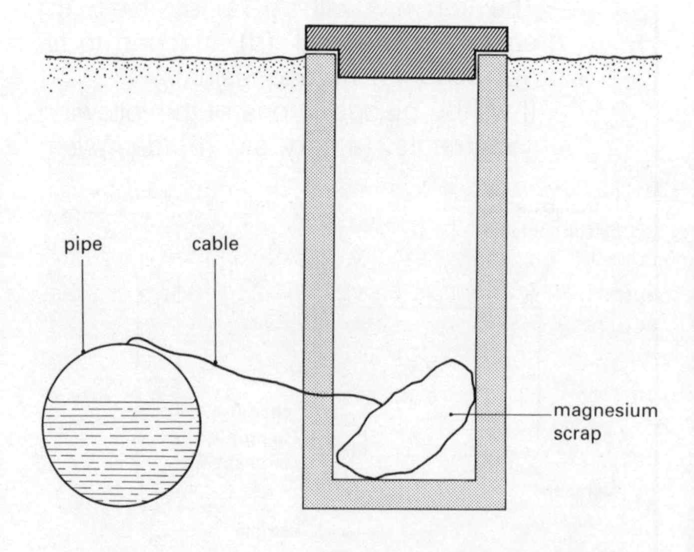

Fig. 9.19 Sacrificial protection of an underground pipe

CATALYSTS

You have already learned that there are substances called catalysts that can change the speed at which a chemical reaction takes place (Unit 7, page 60). The word 'catalyst' was coined in 1835 by the Swedish scientist Berzelius. He found that certain chemical reactions would take place more quickly if another substance (the catalyst) were present. Without the catalyst, the reaction would take place slowly or not at all.

Catalysts have become very important in the manufacture of chemical substances. You have already met the Haber process (page 88) where a catalyst was used to make hydrogen and nitrogen combine to produce ammonia. A catalyst is also used in the manufacture of sulphuric acid (page 54). What is the name of the catalyst used in each case? One of the important uses of catalysts today is in the oil industry. This will be discussed on page 250. It is important to remember that a catalyst will not increase the *amount* of the desired product but the product will be produced more quickly.

Most examples of the use of catalysts are those where the rate at which a reaction takes place is increased. However, there are a few occasions when a catalyst is used to slow down a reaction. The reaction would proceed so fast without a catalyst (called an inhibitor in this case) that it could be dangerous or wasteful of the energy released. One of the most controversial examples of this is the addition of lead (in the form of the compound tetraethyl lead) to petrol. This allows the fuel to be burned more efficiently. But the exhaust gases will contain lead compounds. Why do you think this is undesirable?

In the laboratory, catalysts are often used to increase the speed at which a chemical reaction takes place. The preparation of oxygen is discussed on page 245. A catalyst increases the speed at which the oxygen is produced. Hydrogen peroxide will rapidly decompose in the presence of manganese(IV) oxide to produce water and oxygen. The equation is:

$$2H_2O_2(aq) \rightarrow 2H_2O(l) + O_2(g)$$

Without the manganese(IV) oxide being present, the reaction takes place very slowly.

Assignment – Catalysing the production of oxygen

Oxygen can be prepared from hydrogen peroxide using manganese(IV) oxide as a catalyst (see page 245 for more details).

1 If manganese(IV) oxide is a catalyst it should remain chemically unchanged at the end of the experiment. How would you check that:
   (a) the mass of manganese(IV) oxide was unchanged;
   (b) it was still manganese(IV) oxide (Hint: could it be used again)?
2 Someone suggests that iron(III) chloride could act as the catalyst. What would you do to check this?
3 Someone else suggests that iron(III) chloride is a *better* catalyst for this reaction than manganese(IV) oxide. What tests would you make to check the accuracy of this statement?

How do catalysts work? There is no simple answer, unfortunately. There are a number of theories to explain their action. It is probable that catalysts work in a number of different ways. But they all produce the same end result – a change in the rate of reaction.

One accepted idea is that catalysts can lower the 'activation energy'. For two molecules to react, a certain initial energy (the activation energy) is required. The presence of the catalyst lowers the quantity of energy needed.

Enzymes

In the previous section, the catalysts were simple molecules working in non-living systems. Enzymes are biological catalysts. They are complex molecules working in living systems. All enzymes are proteins and are produced in cells. These substances catalyse the chemical reactions that occur in living things. The action of enzymes is illustrated in Fig. 9.20.

Enzymes are very specific in the way they act. Each type of enzyme will only work on one particular stage of a chemical reaction. For example,

Investigation 9.4
To study the effect of temperature on the action of the enzyme amylase

The first stage of the digestion process is the breakdown of starch in the food to a soluble sugar. The presence (or absence) of starch can be determined by iodine solution. This turns blue-black if starch is present or remains brown if starch is absent.

Collect

1% starch solution
iodine solution
saliva (diluted with distilled water)
test tubes
water bath
ice
teat pipette

What to do

1 Test a small amount of the starch solution with iodine solution to see the colour changes expected if starch is present.
2 It is suggested that four different temperatures are initially investigated:
(a) about 0 °C (beaker of ice cubes);
(b) room temperature;
(c) about 37 °C (water bath necessary);
(d) about 100 °C (beaker with boiling water).
You will be told to do one or more of these.
3 Put 2 cm^3 starch solution into each of four test tubes and 1 cm^3 of saliva solution into each of another four.
4 Set them up in pairs at the four different temperatures.
5 When equilibrium has been reached, mix the saliva and starch solutions.
6 At two minute intervals, remove a small amount of the solution with a teat pipette and test it with iodine solution. Clean the pipette before each test.
7 Record your results in a suitable chart.

Questions

1 Which temperature was the best for the enzyme to work at?
2 What happened with the starch/saliva kept at room temperature?
3 What results would you expect if you used 2 cm^3 of saliva solution instead of 1 cm^3? You might try this out.
4 It is said that enzymes such as amalyse only work efficiently when the solution is neutral (pH = 7). How would you design an experiment to check the accuracy of this statement?

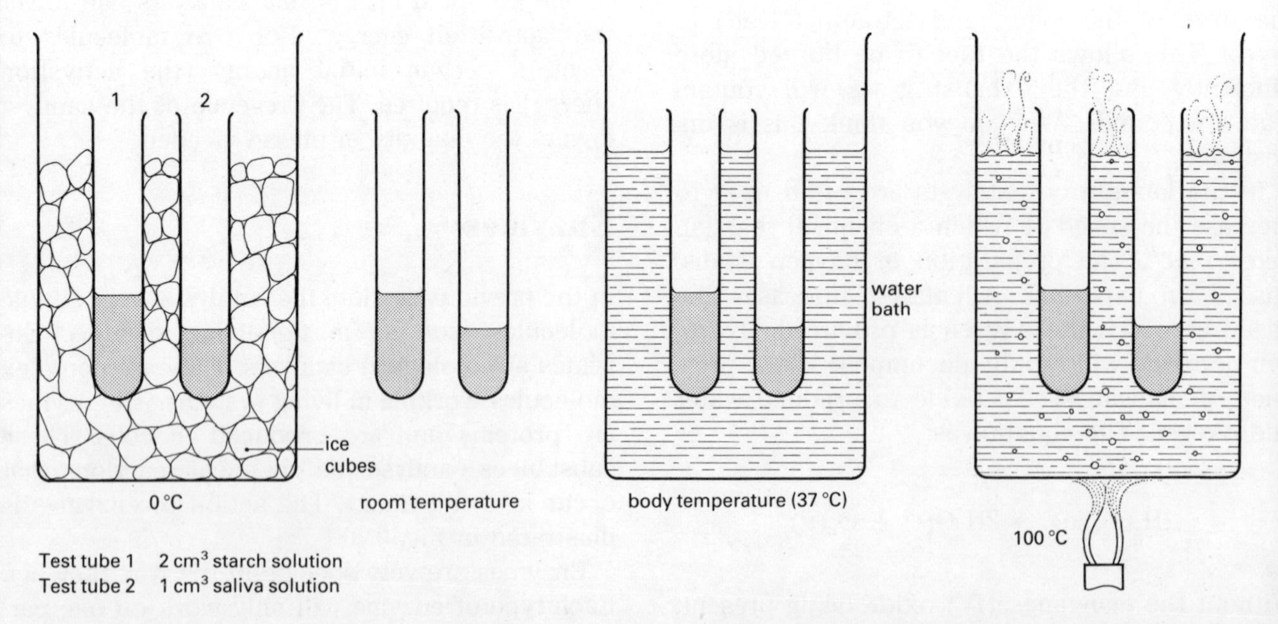

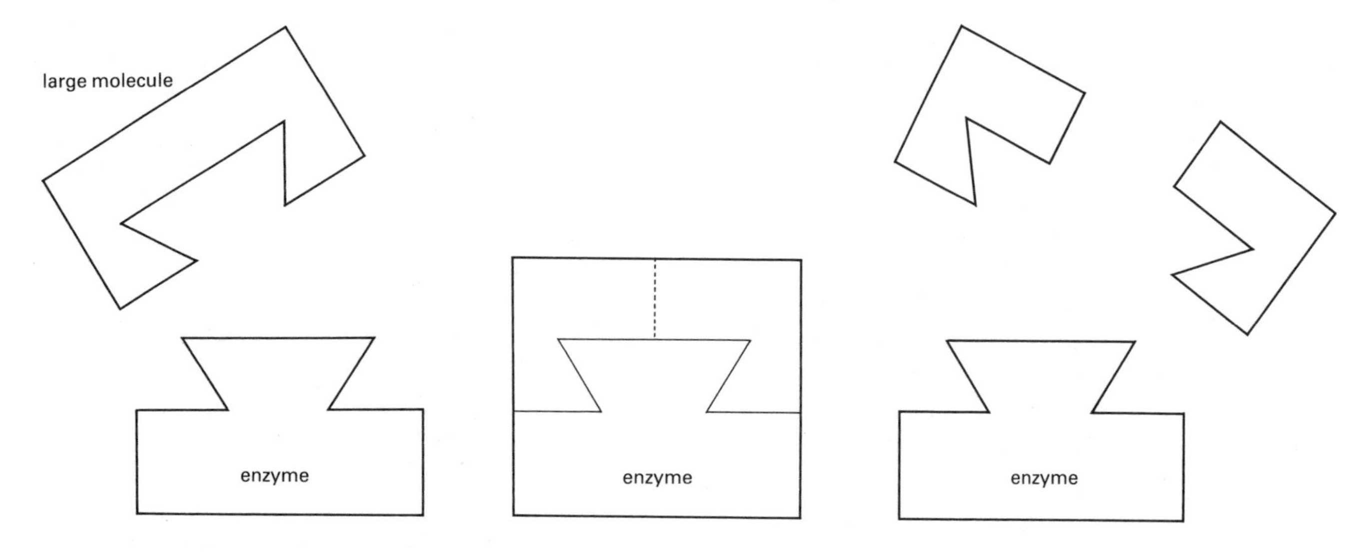

Fig. 9.20 The diagram shows the action of an enzyme that splits a large molecule into two parts

in respiration (see page 217), the breaking down of glucose consists of a large number of different chemical reactions. For each stage, there is one specific enzyme that catalyses the reaction. If that enzyme were missing then the whole chemical process would come to a complete halt.

We will study an enzyme called amylase. This enzyme is present in saliva and acts on the starch you eat to produce a soluble sugar. This is the first stage of the digestion process. Enzymes are usually called after the substance they act on plus 'ase'. Amyl comes from amylum, the Latin for starch. So amylase is the enzyme that acts on starch. Enzymes only work over a small temperature range. The best temperature for enzymes in the human body is 37 °C. If enzymes are raised to a high enough temperature their chemical structure is changed. However, lowering their temperature merely inactivates them. Also, most enzymes work best in neutral solutions (pH = 7). However, in the stomach, the enzymes can cope with the very acidic conditions found there.

Assignment – Using enzymes

Try to find out what part enzymes play in the following processes:
(a) the making of cheese;
(b) the brewing of beer;
(c) the action of biological washing powders.

Reactivity

WHAT YOU SHOULD KNOW

1 The chemical elements can be separated into two groups (metals and non-metals) on the basis of their physical and chemical properties.
2 All metals are crystal structures with fixed positive ions and mobile electrons.
3 The four most widely used metals are iron, aluminium, copper and zinc.
4 The properties of metals can be changed by making alloys. An alloy is a mixture of two or more metals.
5 Metals are a limited resource and must be conserved.
6 The less reactive a metal is the earlier it was used in history.
7 It is possible to produce a reactivity series for metals. This is based on how reactive a metal is with other substances.
8 The higher the position of the metal in the reactivity series the more difficult it is to extract it from its ore.
9 A simple cell produces electrical energy from chemical changes.
10 It is possible to produce an electrochemical series for metals. This is based on the voltage produced between different metals in a solution.
11 The reactivity and electrochemical series are very similar.

12 Because most metals are reactive, they will react with the chemicals around them producing corrosion.
13 The two main methods of corrosion are the formation of oxide layers and electrochemical processes.
14 To prevent iron rusting, it is essential to stop oxygen and water reaching the metal.
15 Catalysts change the speed at which chemical processes take place. They are very important in the manufacture of chemical substances.
16 Enzymes are biological catalysts. Without enzymes, no life would be possible.

QUESTIONS

1 The table shows details of the physical properties of eight elements, A to H.

Element	Melting point/°C	Density/ kg m^{-3}	Electrical conductor	Strength
A	3730	2250	yes	weak
B	98	970	yes	weak
C	−39	13 600	yes	–
D	113	2070	no	weak
E	1535	7860	yes	strong
F	1769	21 400	yes	strong
G	114	4930	no	weak
H	1083	8920	yes	strong

(a) On the basis of the above physical properties, which of these elements are (i) definitely metals; (ii) possibly metals; (iii) non-metals?
(b) The list of elements is: carbon, copper, iodine, iron, mercury, platinum, sodium and sulphur. Can you identify which element is which? Give reasons for your answer in each case.

2 Suggest a reason (or reasons) why each of the following metals has been used in that particular case. (You may need to do research for some of the answers.)
(a) Mercury is used in thermometers.
(b) Iron (or steel) is the most widely used metal.
(c) Sodium is used as a coolant in a nuclear reactor.
(d) Copper is used as a lightning conductor on buildings.
(e) Years ago, lead was commonly used for water pipes but not nowadays.
(f) Platinum is used in jewellery.
(g) Lead 'aprons' are worn by hospital staff working with X-rays or radioactive substances.
(h) Aluminium (or an aluminium alloy) is used for the bodywork of double decker buses.
(i) Gold is often used for the connecting wires to integrated circuits.
(j) Aluminium is used to make kettles and teapots.
(k) Tungsten is used for the filaments of light bulbs.
(l) Chromium is electroplated onto steel car bumpers.
(m) Magnesium is used in distress flares at sea.

3 In a series of displacement reactions to establish the position of a number of metals in the reactivity series, the following results were obtained. 'Y' indicates that a reaction took place, 'N' that no reaction took place.

Solution \ Metal	Copper	Lead	Silver	Zinc
Copper nitrate	–	Y	N	Y
Lead nitrate	N	–	N	Y
Silver nitrate	Y	Y	–	Y
Zinc nitrate	N	N	N	–

(a) Using these results, what is the position of these four metals in the reactivity series?
(b) From the pattern of the results, write down a simple rule to allow a person to establish quickly the position of each metal in the reactivity series.

4 In each of the following examples, say whether or not a reaction would occur. If your answer is 'yes' then write a word equation to describe the reaction.
   (a) Iron filings are added to potassium sulphate solution.
   (b) Zinc metal is added to copper(II) sulphate solution.
   (c) Aluminium powder and iron(III) oxide are heated together.
   (d) Magnesium ribbon is added to sulphuric-(VI) acid.
   (e) Lead metal is added to hydrochloric acid.
   (f) Iron filings are heated with zinc oxide.

5 A student found that, if a zinc plate and a copper plate were inserted into a lemon and connected to a voltmeter (as shown), a voltage was recorded on the voltmeter.

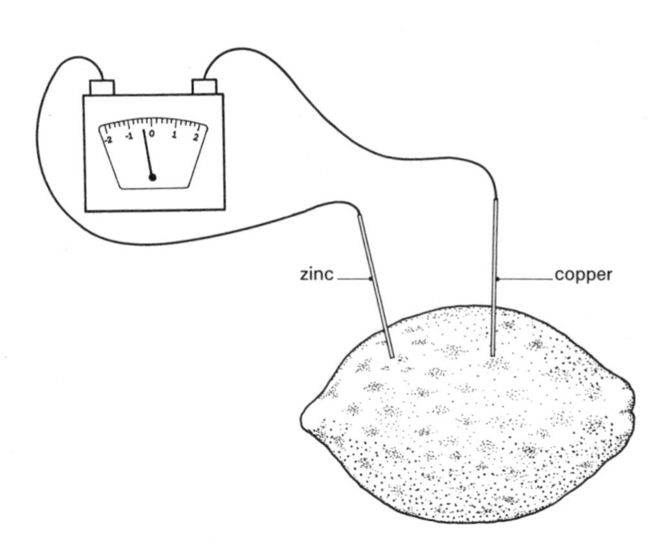

   (a) What would this device be called?
   (b) Using Table 9.5 (page 191), predict a value for this voltage.
   (c) Which metal should be connected to the negative terminal of the voltmeter?
   (d) It was found that the voltage decreased with time. What would you do to get back to the original voltage?
   (e) Do you think the device would work indefinitely? Explain your answer.
   (f) If you had a number of lemons and plates of zinc and copper, how would you produce a bigger voltage?

6 In an experiment on the electrochemical series, the following results were obtained using apparatus similar to that shown. Copper was used as the reference metal.

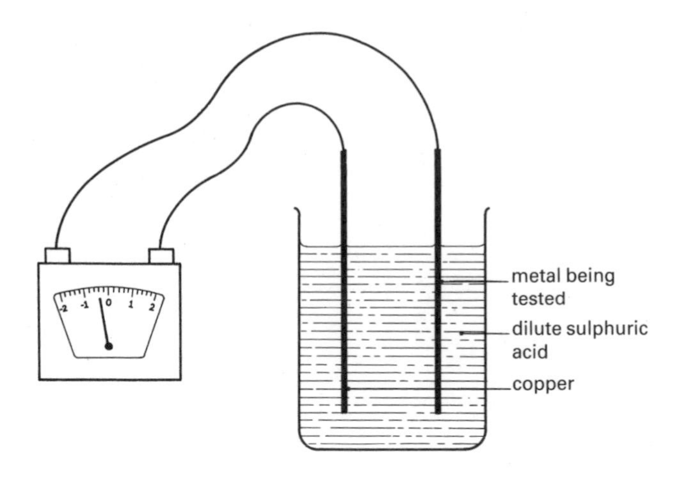

Metal	Voltage/V
Iron	−0.8
Lead	−0.5
Magnesium	−2.7
Silver	+0.5
Zinc	−1.1

   (a) Use these results to establish the position of each metal in the electrochemical series.
   (b) The metal electrode was replaced by a hydrogen electrode and the voltage recorded was −0.30 V. Re-write the results using hydrogen as the reference.
   (c) Describe a practical use of the voltage generated by two different metals.
   (d) Describe a practical problem caused by this effect.

7 Rennin is an enzyme that is produced in the stomach of young mammals. It solidifies the protein in milk. This is called 'clotting'. Assume that you have been asked to investigate how the time taken for the milk to clot is affected by the temperature. You are provided with sufficient rennin solution and milk for all the experiments you do and all necessary laboratory apparatus. Describe how you would carry out this investigation. What sort of results would you expect?

Energy 2

PLANTS AND ENERGY

It may seem strange to move from a topic called 'Reactivity' to one which is going to look at green plants and how they trap energy from the sun. However, in the last section enzymes were introduced. Enzymes play a central part in the chemical reactions which take place in all living things; life would not be possible without them. This section and the next one are concerned with how living things obtain their supplies of energy with which to live, grow and reproduce.

Fig. 9.21 How do plants get their food?

HOW GREEN PLANTS OBTAIN THEIR FOOD

Like other living things, green plants need food in order to live and grow. Green plants do not 'eat' food like most animals, so where does their food come from? In 1692 a Dutchman called Van Helmont carried out an experiment to investigate this. He put 91 kg of soil in a large pot and planted a young willow tree which had a mass of 2.4 kg. For five years he allowed the plant to grow, giving it only water. He then reweighed both the soil and the willow tree. As you would expect, the willow tree had grown considerably and it had gained nearly 75 kg in mass but the soil had lost only 57 g (0.057 kg). From these results Van Helmont realised that the willow had not absorbed much from the soil. He concluded that the increase in mass had come from the water alone.

We now know that as well as absorbing water and other substances from the soil through their roots, plants also absorb simple substances from the air through their leaves. If they are kept in the light, green plants use these simple materials to make their food by the process called PHOTOSYNTHESIS.

To investigate how green plants obtain their food you need to find out what substances green plants contain and where they get them from.

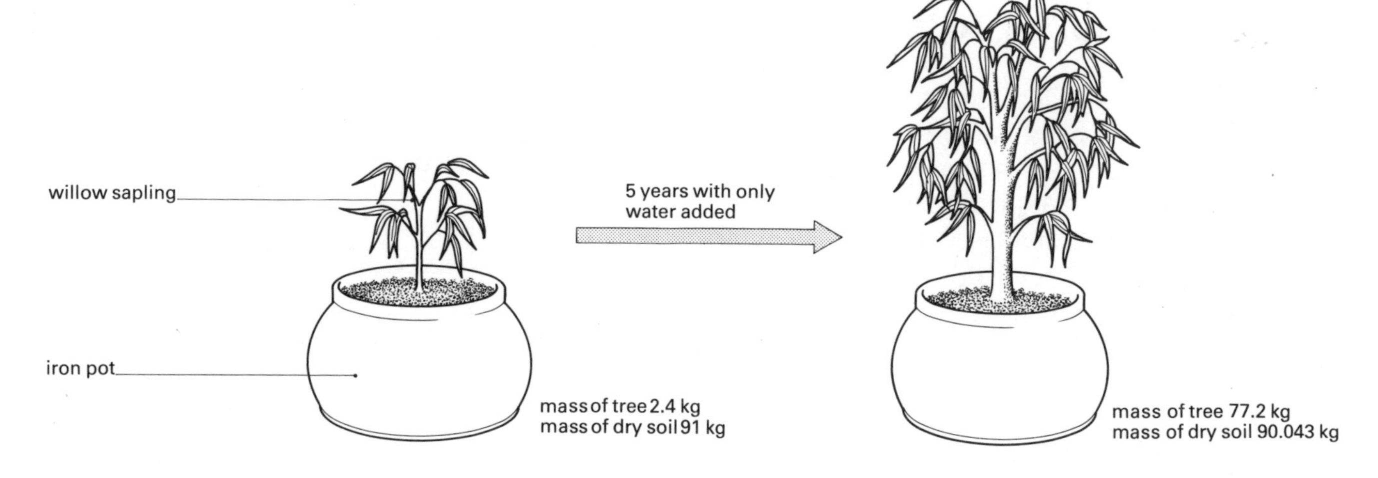

Fig. 9.22 Van Helmont's experiment

Investigation 9.5 To test a leaf for sugar

Most plant leaves contain sugar and starch. Usually, plants convert sugar into starch for storage. The iris plant is an exception, as it stores sugar in its leaves and does not convert it into starch.

Collect

iris leaf
pestle and mortar
sand (a pinch)
filter paper
filter funnel
Benedict's solution
250 cm^3 beaker half full of water
Bunsen burner
tripod
gauze
test tube
dropper pipette

What to do

1 Get ready half a beaker of boiling water.
2 Tear the iris leaf into small pieces and put a few of these into the mortar. Add a pinch of sand and just cover with water.
3 Grind up the leaf with the pestle.
4 Filter the solution into the test tube to a depth of about 1 cm.

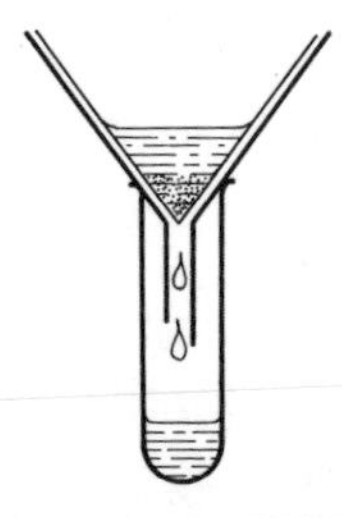

5 Add an equal amount of Benedict's solution to the test tube.
6 Put the test tube into the beaker of boiling water and watch until the contents of the test tube boil.

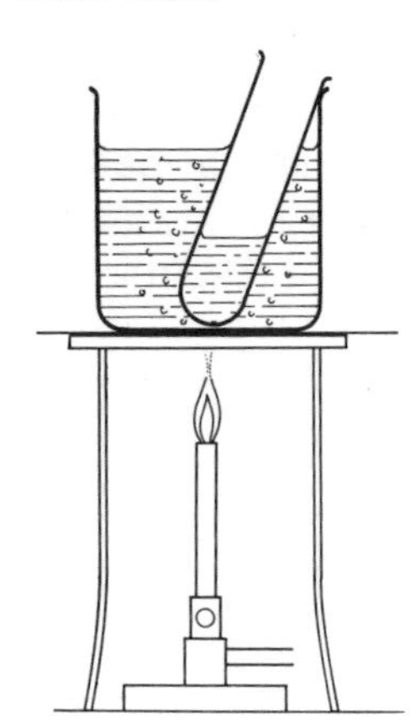

7 If the Benedict's solution turns green, brown or red this means there is some sugar present in the leaf. Record what happens to the solution.

Questions

1 Was there any sugar in the leaf?
2 Do you know where the sugar we eat comes from? In the UK there are two main sources: sugar cane grown in the West Indies and sugar beet grown in the UK. In which parts of these plants is the sugar found?

Investigation 9.6 To test a leaf for starch

If we find that a leaf contains starch, this shows that the plant has been making food during the last 24 hours. Plants do not actually make starch directly; they make simpler carbohydrates like glucose, but these are rapidly converted to starch and are stored as starch in the leaf and other parts of the plant.

Collect

geranium leaf
250 cm^3 beaker half full of water
Bunsen burner
tripod
gauze
forceps
test tube half full of ethanol
white tile
iodine solution
dropper pipette
eye protection

What to do

1 Get ready half a beaker of boiling water.
2 Using forceps, dip the geranium leaf into the boiling water for about 1 minute. This will kill the leaf and soften it.

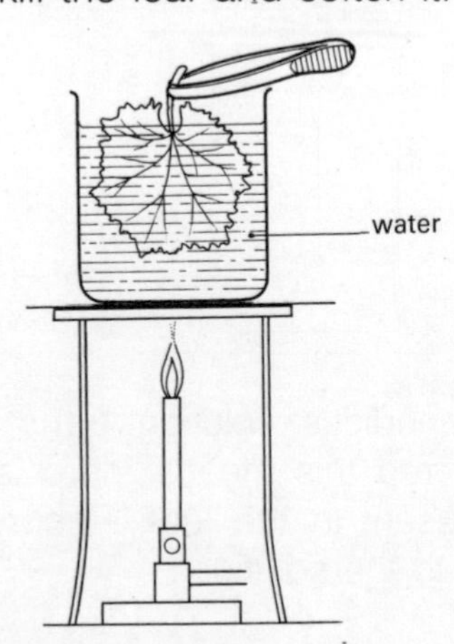

3 Put the leaf into the test tube containing ethanol.

CARE!! Do not continue to boil the water, turn out the Bunsen burner as ethanol is very flammable.

Stand the test tube in the beaker of hot water. Leave it for about 10 minutes. The ethanol should boil and will remove the green pigment (chlorophyll) from the leaf. The leaf will also become very brittle.

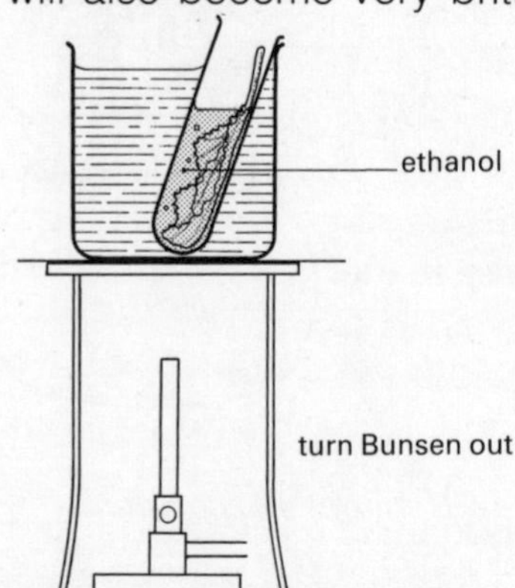

4 Using forceps, remove the leaf from the ethanol and wash it in the hot water to make it softer.

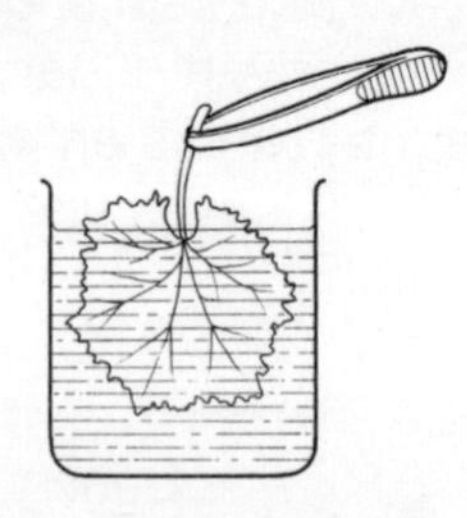

5 Carefully spread the leaf out on the white tile and then cover it with the dilute iodine solution.

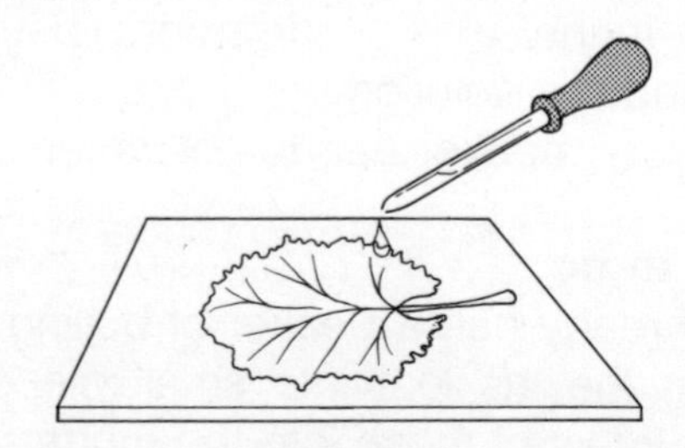

6 If the leaf turns a blue/black colour this shows that there is starch present in the leaf. Record what happens to the leaf when iodine is added.

Questions

1 Is there any starch in your leaf?
2 Which parts of a plant would you expect to contain starch?
3 Which vegetables that you eat contain a large amount of starch? Which parts of the plant are these?

PHOTOSYNTHESIS

This is the process in which energy from sunlight is used to turn simple substances into food. It takes place in green plants, mainly in their leaves.

Certain factors are needed for photosynthesis to take place and you can do some simple experiments to find out what they are. In each of the experiments you will be finding out whether the plant has made starch. Starch is not made in the process of photosynthesis. However, the glucose made during the process of photosynthesis is rapidly converted into starch for storage in the leaf. So the best way of seeing if a plant has been photosynthesising is to look for starch in the leaf.

Because you will be testing for starch it is important to use 'destarched' plants for these experiments. Usually plants have some starch stored in their leaves so this must be removed by keeping the plant in the dark for two or three days. During this time the starch stores will be used up, resulting in 'destarched' plants.

Using controls in experiments

If you do the next three investigations you can find out which substances a plant needs for photosynthesis. In each investigation the plant is given everything it needs except for one substance. Another plant, or another leaf, is used at the same time. This is a CONTROL. The control plant or leaf is given everything it needs, including the substance being tested for. Can you see why it is important to use such a 'control' plant in these experiments?

Investigation 9.7 To find out if light is needed for a plant to make starch

Perhaps you have noticed that plants do not grow very well in poor light. This could be because they need light to make their food: if they do not make much food they will not be able to grow very much.

Collect

geranium plant (destarched)
strip of aluminium foil 2 cm wide
2 paper clips
apparatus for starch test (see Investigation 9.6).

What to do

1 Fasten the strip of foil firmly around one leaf as shown in the diagram below, so that no light can get to either the upper or lower surface.

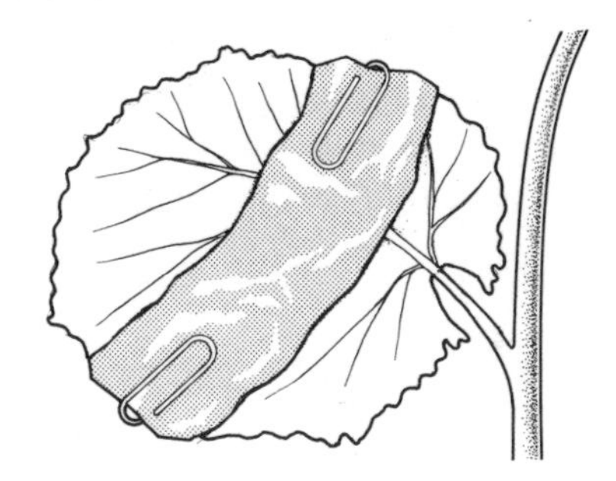

2 Make a drawing of the leaf to show the position of the foil strip.
3 Leave the plant in a well-lit place for 2 or 3 days.
4 Detach the leaf and test it for starch (see Investigation 9.6).
5 Make another drawing of the leaf and mark the brown areas (no starch), and the blue/black areas (starch present.)

Questions

1 Which parts of the leaf contain starch?
2 Compare your two drawings. What does each show?
3 Is light needed to make starch?
4 In this experiment you used only one leaf. Did you have a control? Why?

Investigation 9.8 To find out if carbon dioxide is needed for a plant to make starch

Plants are always surrounded by air. Perhaps they take something from the air to help them make their food?

Collect

2 destarched geranium plants of about the same size
2 large polythene bags
a small dish containing soda lime
2 rubber bands
apparatus for starch test (see Investigation 9.6)

What to do

1 Put the dish of soda lime on the soil beside one of the plants. Soda lime is a mixture of sodium hydroxide and calcium oxide. It will absorb carbon dioxide from the air.
2 Cover both plants with the polythene bags and make an airtight seal using the rubber bands. Make sure no leaves touch the sides of the bags. See the diagram below.

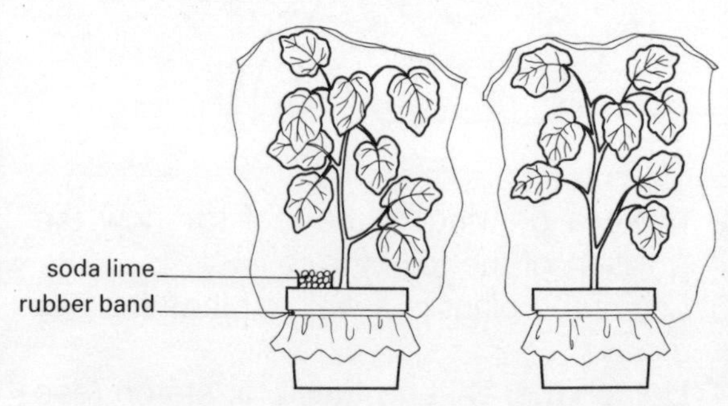

3 Leave both plants in a well-lit place for two or three days.
4 Take one leaf from each plant and test it for starch (see Investigation 9.6).

Questions

1 Which leaf has made starch?
2 Why did you use soda lime with one of the plants?
3 How could you do a similar experiment using only one plant? Would this be a better experiment? Explain your answer.

Investigation 9.9 To find out if chlorophyll is needed for a plant to make starch

Some plants have leaves that are partly white and partly green. These are called variegated leaves. Examples are some ivies, geraniums and spider plants (*Chlorophytum*). The parts which are green contain the green pigment chlorophyll, the parts which are white do not.

Collect

a variegated geranium leaf (from a plant that has been in good light for several days)
apparatus for the starch test (see Investigation 9.6).

What to do

1 Make a drawing of your leaf. Carefully mark the white areas and the green areas.
2 Test the leaf for starch (see Investigation 9.6).
3 Make another drawing of the leaf, mark the brown areas (no starch) and the blue-black areas (starch present).

Questions

1 Which parts of your leaf have made starch?
2 Compare your two drawings. Is chlorophyll needed to make starch?

Assignment – Seaweeds and light

White light is made up of a spectrum of all the colours of the rainbow (see Unit 3). Sea water acts as a filter and screens off some of the light energy from submerged seaweeds. This screening starts at the red end of the spectrum. As light travels downwards through the water, first the red part of light is screened off, then green and finally blue. In *very* clear water, blue light can penetrate up to a maximum depth of 1000 m. Below this all is dark.

The upper layers of the sea contain millions of microscopic organisms called plankton. These are of two types: zooplankton (tiny animals) and phytoplankton (tiny plants). The phytoplankton float near the surface where they receive plenty of light. They are very important as they are the beginning of the food chain for all the animals living in the sea. They also give out oxygen into the water and into the atmosphere.

On rocky shores various seaweeds grow. All these are submerged for part of the day, but some are submerged all the time. Brown and green seaweeds, generally grow on the upper shore. Red seaweeds grow lower down where they are covered with deep water when the tide comes in. All seaweeds make their food by photosynthesis in the same way as all green plants do. The colours of the seaweeds are due to their light-absorbing pigments, not all of which are chlorophyll.

1 Chlorophyll is a green pigment. Which colours of light does it (a) absorb; (b) reflect?
2 What colour light would you expect the pigment of red seaweeds to absorb? Explain your answer.
3 Why are red seaweeds often found lower down the shore than green ones?
4 Why do seaweeds not grow on ocean beds?
5 If there are no seaweeds on ocean beds, what do the animals who live on ocean beds eat?
6 Make up a food chain for organisms living in an ocean miles away from land. (See page 138 for information on food chains.)

If you have done the experiments in Investigations 9.7 to 9.9, you should have found that light, carbon dioxide and chlorophyll are all needed for making starch. Water is also needed, but it is difficult to do an experiment to show this as plants die if you give them no water!

We have seen that starch is produced during photosynthesis, but is anything else produced as well? In 1771 Joseph Priestley carried out the following experiment. He put a burning candle in a sealed bell jar. The candle soon went out. He then put a living plant in the bell jar with the candle and, after a week, when he lit the candle it burned. Priestley was not sure how to explain this, but we now know that the burning candle had used up all the oxygen in the bell jar. Putting the plant in the bell jar had the effect of putting oxygen back into the air so that the candle could burn again. So Priestley had shown that plants give out oxygen.

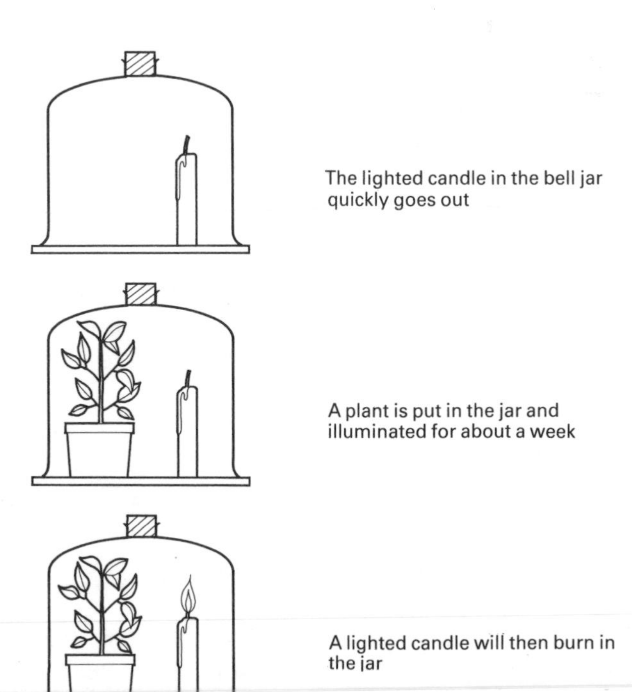

Fig. 9.23 Priestley's experiment

Investigation 9.10 To show that plants give off oxygen

It is difficult to collect oxygen from plants because they give it off into the air. So, for this investigation, you will use a plant that normally lives in water. If the plant gives off any gas you will be able to see bubbles rising in the water.

Collect

a tall beaker and a glass filter funnel which fits inside it
boiling tube or test tube
several pieces of pondweed
matches
wooden splint

What to do

1 Set up the apparatus as shown in the diagram below.

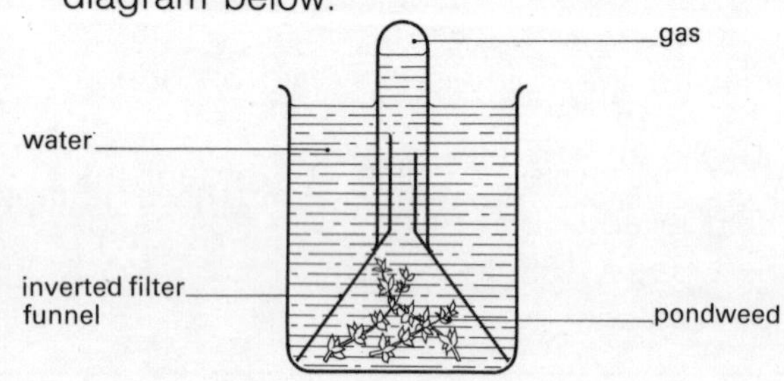

2 Leave for about 1 week in a well lit place. What sort of control plant could you set up at the same time?
3 You should find that the pondweed has given off some gas which will collect at the top of the tube. Test this gas for oxygen. Get ready a glowing splint, quickly remove the test tube and insert the glowing splint. If the splint relights this shows that there is oxygen present. Be careful not to let the splint touch the wet sides of the test tube!

Questions

1 Has the pondweed produced some gas? Was it oxygen?
2 If you set up a control, what results did you get?
3 Why did you use pondweed for this investigation and not a plant such as a geranium?

The process of photosynthesis

Plants need carbon dioxide, water, light and chlorophyll to make food; starch and oxygen are produced. Carbon dioxide and water are the raw materials of photosynthesis. Starch and oxygen are the products. The chemical process of photosynthesis is quite complicated and involves a series of steps. The reactions need energy which comes from sunlight. Chlorophyll absorbs the light so that the plant can use the light energy. So although sunlight and chlorophyll are not 'raw materials' for the chemical reactions, they are needed before the process can happen.

Although starch is the final product it is not the first substance to be made. Glucose is made first and then this is turned into starch.

Photosynthesis is a chemical process. It can be written as the following word equation:

$$\text{carbon dioxide} + \text{water} \xrightarrow[\text{chlorophyll}]{\text{sunlight}} \underset{\text{(food)}}{\text{glucose}} + \text{oxygen}$$

A chemical equation for photosynthesis can also be written:

$$6CO_2 + 6H_2O \xrightarrow[\text{chlorophyll}]{\text{sunlight}} C_6H_{12}O_6 + 6O_2$$

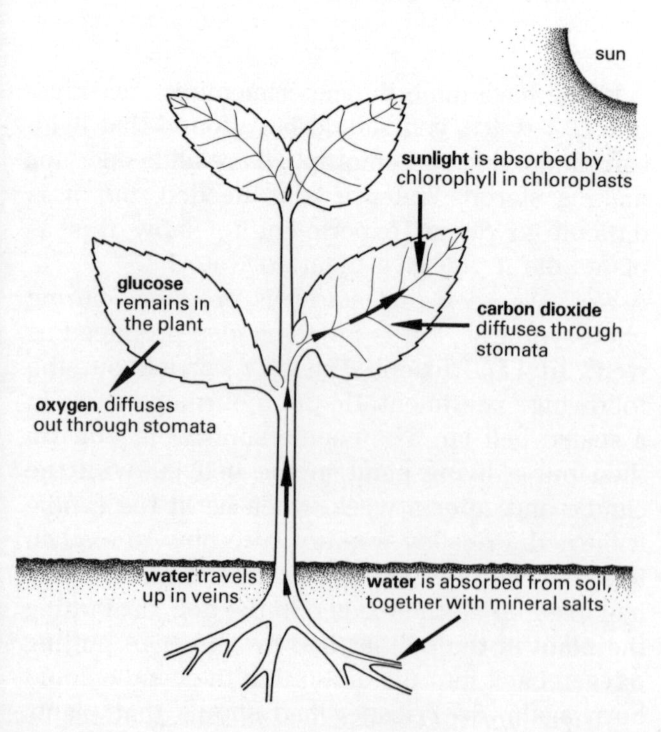

Fig. 9.24 Summary of photosynthesis

Assignment – Sugar content of leaves

The percentage sugar content of the leaves of a green plant was found at different times during one day.

Time	% sugar content
12 midnight	0.5
3 a.m.	0.4
6 a.m.	0.5
9 a.m.	0.8
12 noon	1.6
3 p.m.	1.9
6 p.m.	1.5
12 midnight	0.5

1. Plot the data on graph paper. Put % sugar on the vertical axis.
2. Estimate the sugar content of the leaves at 9 p.m.
3. Describe the changes that occur in the percentage sugar content over the 24 hours.
4. If the maximum light and the highest temperature occur at noon, how would you explain that the highest % sugar content was recorded at 3 p.m.?
5. The data above was collected on a day when the average air temperature was 15 °C and there were 14 hours of daylight. On your graph draw a second line to show what you would expect to find from data collected from the same plants on a day when the average temperature was 6 °C with 12 hours of daylight.

FACTORS AFFECTING THE RATE OF PHOTOSYNTHESIS

Photosynthesis may occur quickly or slowly depending on several factors. The rate of photosynthesis will determine how much food is produced by the plant. This is very important for humans (and for all animals), as we depend on plants either directly or indirectly (see page 213) for our food.

There are four main factors which affect the rate of photosynthesis.

1. Light

In the dark plants cannot photosynthesise at all. In dim light they photosynthesise slowly and, as the light intensity increases, so will the rate of photosynthesis up to a point when the plant is photosynthesising as fast as it can. At this point the plant will not photosynthesise faster even if the light gets brighter.

This is important for gardeners. In order to grow well, vegetables should be able to get plenty of sun. Sometimes bright lights are shone on indoor plants to increase their rate of photosynthesis. However, some plants, such as primroses and ivy, can survive in shady places like woods.

2. Carbon dioxide

The amount of carbon dioxide in the air is usually about 0.03%. However, experiments have shown that an increase in the amount of carbon dioxide in the air increases the rate of photosynthesis up to a certain point.

Extra carbon dioxide is sometimes pumped into commercial greenhouses to increase the rate of photosynthesis and thus the growth rate of the plants in the greenhouses.

3. Temperature

Up to a certain point, the higher the temperature the faster a plant will photosynthesise. For many plants a rise in temperature of 10 °C doubles the rate of photosynthesis. Raising the temperature up to about 40 °C increases the rate of photosynthesis but at higher temperatures photosynthesis slows down and eventually stops altogether. This is because the high temperature destroys enzymes which carry out the necessary chemical reactions.

There are tremendous variations of temperature in the world, both from place to place and at different times of the year. Because of this we find different types of plants growing naturally in different countries and habitats. Plants grown in heated greenhouses generally grow well because of the higher temperature and because a steady temperature is maintained with no 'cold spells'.

Assignment – Rate of photosynthesis

An experiment was carried out to find the rate of photosynthesis of a group of plants at different concentrations of carbon dioxide and at two light intensities. The results are shown in the table below.

CO_2 concentration /% of air	Rate of photosynthesis in arbitrary units	
	low light	high light
0.00	0	0
0.02	20	33
0.04	29	53
0.06	35	68
0.08	39	79
0.10	42	86
0.12	45	89
0.14	46	90
0.16	46	90
0.18	46	90
0.20	46	90

1 Plot these results on a graph.
2 What patterns are shown by the graph?
3 (a) What is the CO_2 concentration of normal air?
(b) What is the rate of photosynthesis in low and high light in normal air?
4 Market gardeners may sometimes add carbon dioxide to the air inside their greenhouses.
(a) What is the advantage of doing this?
(b) Find out *how* they add carbon dioxide to the air.
5 (a) What factors limit the rate of photosynthesis?
(b) How can market gardeners use this information to produce vegetables such as lettuces throughout the winter?
6 Design an experiment to investigate the rate of photosynthesis of pondweed in the laboratory at two different light intensities. (Hint: the 'rate' could be recorded by finding how much oxygen is given off.)

Assignment – Hydrogencarbonate indicator

Hydrogencarbonate indicator solution which has had atmospheric air bubbled through it is orange/red in colour. If carbon dioxide is then passed through it the solution changes to a paler yellow colour. If carbon dioxide is removed from the solution it becomes purple in colour.

The following experiment was set up to investigate the gaseous relationships of plants and animals.

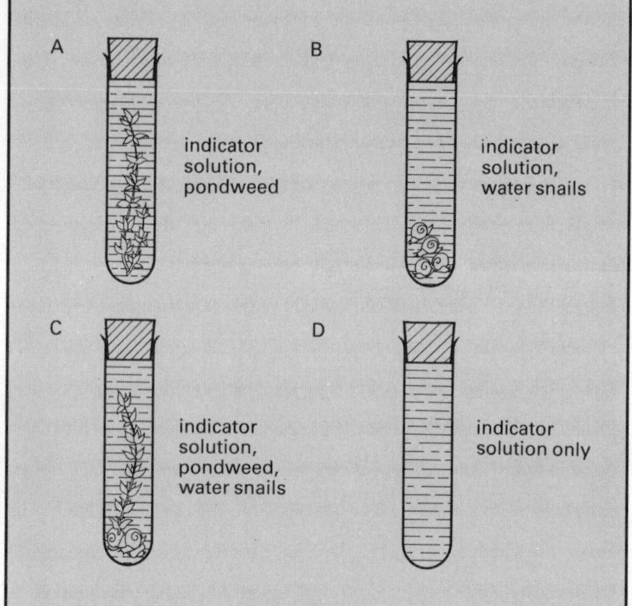

1 Why was it necessary to fill each boiling tube right to the top?
2 Which tube is the control? Why is it necessary in this experiment?
3 If all the test tubes received the same amount of light for 24 hours, what colour changes, if any, would you expect in each of A–D? Explain your answer in each case.
4 If all the test tubes were placed in the dark for 24 hours what would you expect to happen in A–D? Explain your answer in each case.
5 Describe how you would set up an experiment to investigate the gaseous relationships of non-aquatic plants and animals, for example, geraniums and mice. (N.B. hydrogencarbonate indicator would not be suitable.)

4. Water

Plants need water for photosynthesis and if they do not get enough water they will not photosynthesise so quickly.

The best conditions for photosynthesis are found in the tropical rain forests of South America, Central Africa and South East Asia. Lots of sunshine, warmth and a high rainfall mean that the plants can photosynthesise at a high rate and grow quickly. The worst conditions for photosynthesis are dark or dimly lit places, especially if they are cold as well. That is why some indoor plants do not grow well if they are in a dim corner of a cold room.

Although a plant's rate of photosynthesis is affected by light, carbon dioxide concentration, temperature and water supply, these factors do not act separately. In practice they interact and influence each other.

LEAVES AND PHOTOSYNTHESIS

Photosynthesis takes place in most plants in the leaves. However, in some plants photosynthesis may take place in other parts that are green such as stems. Leaves can be thought of as food factories! They are specially adapted to allow photosynthesis to take place as quickly and efficiently as possible.

Figure 9.25 shows the appearance of a typical leaf from the outside.

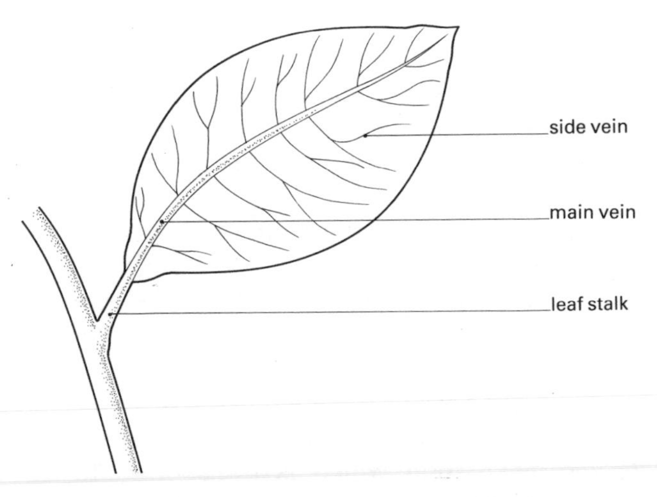

Fig. 9.25 The main parts of a typical leaf

Numerous different shapes and sizes of leaves are found in nature. They are, however, usually flat and thin. Because of this they have a large surface area which makes them good at absorbing carbon dioxide from the air and light energy from the sun. There are usually lots of leaves on a plant or tree and they are positioned in such a way as to get the maximum amount of light. They may fit together so that there are no spaces between for light to get through. That is why it is so dark under many trees and shrubs.

Although a leaf looks thin to our eyes, if you look at a microscope slide of a section through a leaf blade you will see that it is made up of several layers of cells. Leaves are usually less than a millimetre thick. But as leaves are thin they need veins which act as a type of skeleton to prevent them from drooping.

Investigation 9.11
To find the leaf area of a plant

Collect

a large plant which has lots of leaves (or go outside and do this investigation using a shrub or small tree)

What to do

1 Detach one average sized leaf. (How will you decide what is 'average'?)
2 Put the leaf on the squared paper and trace round it with a pencil.
3 Work out the surface area of the leaf in cm^2
4 Count the number of leaves on the plant. Multiply the area of one leaf by the number of leaves.

Questions

1 What is the total surface area of the leaves of the plant?
2 Why is it useful to the plant to have a large leaf area?

The outer layers of the leaf epidermis have tiny pores called stomata. There are usually more on the lower surface of the leaf. They allow carbon dioxide to enter the leaf and oxygen to leave it.

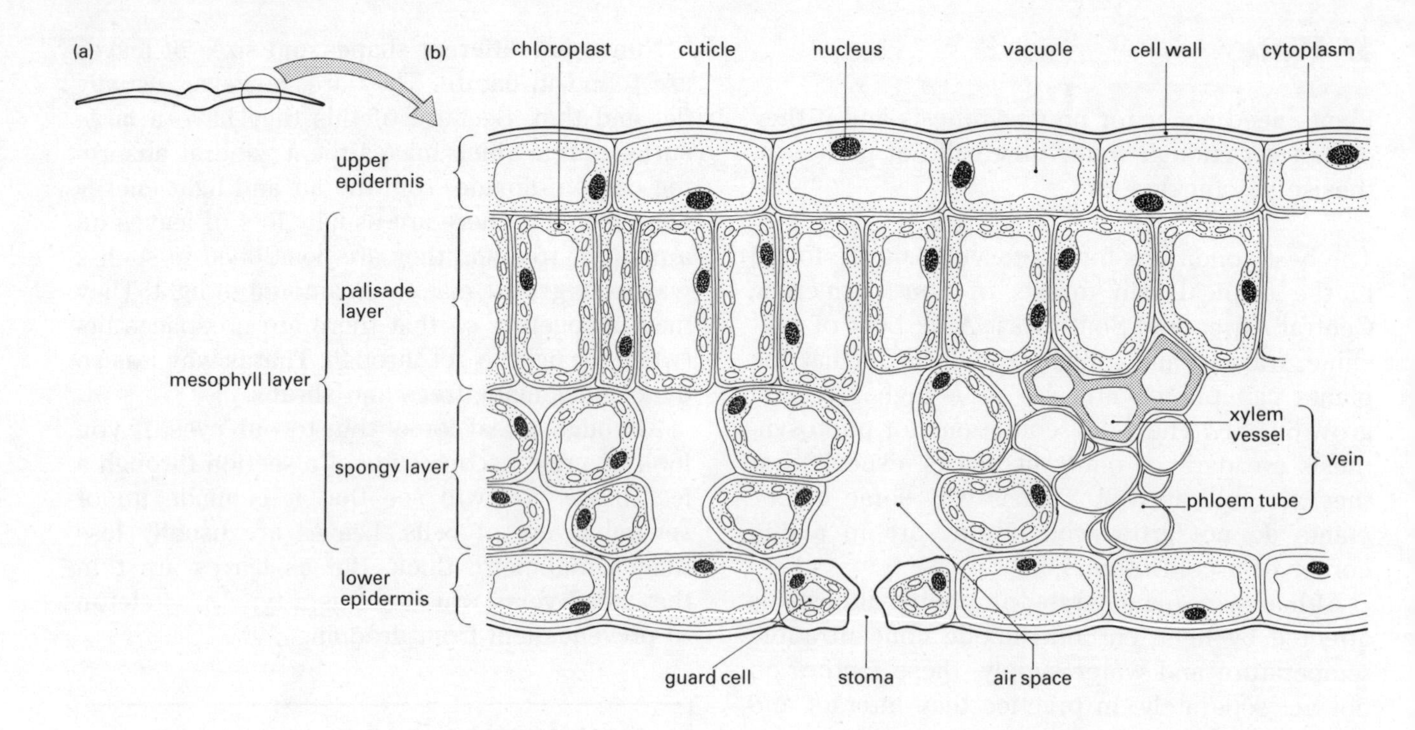

Fig. 9.26 (a) Cross section of a leaf
(b) Surface of a leaf, magnified

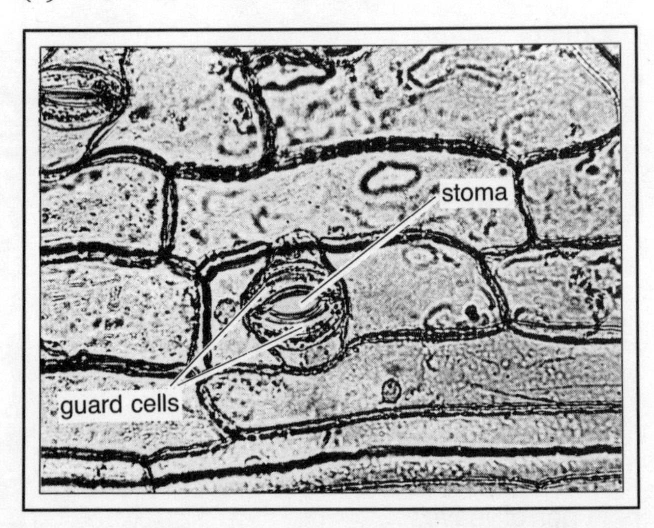

Some leaves have a very large number of stomata, as many as a thousand on every square millimetre.

The cells of the upper epidermis may secrete a waxy substance to form a waxy layer called the *cuticle*. This helps to stop water evaporating from the leaf. Some leaves, like holly leaves, have a thick cuticle.

The middle layers of the leaf are called the *mesophyll*. These cells all contain chloroplasts. There are usually two main layers. The *palisade* has tightly packed cells and the *spongy layer* has loosely arranged cells with air spaces between them. Running through the mesophyll there are also veins. Each vein has thick-walled *xylem* vessels which carry water and smaller thin-walled *phloem* tubes which carry away the food that the leaf has made to other parts of the plant.

The chloroplasts are very important for photosynthesis. They contain the green pigment chlorophyll which traps the light energy. Photosynthesis takes place in the chloroplasts.

Leaves are adapted to photosynthesis because:
1. The palisade and spongy mesophyll layers have chloroplasts and can photosynthesise. Most of the chloroplasts are found in the palisade layer.

The palisade cells are near the upper surface of the leaf. This gets most light and so most photosynthesis takes place here.
2. The cells of the spongy mesophyll are loosely packed with large air spaces. Carbon dioxide can diffuse easily through the stomata into these air spaces and then into the palisade and spongy mesophyll cells.
3. As well as giving strength to the leaf, the veins act as a transport system. The xylem vessels carry water and mineral salts from the roots to the leaves and the phloem tubes carry glucose, sugars and other substances that have been made in the leaves to other parts of the plant. For example, in a carrot or radish plant the leaves make food which is transported to the root to make the carrot or radish grow.

WHAT HAPPENS TO THE GLUCOSE MADE IN PHOTOSYNTHESIS?

Glucose is made during photosynthesis (see the equation on page 206). Several things may happen to this glucose.

1. It may be used to release energy in the leaf. All cells need energy which they obtain by a process called respiration (see page 217).
2. It may be turned into starch and stored in the leaf.

Glucose is a simple sugar. It is soluble, quite reactive and is not a very good substance for storage. The glucose is changed into starch for storage. Starch is a polysaccharide made of many glucose molecules joined together. As starch is such a large molecule it is not very soluble and it is not very reactive. It is made into granules which are stored inside the chloroplasts.

3. It can be used to make other substances.

By adding other elements and compounds to glucose, plants can make other carbohydrates such as sucrose and cellulose, oils and proteins.

Assignment – Fertilisers

Fertilisers contain mineral elements. Mineral elements are needed by all plants for healthy growth. They are added to soil to replace those removed by crops and to increase crop yields.

Although many fertilisers are applied as solids on the top of the soil, they have to be taken up by plant roots in their soluble form (they have to dissolve in soil water).

There are three main mineral elements which are needed by plants.

NITROGEN (found in nitrates). This is used to make plant proteins. Proteins are needed to make new cells so that the plant can grow in size. Plants which have received plenty of nitrates will look healthy, have dark green leaves, provide a good yield and have a long growing season.

PHOSPHORUS (found in phosphates). This helps healthy root growth and encourages fruits to ripen early.

POTASSIUM (found in potash). This helps plants to adjust to extremes of temperature, such as hot days and cold nights. It also helps plants resist drought.

The table gives information about the contents of various fertilisers. (* means that the fertiliser contains a significant amount of the mineral element.)

There are two main categories of fertiliser used by gardeners and farmers:

(a) Inorganic. These are often called 'artificial'; they can be man-made or natural and are quick-acting.

(b) Organic. These are made from plant or animal remains and they usually act slowly.

Fertiliser	N	P	K
Compound fertiliser	*	*	*
Sulphate of ammonia	*		
Nitro chalk	*		
Superphosphate		*	
Sulphate of potash		*	
Farmyard manure	*	*	*
Dried blood	*		
Bone meal	*	*	
Garden compost	*	*	*

1 What do N, P and K stand for in the table?

2 (a) Make a table to show which fertilisers are organic and which are inorganic.
(b) What are the advantages and disadvantages of using organic fertilisers?

3 'Organically grown' vegetables are often sold in health food shops
(a) What do you understand by this term?
(b) Why are such vegetables usually more expensive?
(c) Why do you think people are prepared to pay extra for such vegetables?

4 'Fertilisers are used to increase crop yield.' What does this phrase mean?

5 Go to a shop that sells gardening equipment. Find out what fertilisers are available for (a) houseplants, (b) roses, (c) vegetables, (d) lawns.
Make a table to show which mineral salts each contains.

Assignment – Organic farming

1 (a) What is organic farming?
  (b) How does this differ from conventional farming?
  (c) Is production of organic food likely to increase or decrease? Give reasons for your answer.

2 Why are people in Britain becoming so concerned about food additives?

3 Give three reasons why people may prefer to buy organic food.

4 (a) Why are large supermarket chains not selling much organically grown food?
  (b) Why do you think that organically grown foods tend to be more expensive?

5 If 'organics took up about five minutes of a three-year course' in agriculture and forestry, what do you think the rest of the course would have covered?

Rich harvest for organic farmers

by PAUL LASHMAR

ORGANIC farmers, already enjoying a boom, will benefit still further from the revelation that many additives used in conventional food production can cause adverse effects in humans.

There are now 200 organic farms that meet standards set by the Soil Association – which advocates the growing of crops without chemicals.

Production by organic means accounts for less than 1 per cent of British agriculture but Patrick Holden, chairman of British Organic Farmers, said last week: 'I believe that in 10 years time organic farming will hold at least 20 per cent of the market.'

His view is shared, somewhat cautiously, by major retailers. Safeways, which started stocking organically-grown *mange-tout* beans five years ago, now sells a wide range of vegetables, fruits and salad produce. The firm's view is that the market is 'potentially enormous and almost untapped.'

Sainsbury's too, is investigating the possibility of selling organic food. A spokesman said: 'We haven't taken a decision yet. At this stage it is a matter of getting the quality, quantity and continuity of supply.'

The organic cause will not have been harmed by last week's report of the London Food Commission – set up with Greater London Council funds to provide nutritional advice – which found that more than 40 per cent of the 300 approved food additives in use today can provoke allergic or intolerant reaction. It said that at least 41 approved additives were carcinogens.

Herbie Blake is typical ot the new breed of organic farmer. With his brother John, he runs a 16-acre farm at Flax Drayton, Somerset. The Blake family, descendants of Admiral Blake, have farmed the area for generations.

Mr Blake studied agriculture and forestry at Oxford in the 1970s. 'Organics took up about five minutes of a three-year course. In spite of that I became interested in the philosophy of self-sufficiency and ecology and taught myself organic farming,' he says.

'Flax Drayton farm had been conventionally run and it took us about five years to turn it into a working, economical organic mixed farm. After the soil has got accustomed to being doused with chemicals like nitrate fertilisers it takes years to wean it off. It is a bit like drug addiction; the soil takes time to regain its vitality and balance.'

The Blakes now grow a range of vegetables such as potatoes, spinach, courgettes and cabbage. Their three-quarters of an acre of onions produced a healthy 40,000 lb crop. They also have sheep, hens and bees.

Organic produce tends to be more expensive than ordinary crops although prices are more stable. Mr Blake says: 'For example, one of our 56 lb bags of potatoes costs about £3, compared with £2 for ordinary potatoes.'

The success of organic food is reflected by Dave Cornish of Safeways: 'Our problem has been getting enough organic produce. At the moment we can only supply our 70 shops in the South. The demand is incredible. Customers' reasons for wanting it vary from taste and health to ecological reasons. But we get a lot of people who suffer from allergies and cancer and have been told by their doctors to eat chemical-free food.'

Observer, Sun. 20 Oct. 1985

ENERGY, FOOD CHAINS AND BIOMASS

We have seen that green plants make their own food by a process called photosynthesis. To photosynthesise, plants use light energy from the sun. A plant takes in water and carbon dioxide and uses light energy to convert them into stored chemical energy.

The transfer of energy from the sun to plants and then to animals can be seen in a food chain. (Look back at Unit 8, page 138.) The sun is always at the beginning of a food chain and humans are often at the end.

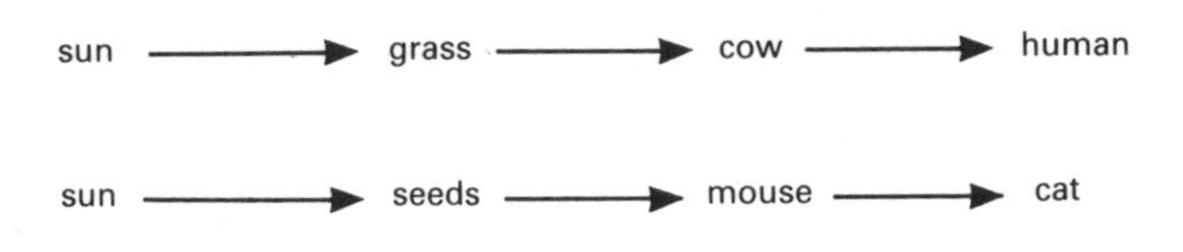

Fig. 9.27 Food chains

Energy flow in food chains

Energy is lost from a food chain at each link. So the energy passed on to the next stage gets less.

Only about 1% of the solar energy reaching a plant is used for photosynthesis. The other 99% is reflected or raises the temperature of the plant.

About 10% of the energy in plants is transferred to the body of a herbivore. Most of the other 90% is changed to heat energy during respiration, excretion and defaecation. The same applies at the next stage: only about 10% of the energy from a herbivore is transferred to a carnivore, or to an omnivore like a human being.

This loss of energy at each level of a food chain puts a limit on the total mass of living matter that can exist at each level. The amount of living matter at each level is termed the BIOMASS. The biomass decreases at each step along a food chain.

Pyramid of biomass

In Unit 8 you saw that the number of organisms gets less the further up the food chain. This is known as the pyramid of numbers (see page 139). A pyramid of biomass follows a similar pattern (Fig. 9.29).

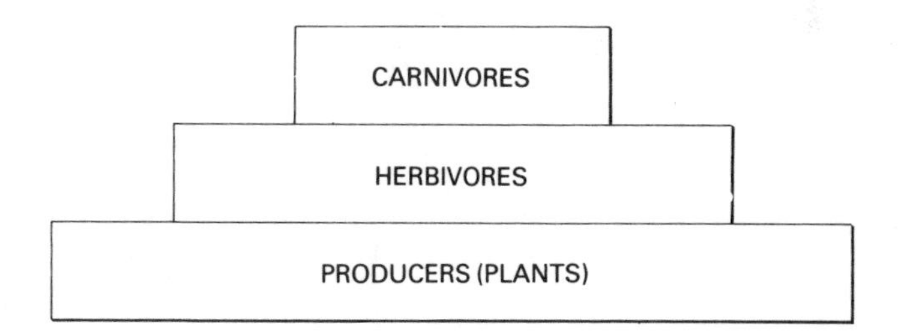

Fig. 9.29 Pyramid of biomass

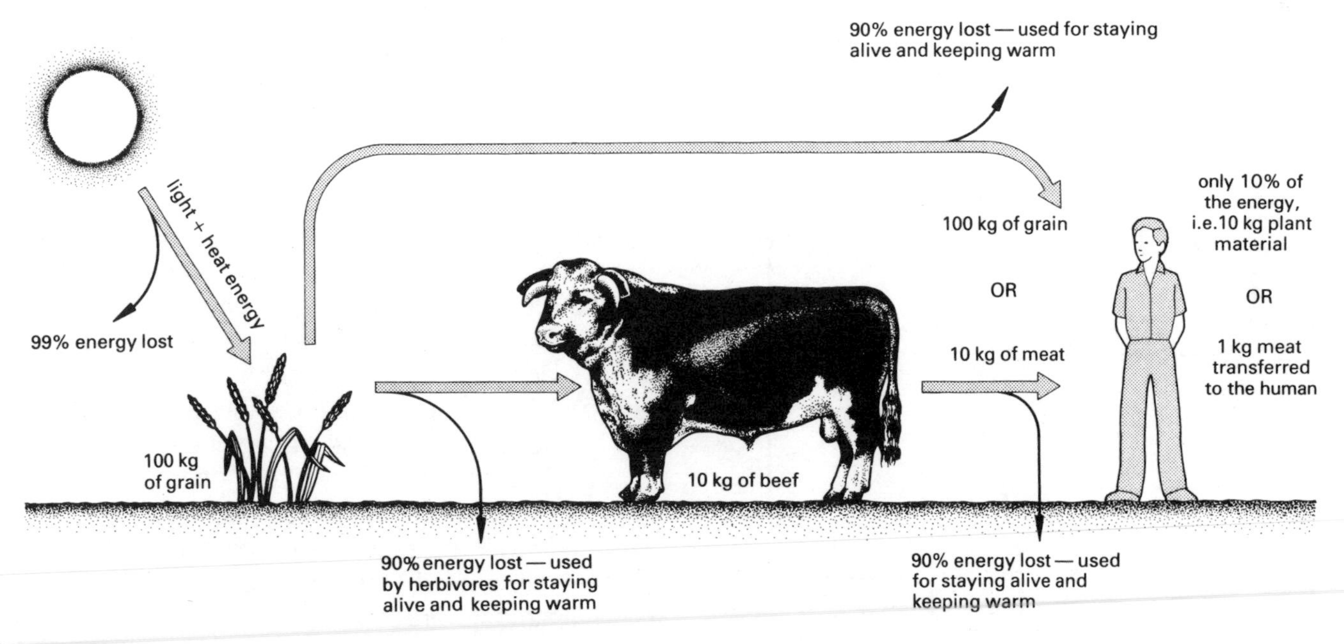

Fig. 9.28

Humans are at the end of many food chains. The size of a human population is limited by the number of links in a food chain. The longer the chain the more energy will be lost, the shorter the chain the more energy will be available. In overpopulated parts of the world where there is a shortage of food the people feed mainly on plant foods. This is because the plant crops produced on a given area of land will provide more energy than the number of animals that could be raised on that area.

A pyramid of biomass will always be roughly the same shape because of the loss of energy at each level. However a 'pyramid' of numbers may not always have traditional pyramid shape. For example, in a tree community various herbivores and carnivores may live in the one tree giving a pyramid like (ii) in Fig. 9.30. When parasites are involved in a food chain this may also result in a different shaped pyramid of numbers. See Fig. 9.30 (iii).

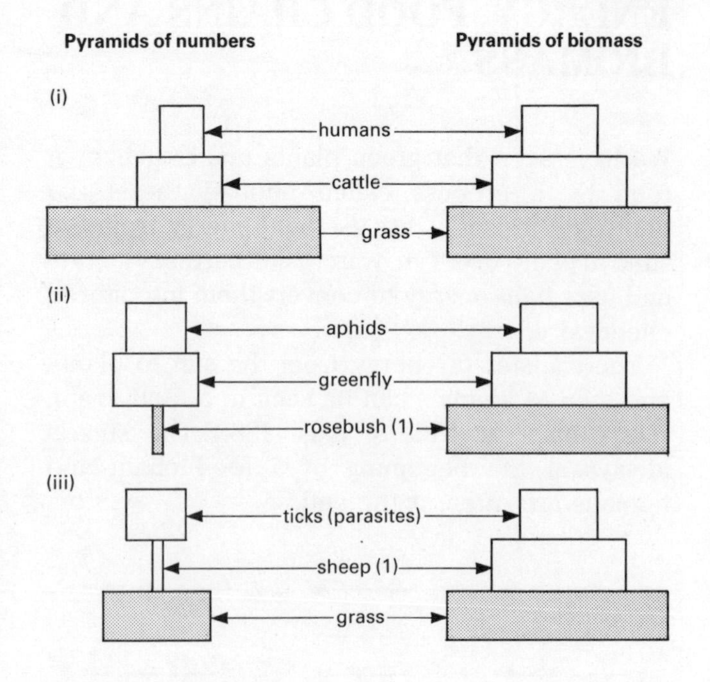

Fig. 9.30 Pyramids of numbers

Assignment – Carbon cycle

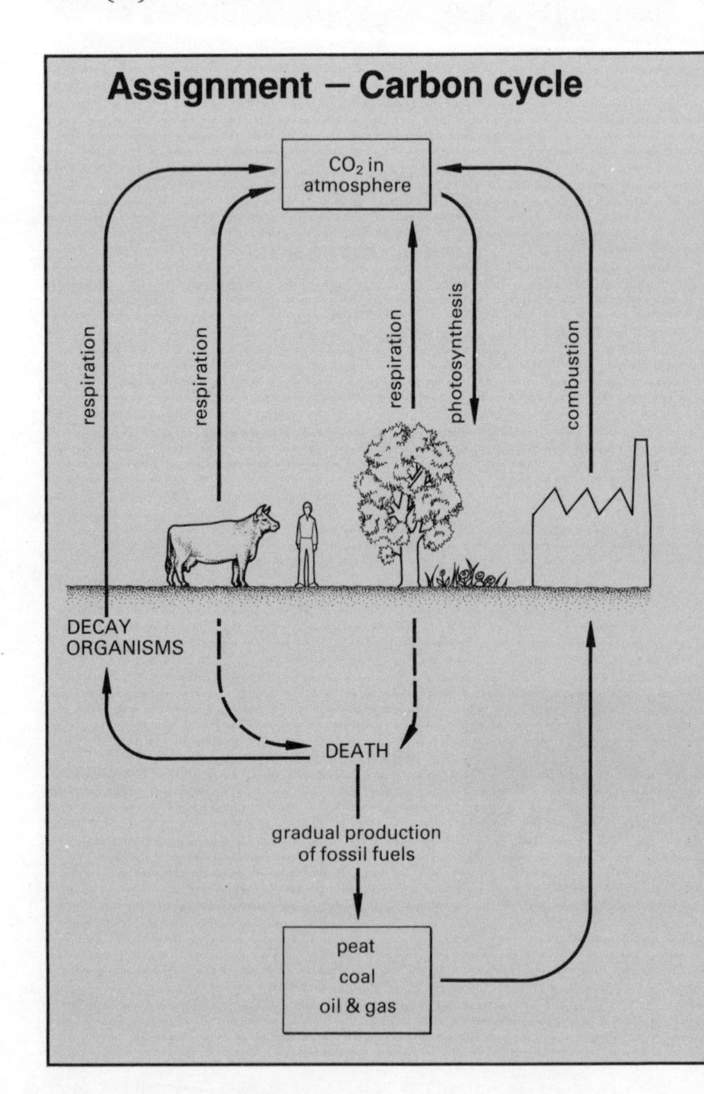

Study the diagram of the carbon cycle and then answer the questions.

1 How is carbon dioxide removed from the atmosphere?

2 Name three processes that add carbon dioxide to the atmosphere.

3 (a) What is the usual amount of carbon dioxide found in inspired and expired air? How can you explain the difference?

(b) Suggest three places where you would expect to find (i) a higher; (ii) a lower percentage of carbon dioxide than in normal atmospheric air.

4 The amount of carbon dioxide in the atmosphere all around the world is thought to be slowly increasing.

(a) Suggest two main reasons to account for this.

(b) What steps could be taken to prevent this increase continuing?

(c) This problem is often called 'the greenhouse effect'. Find out what further problems an increase in the amount of carbon dioxide in the atmosphere would create.

Assignment – Nitrogen cycle

Nitrogen gas is of no use to plants but they need nitrates for making proteins. If nitrates are available in soil water they can be absorbed by the roots of plants.

Green plants can obtain nitrates in four ways:

1. from man-made fertilisers;
2. by means of nitrogen-fixing bacteria. These are found in root nodules of certain plants such as clover and beans. These bacteria are the only living organisms which can convert nitrogen gas into nitrogen compounds;
3. through nitrifying bacteria – these oxidise ammonium compounds to nitrites and then nitrates. To do this these bacteria must have air;
4. through lightning – this causes a little nitrate to fall in rain.

If the soil lacks air, as in waterlogged soil, de-nitrifying bacteria turn nitrates into nitrogen.

When animals and plants die they decay. The bacteria and fungi which cause decay break the protein down into ammonia. Ammonia is also formed from the excretory products of animals.

1 What is the percentage of nitrogen in normal atmospheric air?
2 (a) What are nitrogen fixers?
  (b) Where would you find them?
  (c) What are legumes?
  (d) Find out about crop rotation. How are nitrogen fixers important in crop rotation schemes? Write down two different crop rotation schemes. What are the advantages and disadvantages of each?
  (e) If a gardener wanted a good crop of turnips without using any fertilisers, what would you recommend that he planted in the same place in the previous year?
3 What is the difference between nitrifying and de-nitrifying bacteria?
4 Why do plants grow better in well aerated soil than in waterlogged soil?
5 What factors (a) raise, (b) lower the nitrogen content of the soil?
6 What does 'oxidise ammonium compounds to nitrates and nitrites' mean? Try to write word equations for this process.
7 Explain how garden compost and manure improve the soil for plant growth.

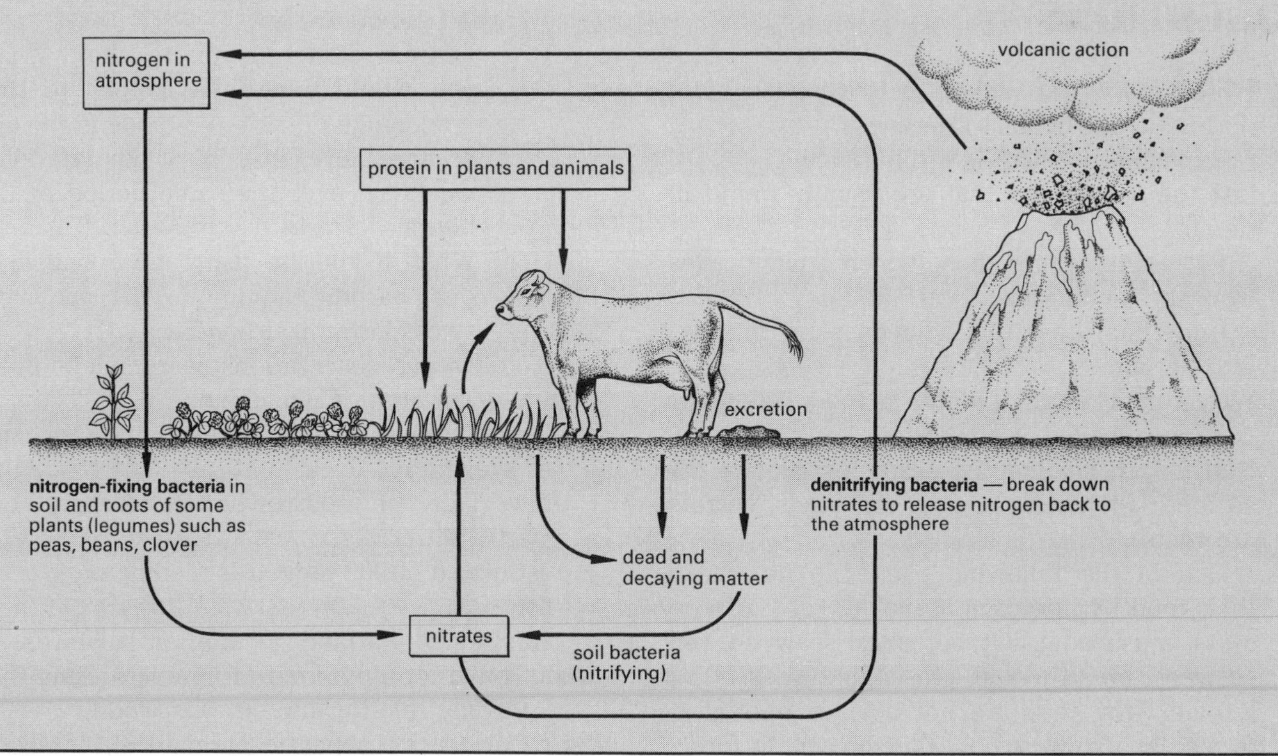

WHAT YOU SHOULD KNOW

1 Plant leaves contain carbohydrates.

2 Green plants make their food by a process called photosynthesis. Carbon dioxide and water are combined to make glucose.

$$\text{carbon dioxide} + \text{water} \xrightarrow[\text{chlorophyll}]{\text{sunlight}} \text{glucose} + \text{oxygen}$$

3 Light energy is needed for photosynthesis to take place. This energy is absorbed by chlorophyll in the chloroplasts of leaves.

4 Oxygen is given off as a waste product of photosynthesis.

5 Factors which affect the rate of photosynthesis are light, carbon dioxide, temperature and water.

6 Plant leaves are adapted for photosynthesis by being broad and thin with many chloroplasts in their cells.

7 The glucose made in photosynthesis may be used for respiration. It may be turned into starch and stored or it may be used to make other substances such as proteins, provided that there is a supply of minerals such as nitrates, phosphate and potassium.

QUESTIONS

1 Which gas(es) are taken in by a green plant during (a) daylight, (b) darkness?
Explain why there is a difference.

2 List the substances that are usually found in the air.
Put a tick against those which a plant takes from the air to make food.
Underline those that a green plant releases into the air.

3 We eat many parts of plants as food. Make four columns headed seeds, stem, root and leaf. Under each heading list examples that we eat as food (for example, peas are seeds, carrots are roots).

4 Which of the following plants do not make their food by photosynthesis?
moss, mushroom, lettuce, green seaweed, red seaweed, bread mould, grass, pine trees.
Explain why they are not able to photosynthesise and suggest how they *do* obtain food.

5 Study the graph below and then answer the questions.

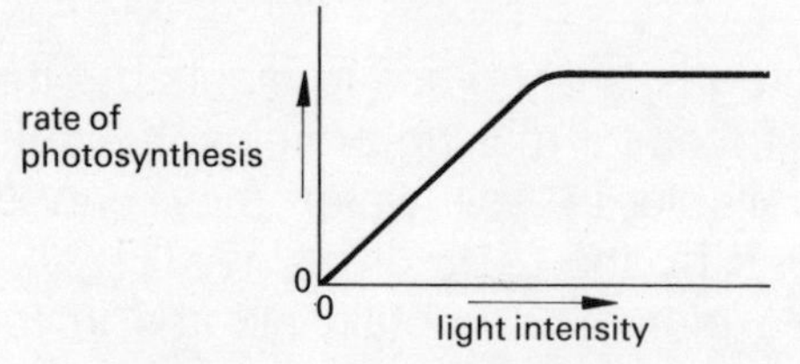

(a) Explain what the graph shows.
(b) Suggest reasons why the curve eventually flattens out.
(c) If the plant was given a greater concentration of carbon dioxide what would you expect to happen to the shape of the graph?

6 The graph below shows the percentage of carbon dioxide in the air immediately above a cornfield during a 24 hour period.

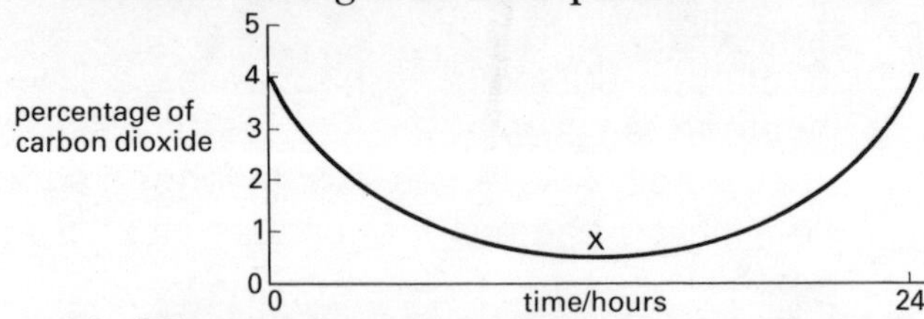

(a) What causes the changes in carbon dioxide concentration?
(b) At what time of day is X likely to be? Explain your answer.
(b) Why would there be such a low level of carbon dioxide at X?
(c) (i) Draw a sketch graph to show what you would expect to happen to the percentage of carbon dioxide in the air at street level in Central London on a weekday, over the same period of 24 hours.
(ii) What would be using up (or giving out) carbon dioxide in the air in a street in Central London?

7 The following observations were made in a study of an area of woodland:
large numbers of greenfly feeding on the sap of sycamore leaves;
one pair of sparrowhawks feeding on warblers;
estimated 50 000 ladybirds feeding on greenfly;
25 pairs of warblers feeding on ladybirds.
(a) Construct a pyramid of numbers using the feeding relationships in this wood.
(b) Construct a separate pyramid of biomass.

Energy 2

RESPIRATION AND ENERGY

In the last topic, you saw how plants made their food by photosynthesis and how animals use plants as food. In this topic we are going to look at how energy is obtained from food.

In how many ways are the cyclists in Fig. 9.31 using energy? Here are a couple of ways to get you started:

1. sending electrical signals around the body to make the muscles contract;
2. expanding and contracting the lungs during breathing.

Fig. 9.31 Using energy

What other ways can you think of? Discuss this with the rest of the class and with your teacher. All the energy has to come from food. The process by which we obtain energy from food is called RESPIRATION.

THE BASIC PROCESS

When certain substances are burned in oxygen, energy is released in the form of heat and light energy. The process is called COMBUSTION. Respiration is *controlled* combustion as it takes place slowly compared with normal combustion.

Fig. 9.32 Burning carbon in oxygen

The complete reaction for the respiration process may be represented as:

$$\underset{\text{glucose}}{C_6H_{12}O_6(aq)} + 6O_2(aq) \rightarrow \underset{\text{carbon dioxide}}{6CO_2(aq)} + \underset{\text{water}}{6H_2O(l)} + \text{energy}$$

The equation may look simple but complete details of the process are extremely complex. To break the glucose down to carbon dioxide, water and energy involves more than twenty separate reactions. Each reaction is catalysed by a *different* enzyme.

Assignment – Respiration and photosynthesis

Below is a list of statements which refer to the processes of respiration and/or photosynthesis. Make two columns headed 'respiration' and 'photosynthesis' and put the appropriate statements under each heading.

uses up energy
does not require chlorophyll
performed only by plants
can happen in the dark
can only happen in light
uses up carbohydrate
requires chlorophyll
takes place in all living things
produces carbohydrates
produces energy
oxygen taken in
gives off oxygen
carbon dioxide given off
carbon dioxide needed

It is useful to look at the differences between inhaled and exhaled air (Fig. 9.33). It can be seen that the amount of oxygen *decreases* while the amounts of carbon dioxide and water vapour *increase*. Is this what you would expect if the equation for respiration is correct? Nitrogen and the rare gases take no part in the respiration process. The exhaled air is also warmer and cleaner than inhaled air. Why do you think this is?

It is easy to show that exhaled air contains more carbon dioxide than inhaled air using the apparatus shown in Fig. 9.34. The inhaled air is bubbled through limewater before being breathed in. The exhaled air is bubbled through limewater as well. Can you explain how the apparatus works? Since limewater turns milky in the presence of carbon dioxide, what differences would you see in the two lots of limewater?

Assignment – Breathing and water vapour

Design an experiment to show that exhaled air contains more water vapour than inhaled air.

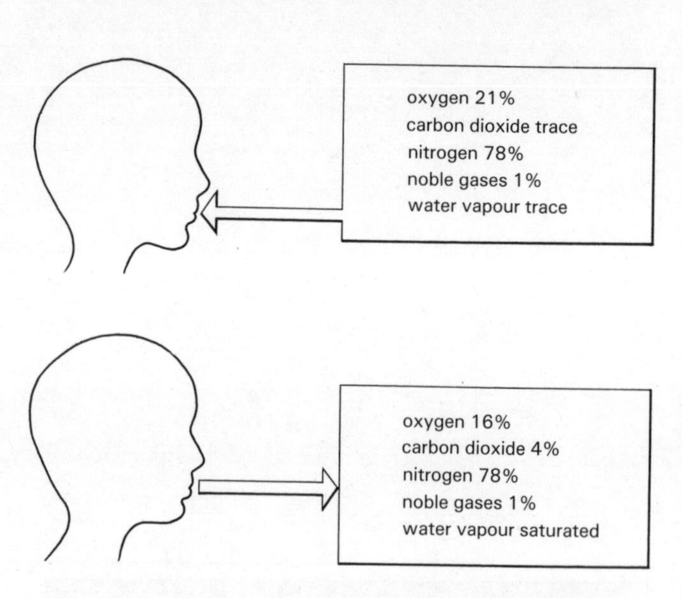

Fig. 9.33 The differences between inhaled and exhaled air

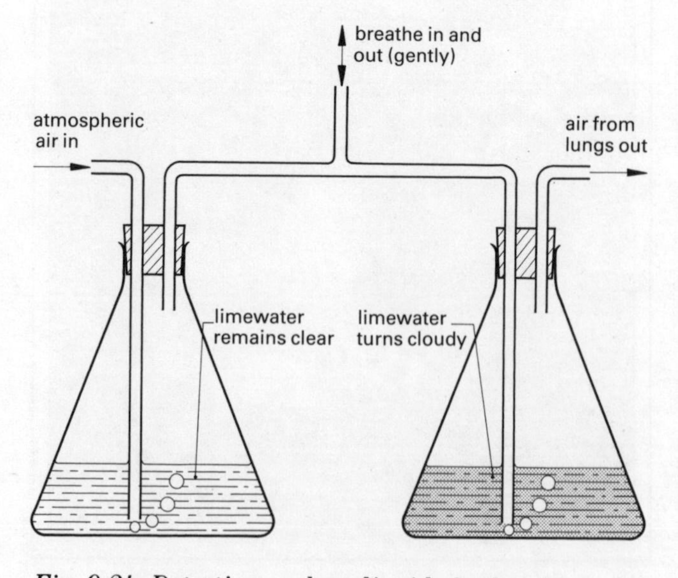

Fig. 9.34 Detecting carbon dioxide in the air

GETTING AIR INTO THE BODY

We shall first look at the process of breathing. This involves getting oxygen into the blood and removing the waste products, such as carbon dioxide and water, from it.

When you breathe in, the air passes down the windpipe (trachea) into the two organs known as the lungs. When you breathe in you can feel (and see) the rib cage moving outwards. At the same time, the diaphragm muscle moves downwards.

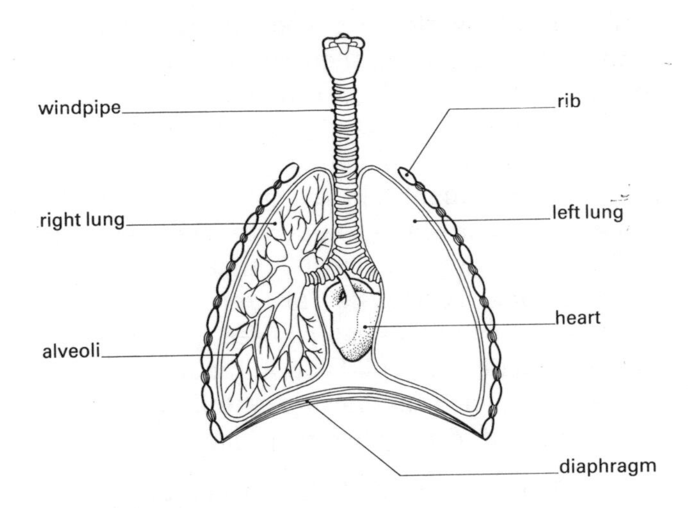

Fig. 9.35 The windpipe and lungs

This increases the size of the lungs and, because the air pressure is lower, air is forced into them as you inhale. The reverse process occurs as you exhale. The rib cage falls and the diaphragm rises. This decreases the size of the lungs, raising the pressure inside. The excess pressure pushes out air. This occurs, on average, twelve times a minute when you are not taking exercise. The complete process can be demonstrated using the model lung shown in Fig. 9.36.

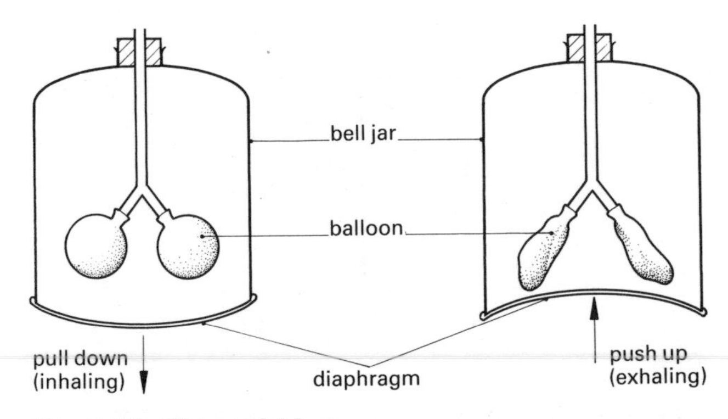

Fig. 9.36 The model lung

From the main windpipe, the air passage divides into two. It then divides continually into smaller and smaller passages. These end up as very tiny air sacs which are called alveoli. These are the regions where oxygen is absorbed into the bloodstream and carbon dioxide and water are removed.

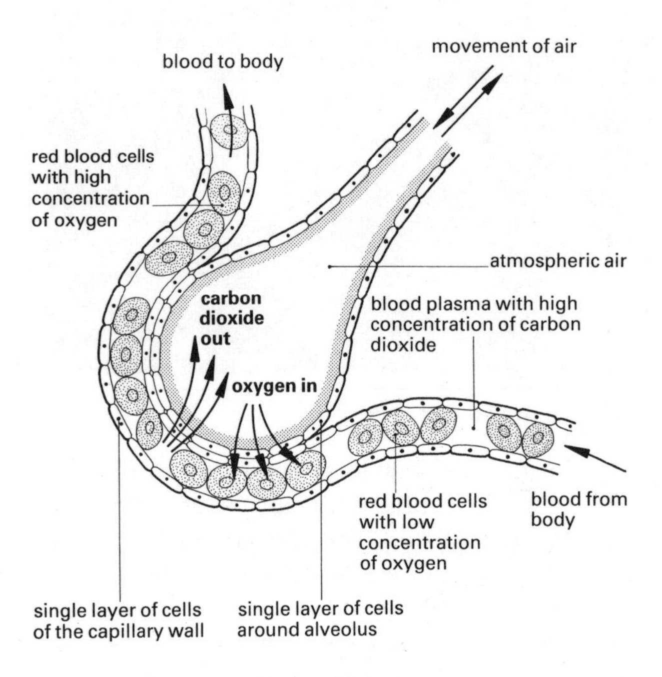

Fig. 9.37 Action of the alveoli

The alveoli are fed with blood capillaries. As the oxygen moves over these, it diffuses into the blood and is then transported through the body's blood system. At the same time, carbon dioxide and water diffuse out of the blood plasma into the alveoli. They are then transported back to the windpipe and eventually exhaled.

HOW IS ENERGY RELEASED?

When the oxygen diffuses into the blood, the molecules become attached to the haemoglobin contained in the red blood cells. These cells (you met them in Unit 2) carry the oxygen around the body to the cells where it is needed for respiration. The body makes the red blood cells in the bone marrow. There can be as many as 5 000 000 cells in each cubic millimetre of blood. (NB: A cubic millimetre is not much larger than a grain of sugar!)

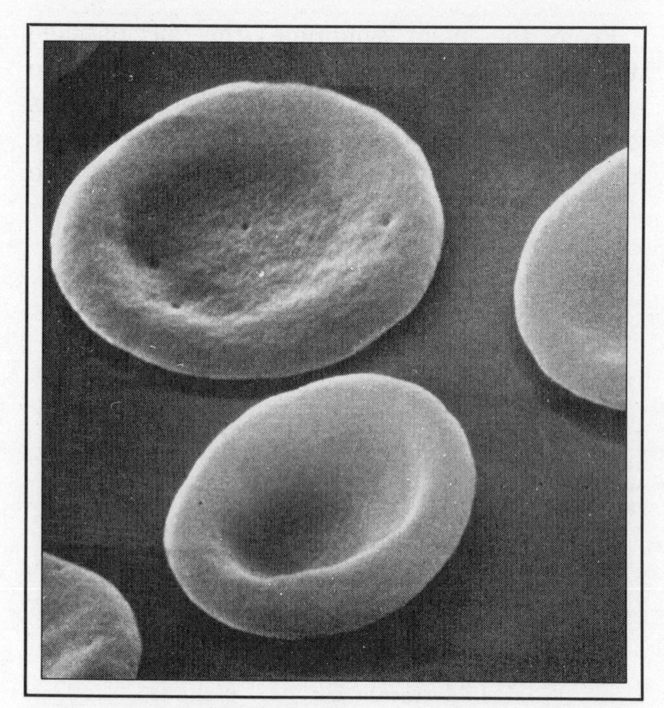

Fig. 9.38 Red blood cells

In the original equation for respiration (page 218), only the initial substances and the final products were listed. But the complete process takes a considerable amount of time because of the number of steps and because enzymes take a while to act. This means that the energy from glucose cannot be released the instant it is needed. The energy has to be stored in a more accessible form. The next section tells us how this is done.

Aerobic respiration

Respiration can occur in two ways: aerobic (with oxygen) and anaerobic (without oxygen). In aerobic respiration, the energy is stored with the help of a chemical found in living cells, called adenosine diphosphate (usually just called ADP). At each stage of aerobic respiration, as the glucose is being broken down, energy is released. This energy is stored by converting ADP to adenosine triphosphate (ATP). It is possible for one molecule of glucose to produce up to 38 molecules of ATP. Energy can very quickly be transferred by the ATP molecules changing back to ADP. So the energy required for all the functions of the body is the energy from the ATP molecules.

It should be mentioned that most plants and animals respire in this way.

Anaerobic respiration

It has been said that obtaining energy from glucose is a fairly slow process and that the breaking down of ATP molecules is very quick. But there are only so many ATP molecules available for energy release at any one time. What happens when they have been broken down to release their energy?

It is possible for the body to produce energy directly from glucose. The process is called anaerobic respiration. Although it is not very efficient at producing energy, it is a useful 'backup' system.

If you have performed a lot of strenuous activity such as running, then you feel 'tired'. This feeling of fatigue is due to the build-up of lactic acid. If all the molecules of ATP have been broken down to produce energy where they are needed (for example, in the muscles), then the 'back-up' system starts to operate. The first stage in the breakdown of glucose is that it is changed to lactic acid ($C_3H_6O_3$). This releases a small amount of energy that can be used quickly. However, the lactic acid builds up in the muscles and tissues, producing tiredness. The following equation can be used to show what happens:

$$\underset{\text{glucose}}{C_6H_{12}O_6} \rightarrow \underset{\text{lactic acid}}{2C_3H_6O_3} + \text{energy}$$

When the exercise stops, some of the lactic acid is broken down to carbon dioxide and water. The energy released is used to convert the rest of the lactic acid present back to glucose. Aerobic respiration then continues as normal.

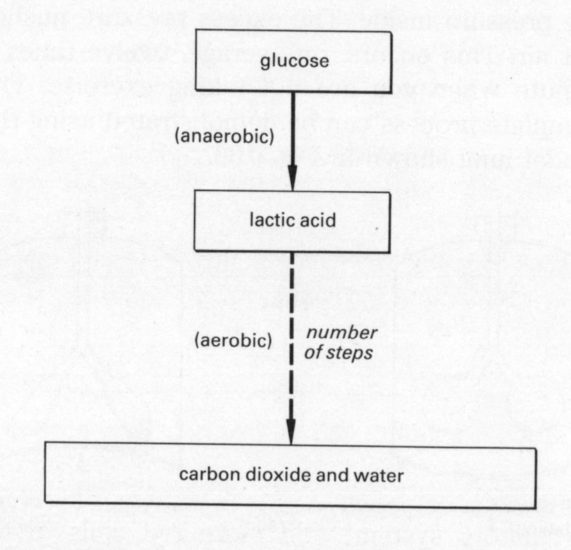

Fig. 9.39 Aerobic and anaerobic respiration

ATHLETES AND RESPIRATION

Athletes are very dependent on both aerobic and anaerobic respiration to cover a given distance in as short a time as possible. When an athlete starts to run, respiration is aerobic. But, depending on the speed at which the athlete moves, all the energy stored in the ATP molecules will be used up. If the athlete is to continue running at the same speed then anaerobic respiration must be used. However, anaerobic respiration can only last for a certain time as the build up of lactic acid rapidly produces fatigue.

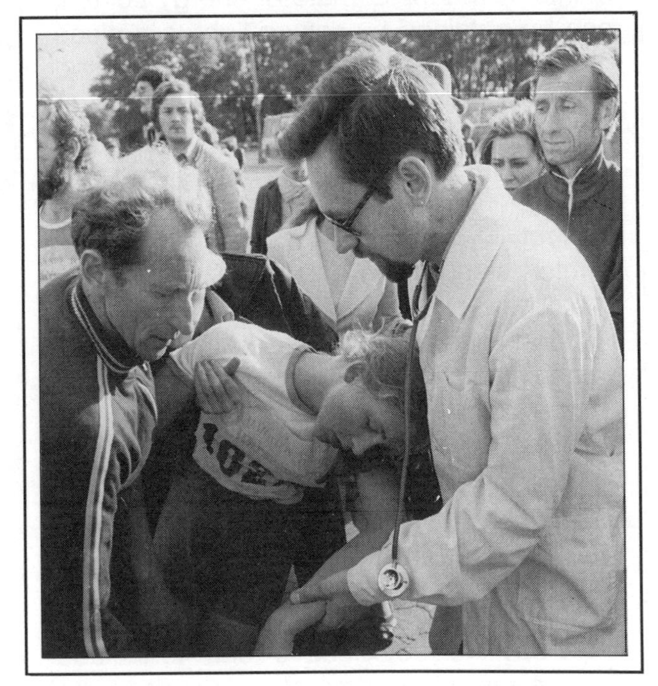

Fig. 9.40 Anaerobic respiration leads to fatigue and collapse

With a sprinter, most of the energy release is due to anaerobic respiration. As the distance to be run increases, so does the amount of aerobic respiration compared with anaerobic. With middle distance runners (e.g. 1500 m), aerobic and anaerobic respiration play roughly equal parts. With long distance runners virtually all the respiration is aerobic.

The athlete uses training methods that will increase the efficiency of the aerobic respiration. Continual training will produce deeper breathing (more oxygen goes into the lungs each time), a stronger heart (pumps blood further through the circulatory system) and more red cells in the blood (more oxygen can be carried).

Assignment – Smoking

Smoking can damage your health. The effects of smoking on the respiratory system and general health are very well known. Smoking can produce cancer, chronic bronchitis, heart disease and other serious ailments. Write an article describing the dangers of taking up smoking when young.

ANAEROBIC RESPIRATION IN PLANTS

This process can also occur in *simple* plants. Complicated plant structures do not need to resort to this process. Perhaps the commonest example of this occurs with yeast, a very simple plant. Yeast will respire aerobically if oxygen is present. But, if there is no oxygen available, then it respires anaerobically. However, unlike the process in mammals, the substance ethanol is produced. The equation that shows the initial and final chemicals is as follows:

$$\underset{\text{glucose}}{C_6H_{12}O_6(aq)} \rightarrow \underset{\text{ethanol}}{2C_2H_5OH(aq)} + \underset{\text{carbon dioxide}}{2CO_2(g)} + \text{energy}$$

Again, the reaction proceeds in a series of steps with enzymes catalysing each separate step. The general name for the process is FERMENTATION. Yeast is used to change the sugars in various substances, such as grapes, to ethanol.

Fig. 9.41 Beer fermenting in a brewery – what causes the froth?

ENERGY FROM FOOD

In the previous topic you found that when you, or other animals, eat food like potatoes or bread, you take in some of the energy captured by the potato or wheat plant from the sun. Plants capture this energy when they make their own food by the process of photosynthesis (page 203). You have also seen that energy is released from food by the process of respiration (page 217).

Energy is measured in joules. One kilojoule is 1000 joules. A kilojoule is the amount of energy needed to raise the temperature of 238 grams of water by 1 °C. Table 9.6 shows the energy needs of various types of people.

Table 9.6 Energy needs of different people

	kJ/day	
	female	male
Newborn baby	1 800	1 800
Child 1 year	3 300	3 300
Child 2–3	5 800	5 800
Child 5–7	7 500	7 500
Child 12–15	9 600	12 000
Adult, light work	9 500	11 500
Adult, heavy work	12 500	15 000
Pregnant woman	10 000	–

From the table you will see that energy requirements vary throughout life. They also vary in adults, depending on the occupation or leisure activities they normally do. See Table 9.7.

Table 9.7 Energy needs of 25 year old man, weighing 65 kg (10 stone)

	kJ/min
Sitting	5.88
Standing	7.14
Washing and dressing	14.7
Walking slowly	12.6
Walking fairly fast	21.0
Walking up and down stairs	37.8
Doing carpentry	15.5
Playing tennis	26.0
Playing football	36.5
Cross-country running	42.0

Investigation 9.12
Measuring the amount of energy in a piece of food

If we want to prescribe a healthy diet for a human, one thing we have to make sure is that it will provide enough energy. To do this, we need to know how much energy there is in the food we eat. Different foods contain different amounts of energy, but finding the precise amount needs complex equipment and very accurate measurements. A rough estimate of energy content can be obtained by setting light to a piece of food. The heat it produces is then used to heat up some water. The rise in temperature will give some idea of the amount of energy in the food.

Collect

retort stand, boss and clamp
boiling tube
thermometer
small measuring cylinder
mounted needle
Bunsen burner
scalpel and tile
samples of nuts and bread (each of mass 1 g)

What to do

1 Copy the table. Leave room to include the results of some other people in the class so that you can calculate an average value for each food you use.
2 Set up the apparatus as shown in the diagram.
3 Put exactly 20 cm^3 of water in the boiling tube.
4 Take the temperature of the water and record it in the table.
5 Take a peanut and cut it until you have a piece with a mass of exactly 1 g. Stick the mounted needle into the peanut.
6 Light your Bunsen burner. Keep it well away from your boiling tube of water. Set fire to the peanut.

▶

Type of food	A Mass of food/g	B Temperature of water before heating/°C	C Temperature of water after heating/°C	D Temperature rise/°C	E Energy produced by 1 g of food/ $kJ\,g^{-1}$

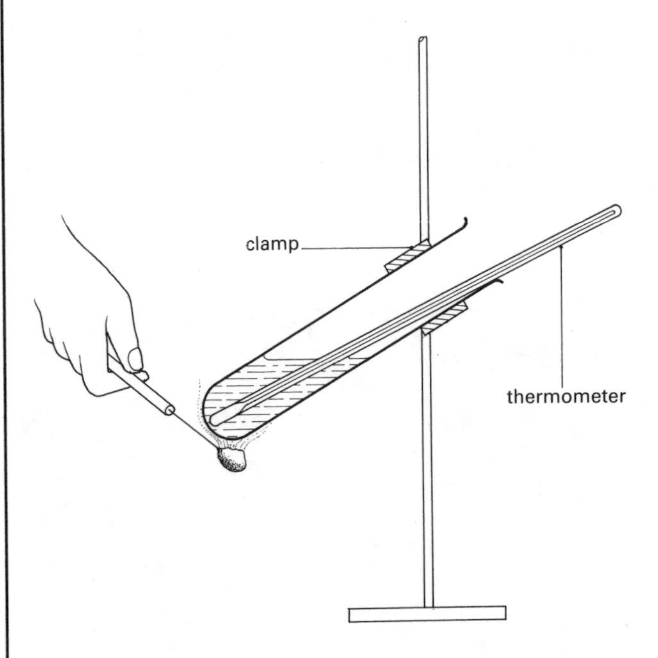

7 As soon as the peanut is alight, hold it under the boiling tube until the nut has stopped burning.
8 Stir the water gently and record the highest temperature reached in your table.
9 Repeat with bread and/or other types of nuts. Use a fresh boiling tube and water each time.
10 Complete column D of your table.
11 4.2 J are required to raise the temperature of 1 g of water by 1 °C.
$1\ cm^3$ of water = 1 g.
Complete column E of your results table like this:

Questions

1 Do some pieces of peanut seem to give more energy than others? Why might this be?
2 Multiply your figure in column E by 100. This gives the amount of energy in a 100 g portion. Look up the energy value of peanuts or other foods you tested in official tables. How does this compare with your figures? Suggest why it may differ.
3 Not all the heat given off by the peanut was used to heat the water. Give two ways in which some of the heat given out will be lost.
4 What substance present in the peanut makes it burn easily? What element is left behind after the peanut has stop burning?
5 Suggest some ways in which the experiment could be improved if you did it again.

$$\text{Energy produced} = \frac{(\text{mass of water used}) \times (\text{temperature rise}) \times 4.2}{1000}\ \text{kJ}$$

$$= \frac{(20) \times (C - B) \times 4.2}{1000}\ \text{kJ}$$

WHY DO WE NEED FOOD?

We have already seen that we need food for energy but it is also needed for other jobs in the body.

Food is needed for:

1. fuel to give us energy and warmth;
2. growth of new tissues; repair and replacement of damaged or worn out tissues;
3. health; keeping our bodies working properly and helping us fight disease.

Our bodies are built from the food we eat. After we have eaten some food it is broken down as it passes through the mouth, stomach and small intestine into simple, soluble substances. (Look back at the work you did in Unit 2.) These soluble substances can then be absorbed into the blood and carried to all parts of the body, where they can be used to make new cells. In some cases, they may be turned into other substances for storage.

DIFFERENT TYPES OF FOOD

Humans have a wide-ranging diet. They eat thousands of different foods that come from other animals (meat, fish, eggs and dairy produce) and from plants (vegetables, cereals and fruit). The food eaten by people in different parts of the world varies considerably . . . for example, the Japanese enjoy raw fish, raw eggs and seaweed; some Arabs like sheeps' eyes; in France many people eat snails and frogs' legs; Australian Aborigines eat insect grubs and some people in

Fig. 9.42 Meals from different parts of the world: (a) curry, (b) Chinese meal, (c) spaghetti, (d) bacon & eggs

England eat tripe, which is part of a cow's stomach, or pigs' brains. Although humans are thought of as omnivores (that is, they eat both animal and plant material), some people choose to be vegetarians or vegans. Some religions forbid their followers to eat certain foods (for example, neither Jews nor Muslims should eat pork).

Whatever the actual food is, it will be made up of one or more of the different food types. Whatever we choose to eat it should include the following: carbohydrates, fats, proteins, vitamins, minerals, water and roughage and in the right proportions. No matter how much food is eaten, a human will become ill and die unless the right amounts of these different food types are eaten regularly.

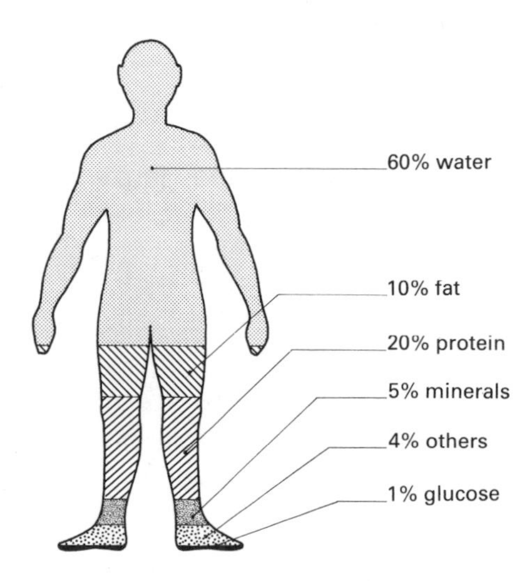

Fig. 9.43 You are what you eat!

Assignment – Vegetarians and vegans

1 What is a vegan?
2 In what ways might such people have to be very careful about their diet?
3 A friend says to you, 'I am a vegetarian because I don't eat meat.'
What other sorts of vegetarian are there?
4 Some people are neither vegan nor vegetarian, but fruitarian. Find out about the diet of such people and say how such a diet might be dangerous.
5 Write a short letter to a friend, explaining either why you are a vegetarian or why you are *not* one.

Carbohydrates

These contain the elements carbon, hydrogen and oxygen. Examples include glucose ($C_6H_{12}O_6$) and sucrose ($C_{12}H_{22}O_{11}$). They are the 'sugary' and 'starchy' foods.

Sugars: There are many different sugars. In fruit the sugar is fructose or glucose, in milk it is lactose. Ordinary table sugar is sucrose, obtained from sugar cane or sugar beet.

Starch: Starch is found in potatoes, rice, flour and many other plant foods.

Cellulose: This forms the cell walls of all plants and is very tough. Plants are often difficult to chew because of their cellulose, and cooking softens them. Fibre or roughage is made up of plant material with a large amount of cellulose. It cannot be digested by humans and so is very helpful in keeping food moving along the gut and preventing constipation.

Carbohydrates have several jobs to do.

1. They provide energy from glucose. 1 g carbohydrate gives 17 kJ of energy.
2. They can be stored. Although glucose cannot be stored in animals it can be converted into glycogen and stored in the liver. In plants, glucose can be converted into starch for storage (see page 211). Both glycogen and starch can be converted back into glucose when needed.
3. Building materials. Cellulose is essential in all plant cells, but is not needed for building in animals.

Fig. 9.44 These foods are rich in carbohydrates.

Fig. 9.45 These foods contain fats.

Fats

These contain carbon, hydrogen and oxygen but the proportion of oxygen is much less than in carbohydrates. For example, mutton fat is $C_{57}H_{110}O_6$.

Most people think of fats, such as butter, lard and fat in meat, as being obtained from animals. These fats are solid at room temperature but change to liquid when heated. But fats can also be obtained from plants (for example, corn oil, olive oil). These are normally in a liquid form at room temperature. Margarine consists mainly of vegetable oils which are turned into solid fat by chemical treatment.

Fats, like carbohydrates, do several jobs.

1. They provide energy. 1 g of fat gives 38 kJ of energy.
2. Fat is stored under the skin. This acts as a fuel reserve and also helps keep the body warm by insulating it.
3. Fat may be found surrounding delicate organs such as the heart and kidneys, where it can be protective.

Investigation 9.13 Finding out what some foods contain

Various tests can be carried out to find out *which* substances are found in different foods. Some of these are outlined below. More complicated experiments can be done to find out *how much* of each chemical substance is found in different foods (see Table 9.8).

Collect

samples of food, e.g. milk, bread, potato, mince, egg, butter
pestle and mortar
test tubes
eye protection

Starch Test:
iodine solution

Glucose (simple sugars) Test:
bunsen burner, tripod and gauze
beaker
white tiles
Benedict's solution

Fat Test:
filter paper

Protein Test (Biuret Test):
dilute sodium hydroxide solution
dilute copper sulphate solution
2 dropper pipettes

What to do

1 Draw a table to record the results of your tests, as shown below.

Food tested	Food contains			
	starch	glucose	fat	protein

√ = yes
× = no

2 Take each food sample and make it as liquid or as 'mushy' as possible. Use the pestle and mortar if this will help. Divide the food into four equal portions in four test tubes.

3 Carry out the four tests using the methods shown in the diagrams below. Record your results in your table.
4 Repeat the tests on several different food samples.

Questions

1 Why was it important to make each food sample as liquid as possible?
2 Which of the tests indicated the *amount* of a food substance in a sample? How did this show?
Compare your results with the data in Table 9.8. Are they the same? If not, suggest a reason.

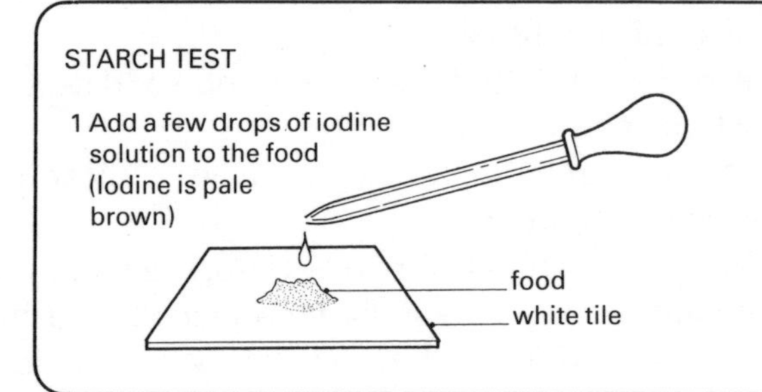

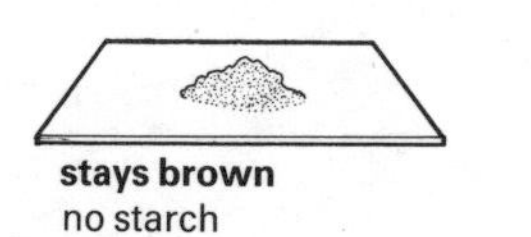

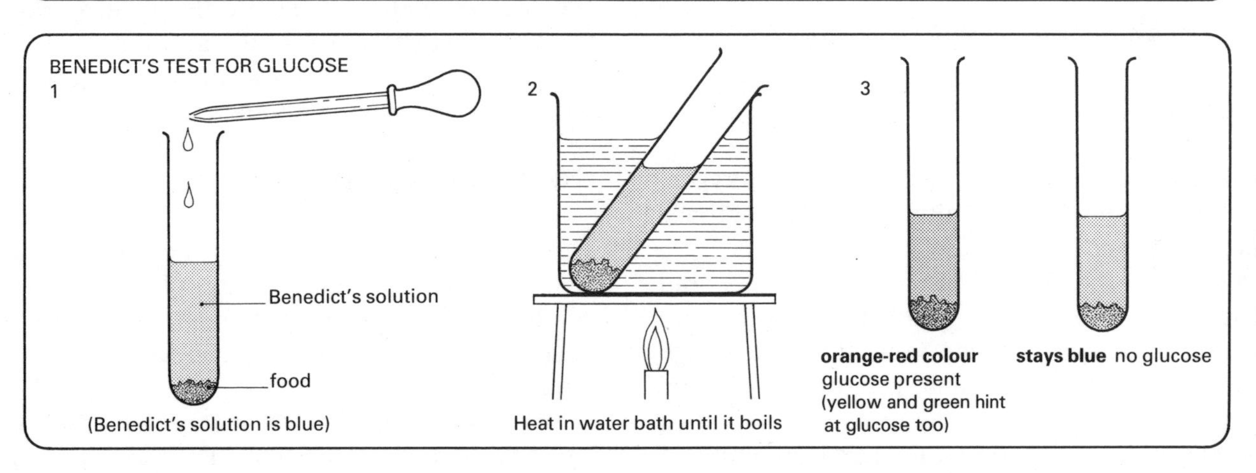

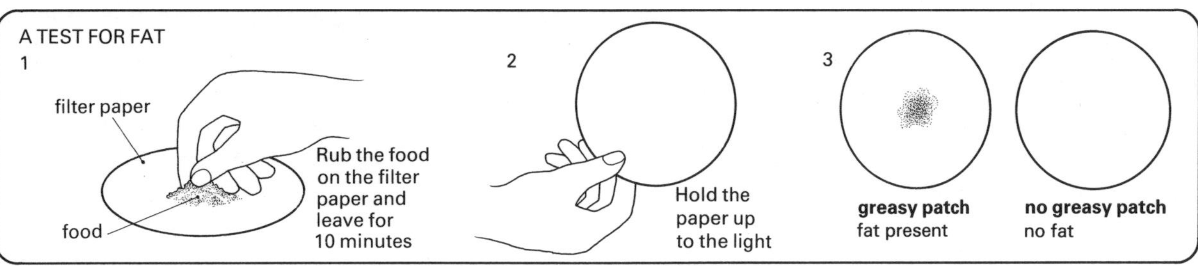

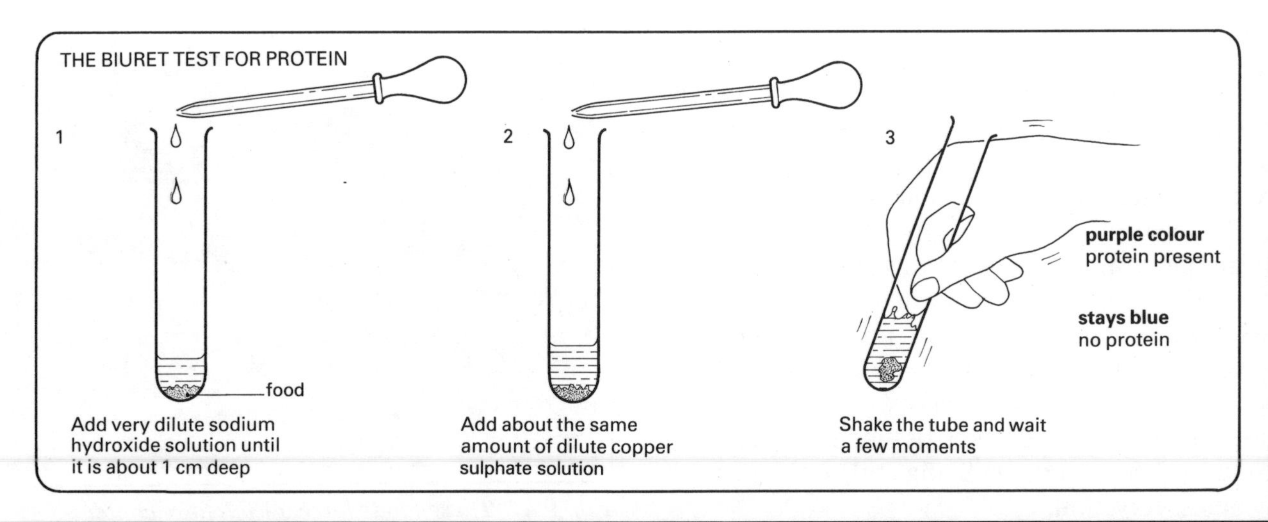

Fig. 9.46 These foods are rich in protein

Proteins

Proteins contain carbon, hydrogen and oxygen. In addition, they contain nitrogen and sometimes phophorus and sulphur too.

Many of the living tissues of animals and plants are made from protein, so a certain amount of protein is found in most foods. A high percentage is found in meat, fish, eggs and milk. Some plant foods such as peas, beans, nuts, cereals and rice also contain large amounts of protein.

Proteins are needed for:

1. Growth: as proteins form most of our tissues, we need them for the growth of new tissues. This is especially important in children.
2. Repair and replacement of worn-out tissues: for example, new skin cells are made to heal a wound and red blood cells, which have a life span of only a few months, are constantly being replaced.
3. To make enzymes: enzymes are made up of proteins. They are very important substances as you saw earlier in this unit (page 195). Enzymes are found in all living cells and are necessary for all chemical processes.
4. Energy: although it is not usually needed for energy if a person has a normal diet containing carbohydrate and fat, 1 g of protein produces 17 kJ of energy. When people are starving and have used up all their reserves of carbohydrate and fat, the protein in their tissues is used up to produce energy to go on living and so the person becomes very thin.

Water

Most foods contain water. Sixty to seventy per cent of our body is made up of water. (See Fig. 9.43 on page 225.)

Water is needed for

1. making up a large part of each cell, for enzymes and for blood;
2. carrying digested food and other dissolved substances around the body;
3. chemical reactions to take place – these can only work in solution.

We lose about 1 litre of water each day in urine and sweat. This must be replaced by drinking or by eating food which has water in it. A shortage of water is usually the main cause of famine; domestic animals die and the crops fail. Humans can survive for weeks without food, but only for a few days without water.

Roughage or fibre

This is all the material which is not digested as it passes through the gut. There is plenty of fibre in whole cereals and in most fresh fruit and vegetables but not in white bread, sugar or processed food.

Fibre adds bulk to the material in the large intestine, stretching its wall. This stimulates the muscles to contract, which pushes the faeces along and keeps them moving. Constipation is often caused by eating over-refined foods which do not contain much fibre. Many doctors also think that plenty of fibre in the diet reduces the risk of cancer of the bowel (large intestine).

Fig. 9.47 These foods have a high fibre content

Assignment – Summary of food tests

Copy the table below and fill it in. The first test has been done for you.

Test for	Reagent	Method	What is seen if test is positive
Starch	iodine solution	add	brown → blue-black
Glucose			
Fat			
Protein			
Vitamin C			

See Investigation 9.14 on page 235 for the test for Vitamin C.

Assignment – What is food made of?

Study Table 9.8.

1 (a) Draw bar charts to show the amount of carbohydrate, fat and protein in 10 g of milk, beef, eggs, bread and rice.

(b) Which of these foods provide a balance of nutrients? Explain your answer.

2 For which main reason would you include each of the following in a 'balanced diet'?

butter, groundnuts, liver, herring and oranges

3 Name two good sources of each of the following:
Vitamin A, Vitamin C, calcium and iron.

Table 9.8 What different foods are made of

10 grams of food	Energy (kilo-joules)	Carbo-hydrate (grams)	Fat (grams)	Protein (grams)	Vitamin A (micro-grams)	Vitamin B (micro-grams)	Vitamin C (micro-grams)	Vitamin D (micro-grams)	Calcium (milli-grams)	Iron (milli-grams)
Milk (whole)	28	0.5	0.4	0.3	0.8	0.4	21	0	12.0	0
Butter	334	0	8.5	0.04	29	0	0	0.01	1.4	0
Groundnuts	252	0.8	4.8	2.8	0	2	0	0	6.0	0.2
Beef	134	0	2.8	1.5	0	0.7	0	0	1.1	0.4
Cheese	177	0	3.5	2.5	8.4	0.4	0	0	81	0.07
Liver	61	0	0.8	1.7	120	3	315	0.01	0.7	1.4
Herring	99	0	1.8	1.7	0.8	0.4	0	0.2	10.2	0.1
Eggs	67	0	1.2	1.2	6.0	1.5	0	0	5.6	0.2
Bread (wholemeal)	97	4.7	0.2	0.8	0	2.1	0	0	2.5	0.28
Rice	150	8.6	0.1	0.6	0	0.7	0	0	0.4	0.04
Potato	37	2.1	0	0.2	0	1.0	210	0	0.7	0.07
Orange	15	0.8	0	0.07	0.4	1.0	492	0	4.2	0.04
Sugar (white)	165	10.0	0	0	0	0	0	0	0	0

1 gram = 1000 milligrams
1 milligram = 1000 micrograms

Assignment – Composition of food

1 Using the information from the bar chart draw up a table to show the percentage composition of carbohydrate, fat, protein and water in the various foods.

2 Which of these foods are the best sources of energy?

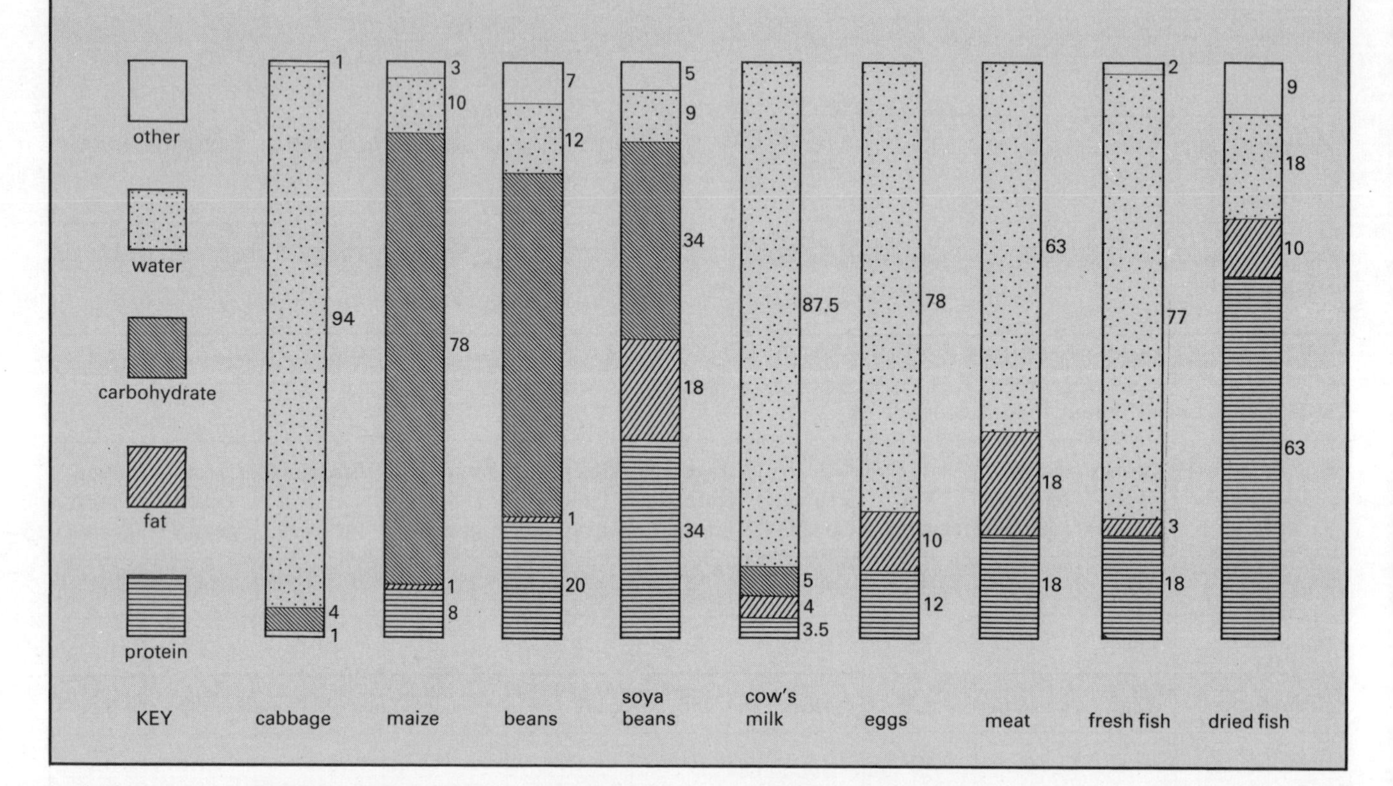

Assignment – Wholefoods and processed foods

Wholefoods are changed little, if at all, before eating. They are free from chemical preservatives, artificial flavourings and colouring. Wholefoods may include nuts, fresh vegetables and fruits, meat, fish, eggs, honey, milk and natural yoghurt.

Food is processed to preserve it and to reduce the time needed to prepare it for eating. Processed foods are altered in many ways before you buy them. Many types of substances may be added during processing, such as colourings, flavourings, preservatives, emulsifiers, antioxidants and stabilisers. Some of these substances are thought to be harmful. Processing removes fibre from fruit and vegetables and bran from wheat grains. Vitamins and minerals are also lost when wheat germ is removed from wheat grains and when foods are canned, frozen or dried.

1 (a) Make two menus for a three-course meal. One menu should be made from wholefoods, the other from processed foods.

(b) Which is the healthier meal? Give your reasons.

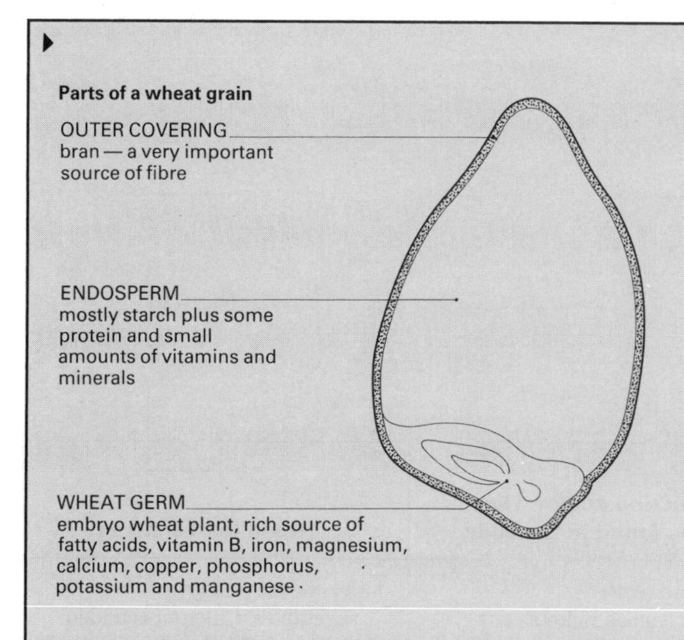

2 (a) Explain the differences between white, brown and wholemeal bread.
(b) What is the difference between white and brown rice? Which is healthier for you?

3 Suggest why people living on remote islands are often healthier than those living in North America or Western Europe.

4 Why are organically grown vegetables classed as wholefoods?

5 Examine the labels of a range of processed foods (you will probably need to visit a supermarket).
(a) All food additives now have a code, e.g. E322, E450. you will probably notice many of these. Fill in a table like the one shown for ten canned or packeted foods.

Food	Additives (just put the code number if these are not specified)

(b) Find examples of foods which contain one or more of the following: colouring, flavouring, emulsifiers, stabilisers, preservatives, antioxidants. Arrange your examples in a table.
(c) What is monosodium glutamate used for? Find three foods that contain it.
(d) Which types of foods have the amounts of vitamins and minerals in their contents list? Are these added or naturally occurring? Explain your answer.

6 (a) Compare the cost of 1 lb of potatoes and 1 lb of potato crisps.
(b) Compare the cost, per 100 g or per lb, of three other foods when fresh and when processed.

7 (a) Find out what drinks such as Coca-cola, lemonade and Seven-up are made of.
(b) Compare the cost of a pint of milk with a pint of Coca-cola. Which is better value for money? Explain your answer. Why are an increasing number of people drinking skimmed milk?

8 Find out the composition of human breast milk and powdered baby milk. What are the main differences?

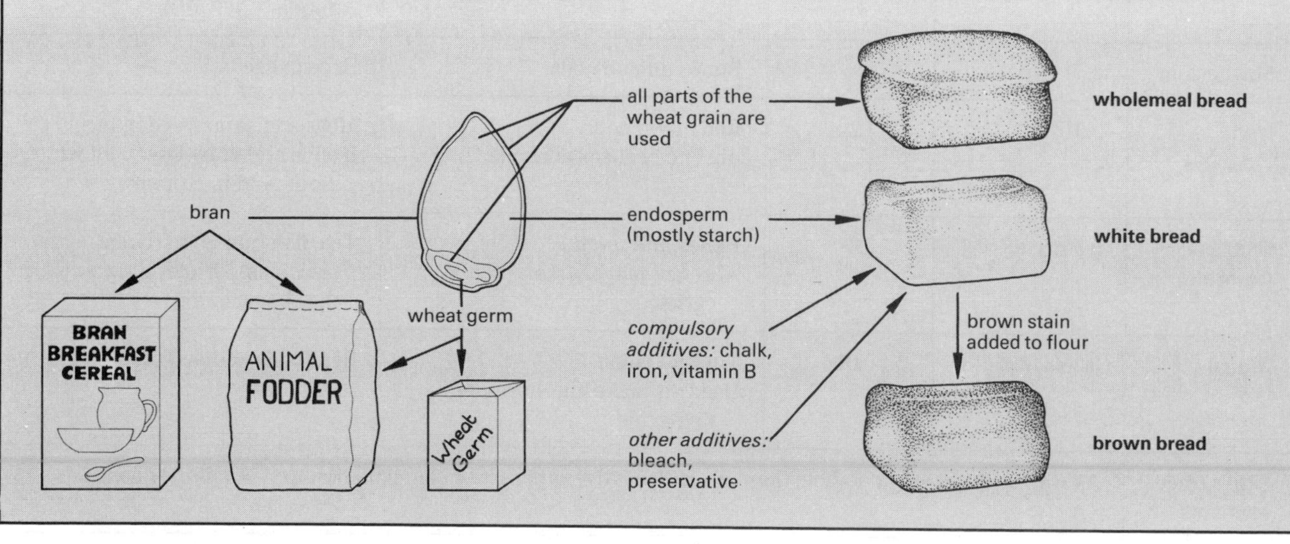

Minerals

These are certain elements which are needed in small quantities to help build up parts of the body and to keep certain processes under control. Some minerals such as iron, iodine and fluorine are needed in extremely small amounts. Calcium, sodium, potassium and sulphur are needed in larger amounts. See Table 9.9 below.

Table 9.9 Minerals

Minerals	Average man intake in grams/day	Total body content/g	Function and/or where it is found in the body	Good food sources
Calcium	1.1	1000	Bones and teeth Absence causes rickets	Milk, cheese and green vegetables. Calcium is added by law to all white flour sold in Britain
Phosphorus	1.4	780	Bones and teeth Essential for energy release in the cells	Present in nearly all foods
Sulphur	0.85	140	Found in muscle, skin etc. Used in making protein	Protein foods
Potassium	3.3	140	An essential mineral in all cells and body fluids Involved in nerve impulses	Vegetables, meat, milk and fruit
Sodium	4.4	100	An essential mineral in all cells and body fluids Used in nerve impulses Lost from body in sweat	Salt, bread, cereals, meat products (e.g. bacon, ham), milk
Chlorine	5.2	95	An essential mineral in all cells and body fluids Lost from body in sweat	Found in the same foods as sodium Added to many manufactured foods
Magnesium	0.34	19	Bones and all cells	Green vegetables
Iron	0.016	4.2	Red blood cells Absence causes anaemia	Meat and some vegetables Iron is added by law to all white flour sold in Britain
Fluorine (fluoride)	0.0018	2.6	Bones and teeth Absence may increase dental caries	Tea, fish bones (in tinned sardines), drinking water (if it has been added)
Iodine	0.0002	0.013	Thyroid gland Used to make the hormone thyroxine	Sea-foods, iodised table salt
Other trace elements	As well as fluorine and iodine, there are at least five other trace elements required in minute quantities			

Assignment – Minerals

Study Table 9.9 and answer these questions.

1 Which minerals are added to which processed foods?
2 Which mineral elements are essential for the formation of healthy bones and teeth?
3 Why might a person be prescribed iron pills?
4 Which minerals are lost from the body daily? Explain your answer.
5 Which mineral elements might vegetarians lack in their diet?

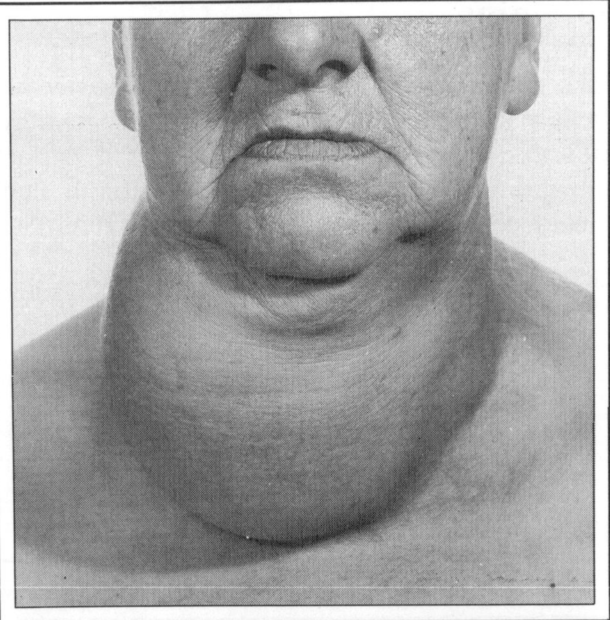

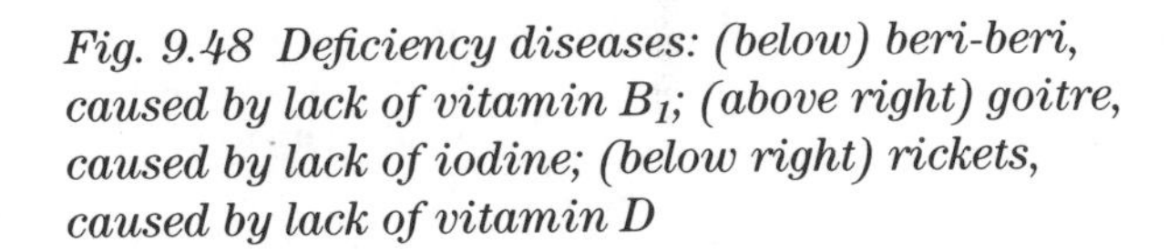

Fig. 9.48 Deficiency diseases: (below) beri-beri, caused by lack of vitamin B_1; (above right) goitre, caused by lack of iodine; (below right) rickets, caused by lack of vitamin D

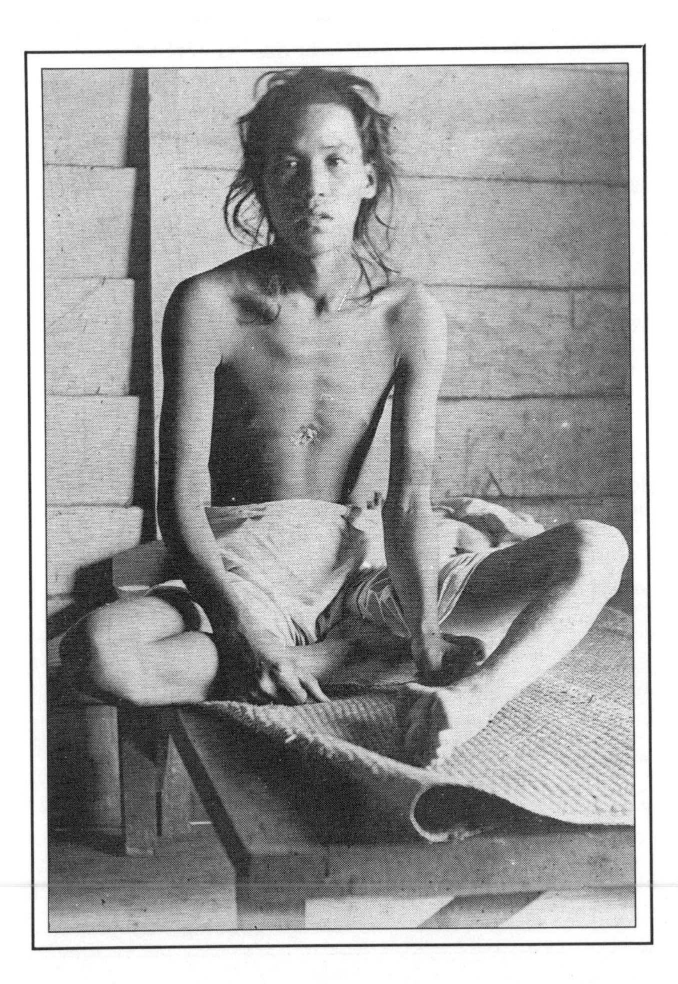

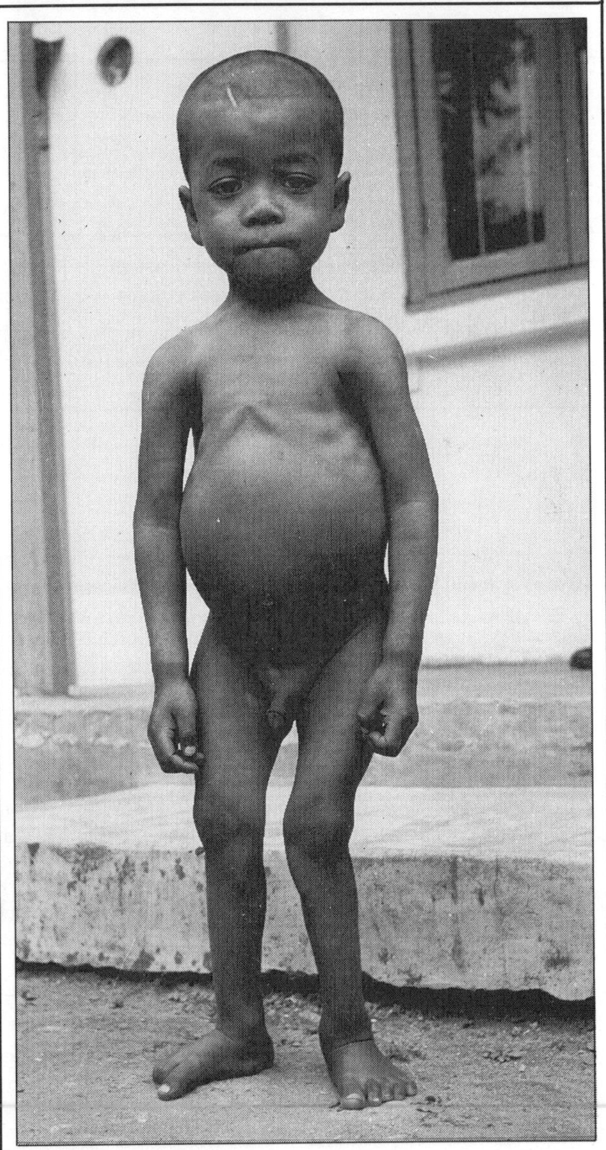

Vitamins

Like minerals, these substances are only needed in very small amounts to keep the body healthy. If a person does not get enough of a certain mineral or vitamin, they may develop a deficiency disease – see Tables 9.9 and 9.10.

Table 9.10 Vitamins

Vitamin	Function in the body	Effect of deficiency	Good food sources
Fat soluble vitamins			
A (retinol)	Essential for night vision Protects eye surfaces	Night blindess Severe eye lesions (xerophthalmia) Complete blindness (keratomalacia)	Fish liver oils, liver, butter, carrots, dark green vegetables Added by law to all margarine sold in Britain
D (calciferol)	Essential for uptake and use of calcium and phosphorus for bone and tooth growth	In children, weak deformed bones which may bend under the weight of the body – this is called rickets	Fish liver oils, dairy products It is made in the skin by the action of sunlight Added by law to all margarine sold in Britain
K	Normal blood clotting	Deficiency is unlikely	Spinach, cabbage, cauliflower peas, cereals It is synthesised (made) by gut bacteria
Water soluble vitamins: There are at least 12 vitamins in the B group. The three listed here are the best known.			
B_1 (thiamin)	Essential for the release of energy from carbohydrate	Beri beri (nervous paralysis and muscle weakness)	Meat, liver, milk, eggs, wholemeal flour Thiamin is added by law to all white flour sold in Britain
B_2 (riboflavin)	Essential for energy utilisation	Restricted growth and poor skin	Milk, meat, liver, eggs Riboflavin is destroyed by UV light
Nicotinic acid (niacin)	Essential for energy utilisation	Skin becomes dark and scaly – this is called pellagra	Meat, cereals, vegetables, milk Niacin is added by law to all white flour sold in Britain
C (ascorbic acid)	Maintaining healthy skin	Slow healing of wounds, bleeding from the gums – scurvy	Fresh fruit and vegetables, especially potatoes, green vegetables and citrus fruits – but vitamin C is easily lost by storage or cooking

Assignment – Vitamins

Study Table 9.10 and the illustration (right).

1. Which vitamins are added to which foods by law in Britain? Why do you think these laws exist?
2. What are the vitamins A, B_1, C and D needed for?
3. In what way is Vitamin D different from the other vitamins? What may happen to a child if it does not have enough Vitamin D in its diet?
4. Why do doctors and health visitors recommend that bottle-fed babies should have vitamin drops added to their feed?
5. Which of the cereals shown would you recommend? Give reasons for your answer.
6. Which vitamin might be lost from milk if it is left on a doorstep all day? Explain your answer.

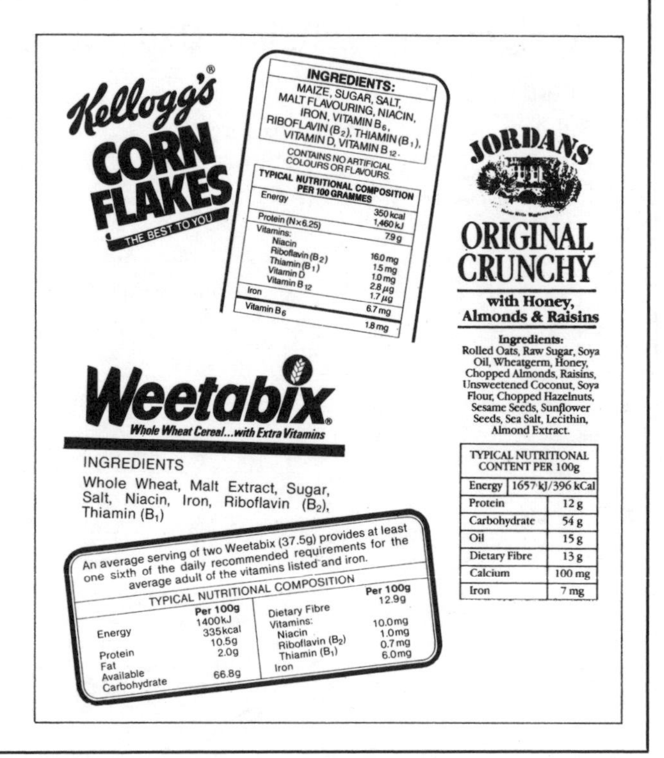

Investigation 9.14 To find out if fruit juice contains Vitamin C

Most people think that fruit juice contains Vitamin C. But do all fruit juices contain the same amount? You can compare various fruit squashes, fruit juices (packets, cans or bottles) and freshly squeezed fruit juice using the method below.

The test uses a blue dye, DCPIP (2,6-dichlorophenolindophenol). Vitamin C has a bleaching property: this means it can decolorise the blue DCPIP.

Collect

samples of at least three of the following:
- squash – orange or grapefruit
- fruit juice (packet, can or bottle) – apple, orange or grapefruit
- freshly squeezed juice – lemon, orange or grapefruit.

syringe (2 cm or 5 cm)
test tubes and test tube rack
DCPIP solution (0.1%)
dropper pipettes, one for each sample of juice
Bunsen burner and test tube holder

What to do

1. Use a syringe to measure 2 cm^3 of DCPIP into a test tube.
2. Add the first sample of fruit juice one drop at a time. Record how many drops are needed to decolorise the dye.
3. Repeat with at least two other samples.
4. Boil a sample of freshly squeezed juice. Repeat the test on this sample.
5. Record all your results in a table.

Questions

1. For each juice, what was the least number of drops needed to decolorise the DCPIP? What does this tell you about the Vitamin C content of this fruit juice?
2. Put the juices you tested in rank order according to Vitamin C content. Start with the one containing most Vitamin C.
3. How did the boiled fruit juice compare with the fresh sample of the same type? What does boiling do to Vitamin C?
4. Would this type of test be suitable to use with all fruit juices, such as blackcurrant, tomato and grape? Explain your answer.

Assignment – Vitamin C in potatoes

Vitamin C and the B vitamins are water soluble. Most of them are destroyed if they are heated for longer than a few minutes. Study the graph and the illustration.

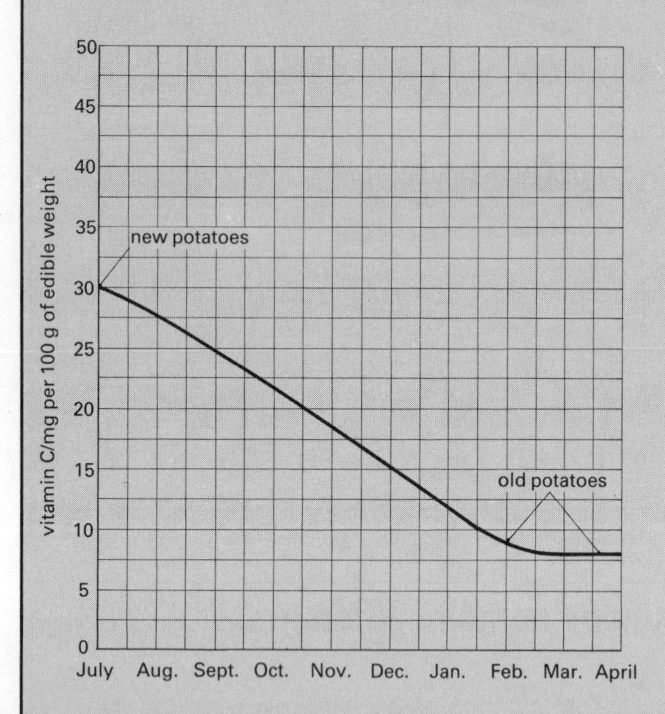

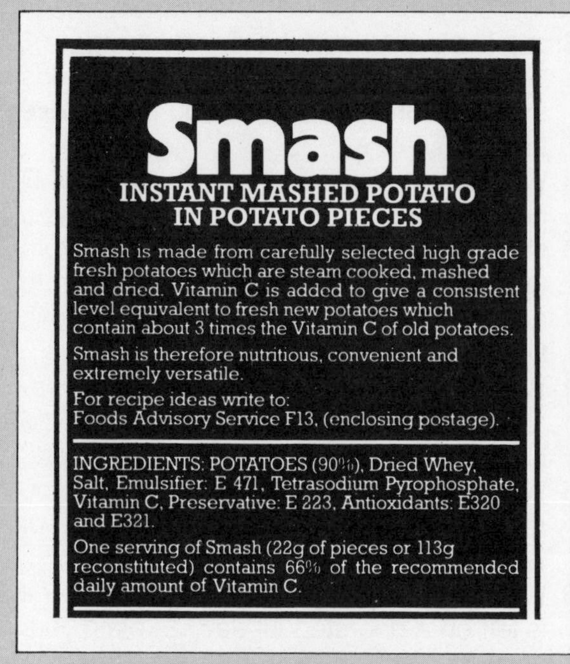

1 (a) Why is it best to cook vegetables in a small amount of water rather than a large amount?
(b) Why is it best to cook vegetables for only a short time, until they are just cooked?

2 Why is it a good idea to use the water vegetables have been cooked in for making gravy or other sauces instead of throwing it away?

3 (a) What is the Vitamin C content of new potatoes?
(b) What is the Vitamin C content of old potatoes in April?
(c) Describe what happens to the Vitamin C content of potatoes between July and the following April.

4 (a) Why is Vitamin C added to Smash?
(b) How does the amount of Vitamin C in Smash compare with (i) new potatoes, (ii) old potatoes?

5 List the advantages and disadvantages of using Smash instead of fresh potatoes.
Which would you recommend cooks to use? Give reasons for your answer.

BALANCED DIET

You eat food. All the food you eat over a length of time is called your DIET. To have a balanced diet you must have the right types of food in the right amounts. The diet must consist of carbohydrates, fats, proteins, vitamins, minerals, roughage and water in the right proportions. A good diet for one person is not necessarily a good diet for someone else. What makes up a *balanced* diet depends on several factors such as your age, size, sex, occupation and leisure activities. Special needs, such as pregnancy or illness, also make a difference.

A balanced diet should provide

(a) enough energy;
(b) enough protein for growth, repair and replacement;
(c) some fat (but not too much as it may lead to obesity; see page 238);
(d) a mixture of different foods to provide vitamins and minerals;
(e) foods containing roughage.

Assignment – Food and religion

The table shows some foods not eaten by people of certain religions or beliefs. However, there may be certain variations within each group. For example, some Hindus, especially those who have come from East Africa, may eat animal protein.

1 What is the difference between a vegan and a vegetarian?
2 Which religions forbid
 (a) alcohol, (b) pork, (c) beef, (d) lamb, (e) shellfish?
3 Which religion(s) have a vegetarian diet?
4 (a) Which type of diet may not contain enough protein?
 (b) What would you recommend as good sources of protein that could be eaten by people of any of the groups shown in the table?
5 The best sources of Vitamin D are fish, dairy products and margarine.
 (a) Which of the groups shown in the table may not have enough Vitamin D in the diet?
 (b) What problems may a lack of Vitamin D in the diet cause? Apart from eating foods containing Vitamin D, how else can a person get it?
6 Find out what the terms 'kosher' and 'hallal' mean.
7 Find out what Muslims eat during the month of Ramadan.

			Christian					
	Vegetarian	Vegan	RC	Mormon	Jew	Sikh	Muslim	Hindu
Alcohol	√	√	√	×	√	√	×	×
Tea, coffee, cocoa, Coke	√	√	√	×	√	√	√	√
Eggs	√	×	√	√	√	√	√	√*
Milk and yoghurt	√	×	√	√	√	√	√	√
Cheese	√	×	√	√	√	√	√	√
Chicken	×	×	some prefer to avoid meat on Friday	√	kosher	√	hallal	×
Mutton	×	×		√	kosher	√	hallal	×
Beef	×	×		√	kosher	×	hallal	×
Pork	×	×		√	×	rarely eaten	×	×
Fish	×	×	√	√	√	√	√	×
Shellfish	×	×	√	√	×	√	√	×
Nuts	√	√	√	√	√	√	√	√
Pulses	√	√	√	√	√	√	√	√
Animal fat (lard and some margarine)	√	×	√	√	kosher	√	hallal	×
Butter/ghee	√	×	√	√	√	√	√	√

* Some Hindus do not eat eggs

Unbalanced diets

Various health problems can occur if a person does not have a balanced diet. Two of these are described below.

Obesity

An obese person is one who has too much body fat. Usually this is caused by eating too much of the wrong foods, especially during the first two years of life. If more food is eaten than is needed to provide energy, then the extra will be converted into fat.

Many people in the western world are obese. Poor people often eat a fattening diet which consists mainly of fats and carbohydrates. Foods rich in carbohydrates are usually much cheaper than those rich in protein. Compare the cost of a pound of potatoes with a pound of steak, or the price of 250 g of biscuits with 250 g of cheese.

Anorexia

When a person carries slimming to extremes, they may suffer from anorexia. There are two forms of this disease.

Anorexia nervosa This is when slimming becomes such an obsession that the person becomes undernourished. The sufferers lose their appetite and eat virtually nothing. They become very concerned with their appearance, seeing themselves as very overweight even when they are extremely thin. This type of anorexia is often found in teenage girls. In early stages it can be easily treated, but if it continues too long hospital treatment and psychiatric help are often needed.

Bulimia nervosa In this condition the sufferer appears to eat well, in some cases eating very large meals. Later, in private, the sufferer makes themself vomit before any of the food can be digested. This disease is often more difficult to detect as the person appears to be eating normally. As with anorexia nervosa, sufferers are often obsessed by their appearance and wanting to be even thinner.

Assignment – Case study of a British family

Children	*Born in*
A. boy	1984
B. girl	1982
C. girl	1981
D. boy	1975
E. girl	1973

D and E were born in India where they lived until 1980. The other three children were born in England.

When C was 2½ years old her teeth started to blacken at the roots, some fell out and the others became brittle and chipped. B's teeth had similar problems at 2½. As yet A has few teeth and they are all healthy.

The mother of these children said that there were no problems with the teeth of the two older children. She said that she gave vitamin drops to the children born in England until they were nearly two years old.

1 (a) What would you suggest was the cause of the problems with the childrens' teeth?
  (b) What is the disease called?

2 If all five children had the same food to eat why do you think the children born in India had no problems with their teeth?

3 (a) Why do you think that A's teeth are healthy but B's and C's are not?
  (b) What would you advise the mother to do to ensure that A never has this sort of problem with her teeth?

4 What other health problems might B and C have? Which other parts of their bodies might be affected?

5 (a) What vitamins are contained in the vitamin drops given to babies?
  (b) Why are these needed as well as milk and other food?
  (c) How much does a bottle of vitamin drops cost (find out from the chemist or local child clinic).

Assignment – Diet and heart disease

Heart disease causes more deaths than any other illness.

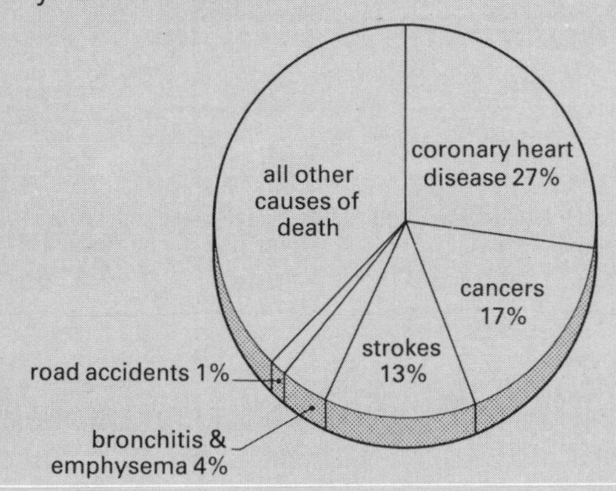

1 (a) Put the information in the pie chart into a table.
(b) Give examples of 'all other causes of death'.

The main cause of heart disease is the build-up of a fatty substance called *cholesterol* on the inside of arteries.

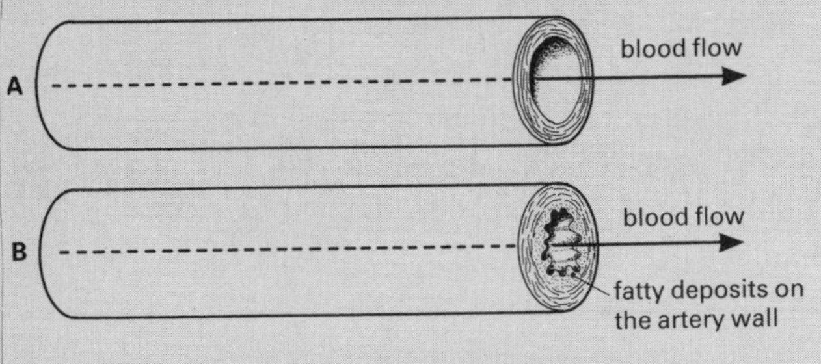

2 (a) Which of the arteries shown in the diagram will allow the most blood to pass through in a given time? (You can assume that the blood pressure is the same in both.)
(b) Blood carries food, oxygen and other materials to all parts of the body. If artery B is carrying blood to the arm, how will the amount of food and oxygen reaching the arm be affected?

The arteries supplying the heart with food and oxygen are called the coronary arteries. If cholesterol clogs the coronary arteries the heart will become weaker and may stop beating properly (a heart attack). The blood supply to the brain may stop, causing the person to die. Some people survive one or two heart attacks but are killed by a later one.

If the arteries leading to the brain become clogged by cholesterol this may cause a stroke when the brain does not get enough oxygen. The person will collapse and may die. If they recover, the brain, or parts of it, are often permanently damaged so that the person may not be ble to control certain parts of the body, may not be able to speak properly or may not be able to think as they did before.

3 Find out what type of foods contain the most cholesterol.

4 (a) What is the difference between a stroke and a heart attack?
(b) What do the terms 'cerebral thrombosis' and 'coronary thrombosis' mean?

5 (a) Find out what different parts of the brain do.
(b) If as a result of a stroke a person could not speak or walk, which part(s) of the brain would have been damaged?

Coronary heart disease is much more common in smokers than in non-smokers. Nearly all people with arterial disease in the legs are smokers. For these people walking becomes painful and they may get gangrene. Gangrene is caused when not enough food and oxygen is supplied to the cells, in this case because of blocked arteries. Gangrene is almost impossible to cure so one or both legs have to be amputated (cut off).

A number of factors seem to contribute to the risk of getting heart disease: smoking, high blood pressure, obesity, high levels of cholesterol in the diet, lack of exercise, stress (worry) and old age.

6 List at least four things that you can do now, and for the rest of your life, to reduce your risk of getting heart disease.

Assignment – Milk

What a pint of milk provides

	Pasteurised	Homogenised	Channel Islands	Semi-skimmed	Skimmed	Sterilised
Calories (energy)	380	380	445	280	195	380
Fat	3.8%	3.8%	4.8%	1.8%	0.1%	3.8%
Protein	19.3 g	19.3 g	21.1 g	19.5 g	19.9 g	19.3 g
Carbohydrate (lactose)	27.5 g	27.5 g	27.5 g	28.4 g	29.3 g	27.5 g
Calcium	702 mg	702 mg	702 mg	729 mg	761 mg	702 mg

Source: The National Dairy Council

1 For each type of milk find out the
(a) cost per pint,
(b) colours of the bottle tops.

2 What are the differences between
(a) pasteurised and skimmed milk,
(b) pasteurised and Channel Island milk?

3 Which type of mik would you recommend for
(a) young children,
(b) obese people,
(c) undernourished adults?
Give your reasons in each case.

4 What has happened to the following types of milk before it is put in bottles:
(a) pasteurised;
(b) homogenised;
(c) sterilised?

5 Long-life milk can be bought, in packets, in shops. This milk will have had UHT treatment.
(a) What is UHT treatment?
(b) Why is this milk 'long-life'?

6 (a) Find out the costs of 1 pt of (i) milk (pasteurised), (ii) Coke, (iii) orange juice. Find out what food types each of these contains. Arrange all your information in a table, similar to the one above.
(b) Comment on the usefulness of these three drinks in the diet of schoolchildren. Which are the most useful and which are not at all necessary? Give reasons for your answer.

Respiration and energy

WHAT YOU SHOULD KNOW

1 All our energy comes from the food we eat.
2 The process by which the energy is released is called respiration. Respiration is a form of controlled combustion.
3 During respiration glucose is broken down into carbon dioxide and water.
4 As a result of respiration, exhaled air contains more carbon dioxide and less oxygen than inhaled air.
5 Inhaled air passes into the lungs and to the alveoli, where gaseous exchanges take place.
6 Red blood cells carry oxygen from the alveoli and blood plasma carries carbon dioxide back to them.
7 Normal respiration involving oxygen is called aerobic respiration. The energy released is stored in molecules of adenosine triphosphate.
8 If the energy requirements are greater than the amount available from aerobic respiration, then anaerobic respiration takes place.
9 Simple plants (e.g. yeast) can obtain their energy from anaerobic respiration.
10 Foods can be grouped according to their reaction to various chemical tests.
11 Food is needed for energy; growth, repair and replacement of tissues; storage and for keeping the body healthy.
12 A balanced diet must contain enough energy, carbohydrates, fats, proteins, vitamins, mineral salts, water and roughage (fibre) in suitable proportions.

QUESTIONS

1 In a classroom containing a large number of students and with the windows closed it is observed that (a) the air temperature increases, (b) there is a rise in humidity and (c) breathing becomes uncomfortable. Use your knowledge of respiration to explain each of these changes.

2 The spirometer is a device that allows the volume changes in the lungs to be investigated. It contains a chamber filled with pure oxygen. The oxygen is breathed in by the person. On breathing out, the expired gases are passed through a substance that absorbs carbon dioxide before returning to the chamber. The changes in volume of the oxygen in the chamber are monitored by a pen moving over a calibrated sheet. The diagram shows some results obtained.

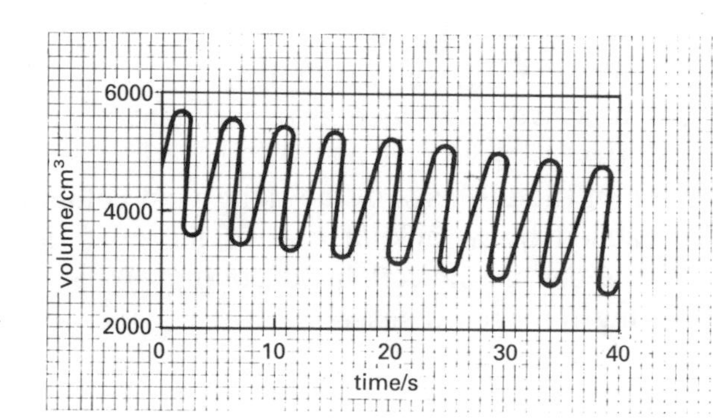

(a) What average volume of oxygen was breathed in?
(b) What was the rate of breathing?
(c) Explain the shape of the volume changes.
(d) What differences would you notice to the trace if the person was attached to the spirometer after a period of vigorous exercise?

3 Design experiments to show whether each of the following is true.
(a) Exhaled air is hotter than inhaled air.
(b) Germinating seeds undergo respiration.
(c) Yeast respires anaerobically.

4 Two students were investigating the effect of exercise on their breathing rates. They measured their normal breathing rate and then both did the same amount of exercise. After this, they measured their breathing rate every two minutes. The results are given in the table.

	Breathing rate before exercise	Breathing rate at 2 minute intervals after exercise							
		0	2	4	6	8	10	12	14
Student X	14	38	33	28	24	20	17	14	14
Student Y	12	34	29	24	20	16	12	12	12

(a) Suggest a reason why the two breathing rates before exercise are not the same.
(b) Why does the breathing rate increase when exercise is taken?
(c) Why does the breathing rate stay high for a time even after exercise has finished?
(d) Which one of the students is more likely to have done a lot of athletic training? Explain your answer.

5 Look at the photograph of the child who has never had enough food to eat.
(a) Make a list of the features which you think are caused by starvation.
(b) How do you think each of the following have been affected by the child's poor diet: energy, growth, resistance to disease, ability to heal wounds?
(c) If you were able to give this child food what types would you give him? Explain your reasons in each case.

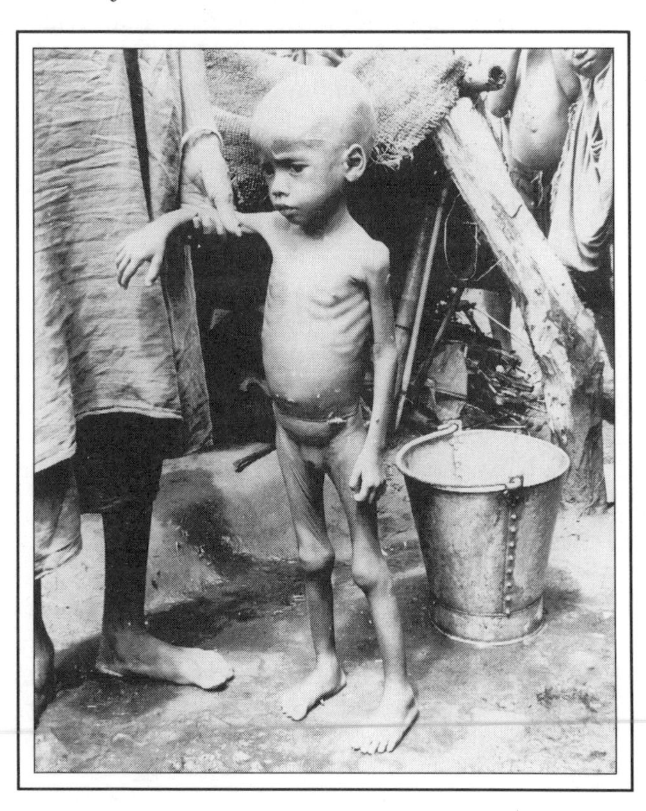

6

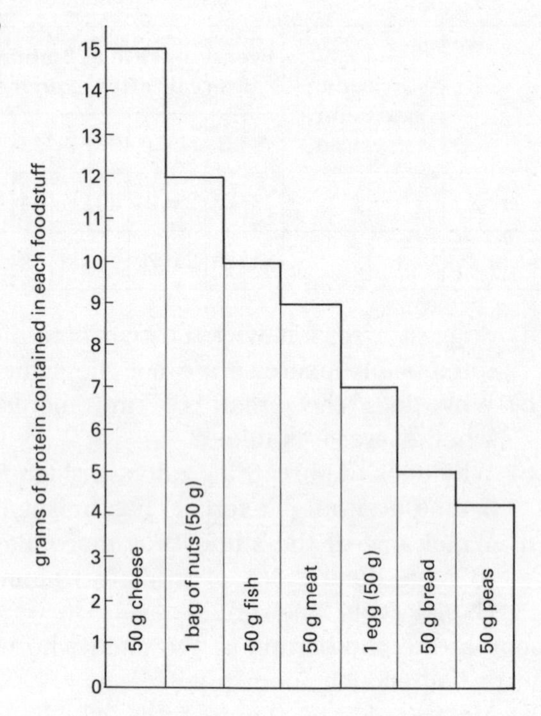

Study the data above.

(a) What is the mass of protein in 100 g of (a) bread, (b) cheese, (c) egg, (d) nuts, (e) meat?

(b) In one day a child eats 200 g bread, 100 g cheese, 2 eggs, a bag of nuts and 100 g of meat.

(i) How much protein does this provide?

(ii) Study the table below. For which types of person would this be about the right amount of protein?

Table to show protein requirements per day

Age/years	Boys/g	Girls/g
7–9	68	68
10–12	86	86
13–15	110	96
16–20	119	88

(c) Proteins are found in plants and animals. Nitrogen is one of the elements found in ALL proteins.

(i) Give three different ways that a plant can get nitrogen.

(ii) How do animals get their nitrogen?

(iii) Why is it more economic to produce protein for humans to eat from plants (e.g. peas, beans, nuts) than as meat?

7 In 1912, before anyone knew about vitamins, a scientist performed two experiments using two groups of young rats.

1st experiment

Group A: rats were fed highly purified protein, carbohydrate, fat, minerals and water. The food contained no vitamins although the scientists did not know this.

Group B: rats were fed the same diet but with a few drops of milk added.

Group A rats stopped growing and lost weight. Group B rats grew steadily and gained weight.

2nd experiment

The milk was removed from Group B's diet and added to Group A's diet. Group A now gained weight and Group B rats stopped growing and lost weight.

These results told the scientists two important things about milk.

(a) What are these two things?

(b) Why was the second experiment necessary?

8 Each of the diseases in the right hand column is caused by a lack of one or more of the substances in the left hand column. What is lacking for each disease?

iron	rickets
Vitamin A	anaemia
iodine	night-blindness
Vitamin C	goitre
Vitamin D	scurvy

9 Explain the scientific basis for the following statements.

(a) Carrots help you see in the dark.

(b) Young children are often given orange juice and/or cod liver oil.

(c) 'An apple a day keeps the doctor away.'

(d) Asians who come to live in great Britain are particularly prone to rickets.

(e) Pregnant women sometimes become anaemic for the duration of the pregnancy.

(f) Bottle-fed babies should have vitamin drops added to their milk. Breast-fed babies do not need additional vitamin drops.

10 An experiment was carried out to find out whether heating cabbage affected the amount of Vitamin C in it. The cabbage was put in a saucepan, covered with water and brought to the boil. Samples of the cabbage were tested at 10 minute intervals to find their Vitamin C content. The table below shows the results.

Time after boiling/ min	Vit. C content/ mg per 100 g
0	50
10	50
20	50
30	50
40	50
50	50
60	50
70	30
75	5
80	0

(a) Plot these results on a piece of graph paper.
(b) What does the graph show? Suggest reasons for this.
(c) What advice would you give to people cooking cabbage?

Energy 2

FUELS AND COMBUSTION

WHAT IS IN THE AIR?

Having looked at how plants and animals obtain their food and energy, we will now look in more detail at the gas that surrounds all living things and without which life would not exist as we know it – the air. We are at the bottom of a vast 'ocean' of gas held to the earth's surface by the force of gravity. Although the pressure due to the gas is very high ($10^5\,N\,m^{-2}$), we are not very often aware of its presence. If the pressure changes (for example, if you go up in an aeroplane or through a tunnel), then we can detect it through its effects on our ears. If the air is set into motion (i.e. a wind), we can feel and see its effects. When air is in motion, the effects can cause widespread damage – as with a cyclone.

In 1773, it was discovered that there were two different gases in air. One of these gases was an active gas that helped things burn and the other gas was inactive. The gases are called oxygen and nitrogen. As you can see from Table 9.11, oxygen and nitrogen form 99% of the atmosphere. But it was found that air is a mixture of many gases and eventually all the other gases were discovered.

Fig. 9.49 Air in motion can cause an enormous amount of damage: the result of a cyclone in Australia

Table 9.11 Gases in the air

Gas	Percentage by volume
Nitrogen	78
Oxygen	21
Argon	0.9
Carbon dioxide Neon Helium Krypton Xenon	0.1

Where do you think the carbon dioxide gas comes from? The other gases (argon, neon, helium, etc.) are usually called inert or 'noble' gases. The word 'noble' indicates that the gases are very unreactive. Air is a mixture of gases and the percentage composition can change. Water vapour has not been included in the list.

Because it is used in respiration, oxygen is probably the most important gas in the atmosphere. Oxygen is essential to the process called combustion (see page 246). For plants the presence of carbon dioxide is essential. Why is this? It should be remembered that the whole 'ocean' of gases protects us from much of the electromagnetic radiation and radioactive particles that would otherwise hit the earth's surface.

Assignment – Air

1 Air is a mixture of gases. What are the differences between a mixture and a compound?
2 What differences would you expect in the percentage composition of the air from the following regions: (a) a forest, (b) an industrial area and (c) an island in the middle of an ocean?
3 Try to find out uses of the following 'noble' gases: argon, neon and helium.
4 The percentage of oxygen is normally 21%, and of nitrogen 78%. What differences would you observe if the percentages were reversed, i.e. oxygen 78% and nitrogen 21%?

Making oxygen in the laboratory

It is easy to prepare oxygen in the laboratory using hydrogen peroxide, H_2O_2. This compound is unstable and can be converted to water and oxygen using the apparatus shown in Fig. 9.50

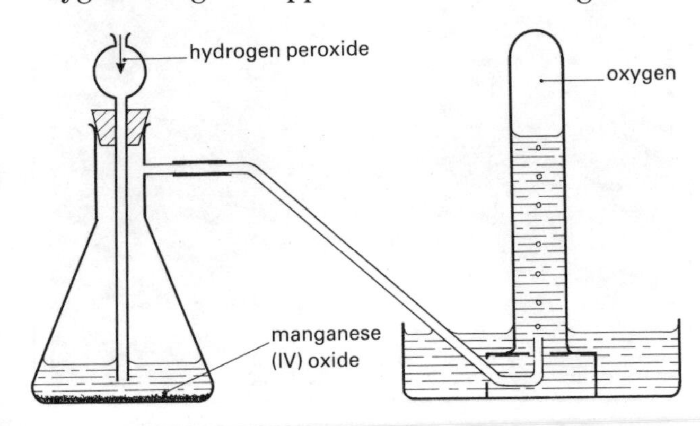

Fig. 9.50 Preparing oxygen

The equation is:

$$2H_2O_2(aq) \rightarrow 2H_2O(l) + O_2(g)$$

A catalyst is used to speed up the chemical reaction. It is manganese(IV) oxide. The gas can be collected by bubbling it through water as oxygen is only slightly soluble in water. After adding the hydrogen peroxide to the flask, it is usual to allow the gas to escape for a few seconds before starting to collect it. Why do you think this is done? How would you test the gas to make sure it was oxygen?

Making oxygen industrially

Large amounts of oxygen gas are needed and the largest supply available is, of course, the atmosphere. Because the air is a mixture of gases, it is possible to separate the constituents by using physical means rather than chemical means. The physical property used is the different boiling points of the gases.

You can demonstrate the way in which the industrial process works by using a bicycle pump. With your finger over the outlet, compress the air inside the pump. It will become warm. Allow the gas to cool back to room temperature. Then let the air escape through the outlet. It will be colder. This principle is used to separate gases in the air. The air is compressed and heats up. It is allowed to cool down to its original temperature. It then expands through a small hole causing a rapid drop in temperature.

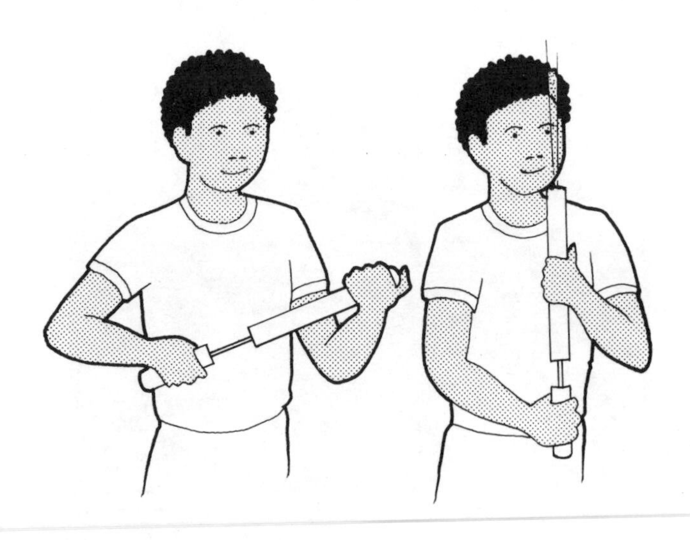

Fig. 9.51 Cooling air

In the industrial manufacture of oxygen, water vapour and carbon dioxide must first be removed. How could this be done? The boiling point of oxygen is −183 °C and that of nitrogen is −196 °C. The air is compressed, cooled and then allowed to expand. This has to be repeated a number of times before the air condenses to form a liquid. 'Liquid air' is obtained *below* −196 °C. If the liquid air is allowed to warm up then the nitrogen will boil off as the temperature reaches its boiling point, −196 °C. By controlling the temperature between −196 °C and −183 °C, the nitrogen is removed leaving the liquid oxygen behind.

Liquid oxygen can be transported around in special refrigerated containers. Why is it an advantage to be able to transport oxygen as a liquid rather than as a gas? Where the oxygen cannot be kept at such a low temperature, metal cylinders are used to contain the gas under very high pressure.

You may have used such a cylinder in the laboratory. When the cylinder is in use, gauges are fixed to it that tell you the pressure of the gas inside. Your teacher may say that the oxygen cylinder is 'empty'. What do you think the word 'empty' means here?

Fig. 9.52 Transportation of oxygen in high-pressure cylinders

AIR AND COMBUSTION

Combustion, often called burning, means the reaction of a substance with oxygen. The reaction releases energy, usually in the form of heat and light. When combustion occurs, nitrogen and the other gases play no part in the process. But it is very fortunate that nitrogen forms 78% of the air. If you have seen the difference between a substance burning in air (21% is oxygen) and burning in pure oxygen, you will understand why.

Fig. 9.53 Combustion of wire wool (above) in air; (below) in oxygen

In the following sections we will be interested in rapid combustion where the reaction takes place quickly. But we have already met two slow forms of combustion: corrosion and respiration. The majority of substances to be studied in the following sections are called hydrocarbons. These are compounds that contain only the elements carbon and hydrogen. Do you remember the alkanes in Unit 6, page 40? These are the most important group of hydrocarbons with respect to combustion.

Some of the combustion reactions of hydrocarbons are quite violent but the combustion of a candle is less so. Candle wax is made of solid hydrocarbons. The products of combustion can be demonstrated with the apparatus shown in Fig. 9.54.

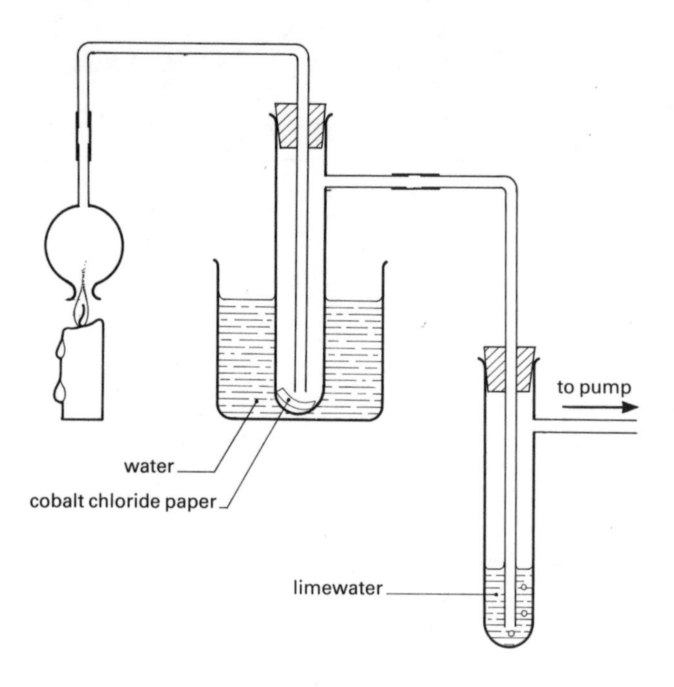

Fig. 9.54 Combustion products of a candle

The pump pulls the air and the gaseous products of combustion through the apparatus. The cobalt chloride is blue when dry but turns pink in the presence of water. Limewater turns cloudy in the presence of carbon dioxide. Both tests show positive results. Can you see the connection with respiration?

It is not possible to write a simple equation for the burning of candle wax as it consists of a number of different hydrocarbons. We can say that, when a hydrocarbon is burned in oxygen:

(a) carbon atoms produce carbon dioxide;
(b) hydrogen atoms produce water.

The simplest alkane is methane, CH_4. The combustion of methane is represented by the following equation.

$$CH_4(g) + 2O_2(g) \rightarrow CO_2(g) + 2H_2O(g)$$

If there is not enough oxygen, then carbon monoxide, CO, may be formed instead of CO_2.

FUELS

What is a fuel?

A fuel is usually classed as a substance that reacts rapidly with oxygen to change chemical energy into other forms of energy. The energy released is most often in the form of heat. Often light energy is produced as well. The energy from a fuel is used in many different situations. Can you identify the fuel and the main energy form produced in the situations shown in Fig. 9.55?

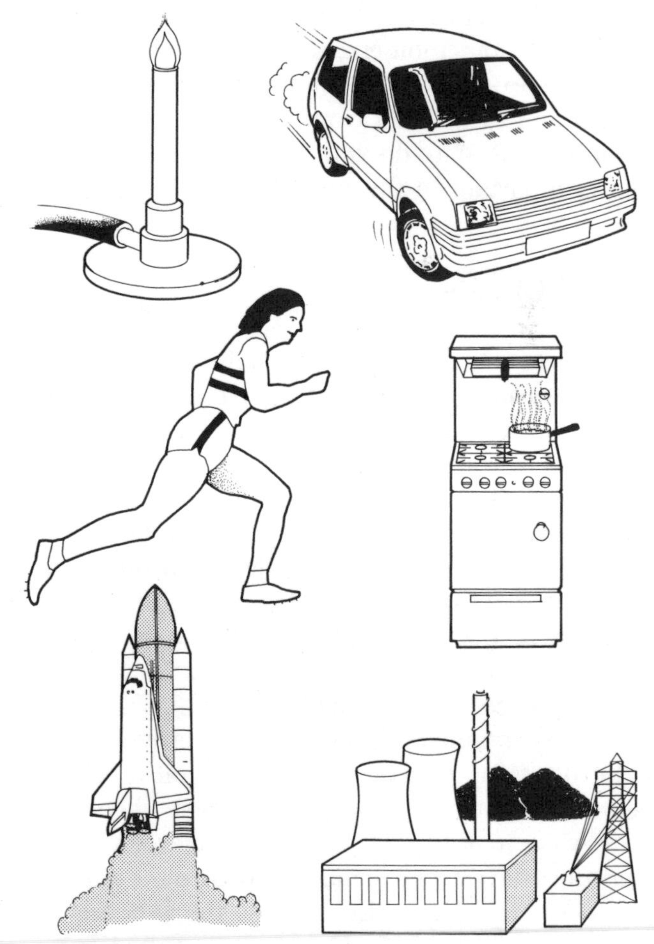

Fig. 9.55 Using fuels

Most fuels used today are hydrocarbons. It is possible to classify them under three headings: solid, liquid and gaseous fuels.

Solid fuels

The most important solid fuel in this country is coal. In Unit 1, you studied the formation of coal. Until recently it was the most important fuel that we had. Although oil and natural gas have assumed greater importance, they are not going to last much longer. Perhaps coal may regain its former importance, as estimated reserves will last for 200–300 years.

Although coal is mainly carbon, it does contain many other substances. When burning coal as a fuel, only the heat produced is of importance. However, if coal is heated in the absence of air (this is called destructive distillation) then many other substances can be produced. Some examples are fertilisers, insecticides, detergents, explosives, dyes and plastics. An added bonus is that the material left behind is practically pure carbon. At the moment, it is cheaper to obtain these by-products using oil. It is possible to make petrol and diesel fuel from coal but the process is little used. Why do you think this will become more important in the future?

Fig. 9.56 Oil production platform in the North Sea

Liquid fuels

The most important liquid fuels are obtained from petroleum. It is usually just called 'oil'. In Unit 1, you learned about the way that oil was probably formed. The original oil wells were all drilled on land. As technology has developed, it has become possible to obtain oil from under the sea. Great Britain is 'self-sufficient' in oil at the moment due to the oil fields found under the North Sea. 'Self-sufficient' means that enough oil can be produced for our own needs. But this oil is being used up very quickly.

Oil as obtained from under the ground is of little use. It is usually called crude oil or petroleum. Oil consists of many different chemicals (the vast majority of these are hydrocarbons) and they have to be separated before the oil can be used. You have already met the process by which this is done in Unit 5. It is called FRACTIONAL DISTILLATION. It relies on the fact that the different parts (fractions) of oil have different boiling points. The lightest fraction (smallest molecules) has the lowest boiling point and the heaviest fraction (largest molecules) has the highest boiling point. As the carbon atoms have a much greater mass than the hydrogen atoms, the number of carbon atoms is a good indicator of the molecular mass. The separation of oil into its different fractions is called REFINING. Figure 9.58 illustrates how the process takes place.

Each fraction contains molecules with a range of molecular masses. Further purification is usually needed before they are useful for practical purposes. Table 9.12 gives some information about the different fractions of oil.

One of the problems with crude oil is that the amount of each fraction does not correspond with the demand for it. For example, the amount of petrol required greatly exceeds the amount found in the crude oil. To overcome these difficulties, larger molecules are broken down into smaller, more useful molecules. This process is

Fig. 9.57 Oil refinery towers

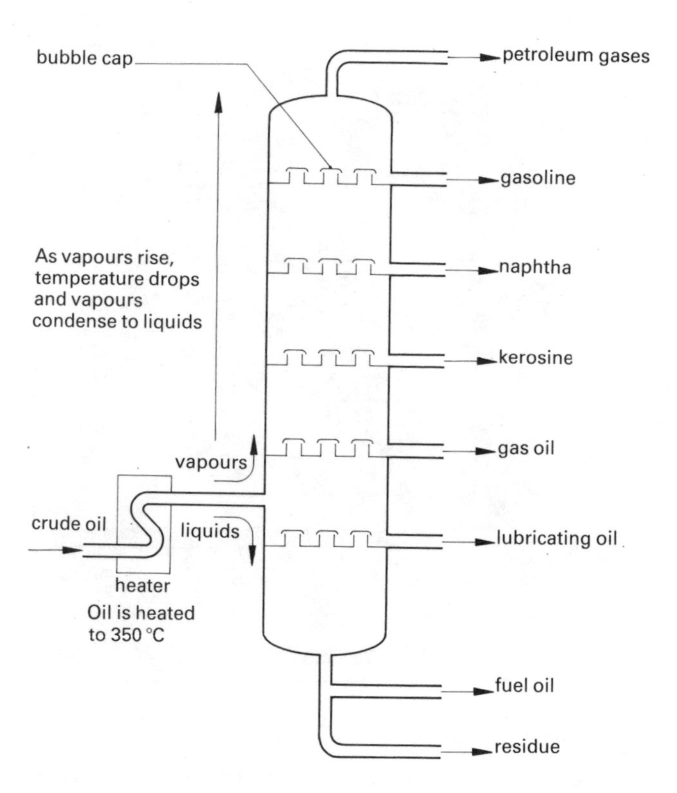

Fig. 9.58 Fractional distillation of oil – this is a simplified diagram of the refinery shown in Fig. 9.57

Table 9.12 The fractions of oil

Name of fraction	Approx. range of boiling points	Number of carbon atoms	Some uses
Petroleum gases	less than 20 °C	1 to 4	bottled gas; making petrochemicals
Gasoline	20 °C to 60 °C	5 to 6	petrol for cars
Naphtha	60 °C to 180 °C	6 to 10	most important source for petrochemicals; made into petrol and aviation fuel
Kerosine	180 °C to 260 °C	10 to 14	aviation fuel; home heating; paraffin
Gas oil	260 °C to 340 °C	14 to 20	diesel fuel; waxes; petrochemicals
Lubricating oil	over 340 °C	20 upwards	lubricants; grease
Fuel oil	over 500 °C	very large	fuel for ships and power stations
Residue (bitumen)	–	very large	road surfaces; waterproofing materials

Fig. 9.59 Catalytic (or 'cat') cracker

called *cracking*. To break the molecules down they are heated and then passed over a suitable catalyst. This is carried out at the refinery using what are usually called 'cat crackers'.

It is possible to demonstrate this in the laboratory. Liquid paraffin can be broken down into smaller gaseous molecules by strongly heating it and allowing it to pass over heated pieces of china. Strictly speaking, this is called thermal cracking, as the pieces of china do not act as an active catalyst. The gas can be collected over water.

In refineries, it is also possible to build up larger molecules from smaller ones. This process is called *polymerisation*. Other processes can change the shape of the molecules (for example, to make them better for burning as fuels). This is called *reforming*.

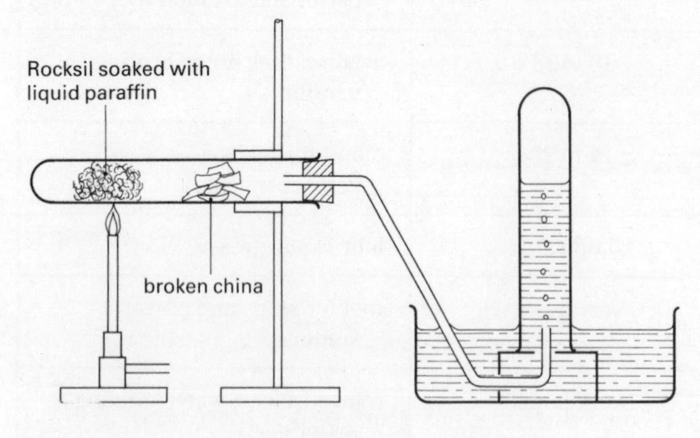

Fig. 9.60 Cracking of paraffin

One of the fastest growing industries at present is the manufacture of petrochemicals. This is the name for substances that can be made from oil or natural gas. Figure 9.61 shows a few of the many petrochemicals in use today.

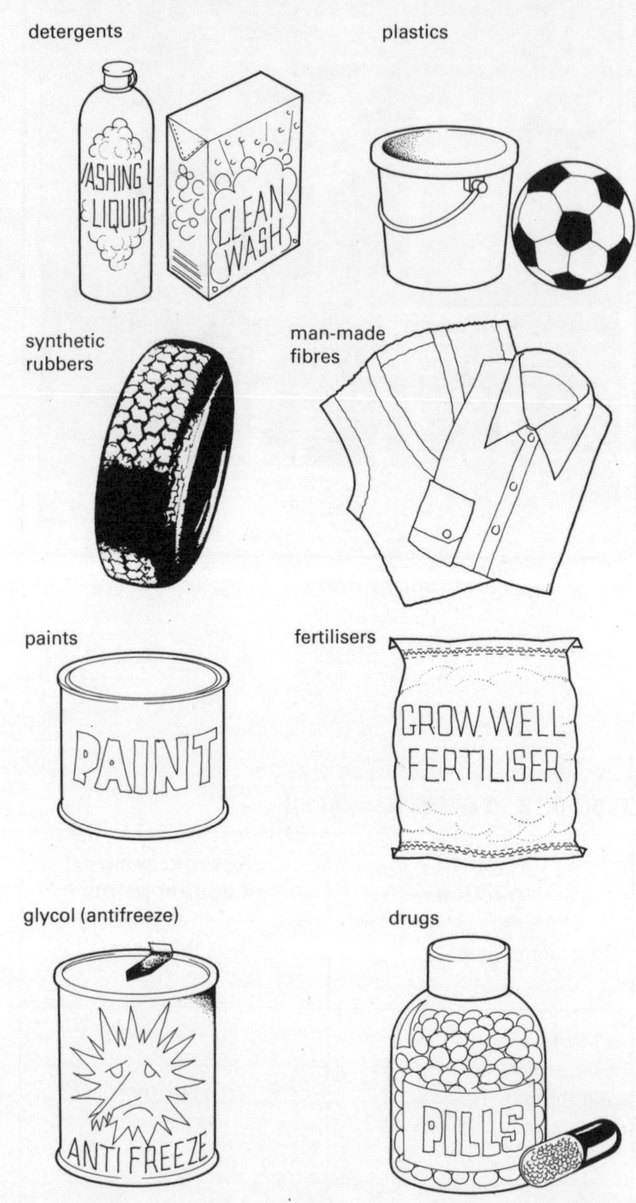

Fig. 9.61 Some uses of petrochemicals

Assignment — Petrochemicals

Read the Shell book *Petrochemicals*. Then write a short account suitable for a person with no scientific training on one or more of the following areas of petrochemical use: (a) solvents; (b) detergents; (c) industrial chemicals; (d) agricultural chemicals.

Gaseous fuels

There are three gases that are normally used as fuels. Two of the gases are present in the lightest fraction produced in the distillation of oil. They are propane (C_3H_8) and butane (C_4H_{10}). It is possible to liquefy these gases by compressing them. These fuels are used in situations where there is no piped gas available. The gas is sold in containers of various sizes – depending on the requirements. You may have used these gases for portable cookers when camping. Or you may have seen heaters that use these gases.

Fig. 9.62 Butane is used as a fuel in camping stoves

Whatever the use, the gas is allowed to escape through a valve. As it leaves the container, it becomes a gas. The gas is mixed with air and then burned.

The third and most important of the gaseous fuels is natural gas. This mixture of gases is found most often with oil but occasionally on its own. Its formation is thought to be similar to that of oil. But as the name shows, its constituents are all gases. The actual proportions vary according to where the gas is found. The main constituent is methane (CH_4). In natural gas from the North Sea, methane makes up over 90% of the total. Mixed with the methane are small quantities of ethane, propane and butane. You may remember that these are the first four members of the alkanes (Unit 6, page 40). In addition to these gases, there is usually a small quantity of non-combustible gas.

Until the 1960s, the only convenient sources of natural gas were to be found under the Sahara Desert in North Africa. This was sent through pipes to the coast. Here it was liquefied (by cooling it to −160 °C) before transportation by ship to the UK. In this state it was called liquefied natural gas, or LNG. At this time, the gas that was used in homes and industry was called town gas. It was manufactured from coal originally and then from oil as well. A network of pipes was built to deliver gas to the home. The natural gas from the Sahara could produce twice as much heat energy as town gas and was used to 'enrich' the town gas.

In the 1960s, vast quantities of natural gas were discovered under the North Sea. In spite of the depth of the sources (down to 3000 m), it was thought economical to recover it. Wells had to be dug to bring the gas to the surface. The gas then had to be transported by pipe to the nearest shore and piped around the country. Because there was so much gas available, the complete UK mainland gas system was adapted to run on natural gas.

A large amount of energy is used in the home. Electrical energy is produced using heat energy from fuels. The best way to use this is for devices that can only run on electricity, such as radios, televisions, refrigerators, washing machines etc. It is cheaper to use a gaseous fuel for the purpose of heating and cooking rather than change electrical energy into heat energy. Why do you think this is?

What makes a good fuel?

It might be thought that the amount of heat energy produced by burning the fuel is the most important factor. This is usually called the CALORIFIC VALUE of the fuel. In Table 9.13 hydrogen has the largest calorific value.

If 1 g of hydrogen is completely burned in oxygen, then 120 000 J (1.2×10^5 J) of heat energy is released. But does that make hydrogen a good fuel? You have only to look at the picture of the

Table 9.13 Calorific value of various fuels

Fuel	Calorific value ($J g^{-1}$)
Hydrogen	120 000
Natural gas	55 000
Butane	50 000
Petrol	47 000
Diesel	45 000
Coal	34 000
Ethanol	28 000
Wood	22 000
Bread	11 000
Potatoes	3 500

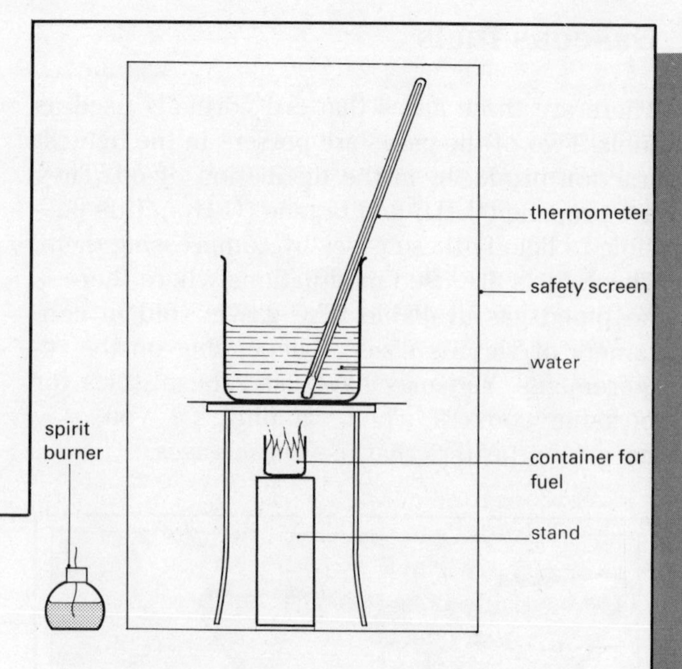

Investigation 9.14
To study the burning of fuels

This investigation will enable you to calculate the approximate calorific value for a number of different fuels and to suggest whether they are good fuels or not.

Collect

100 cm^3 beaker
thermometer (0 to 100 °C)
tripod, gauze and mat
stand for fuel
containers for fuel
various fuels (e.g. paper, ethanol, candle wax, wood shavings, coal, sugar)
eye protection
safety screen

WARNING – 1. Always wear eye protection when burning the fuels.
2. Always keep the safety screen in position when burning the fuels.
3. Allow the fuel to cool down after burning if it has to be re-weighed.

What to do

1 Set up the apparatus as shown below. The water should be at room temperature. Measure this temperature.
2 Collect the fuel and find its mass.
3 Ignite the fuel and place it in position. For some fuels you will be using spirit burners.
4 When the fuel has been completely burned (or the spirit burner has been turned off), stir the water and measure the temperature rise. If not all the fuel was burned then you will have to find the mass used.
5 Calculate the heat energy supplied to the water. This is calculated by the formula: mass of water (kg) × specific heat capacity × temperature change (°C). The specific heat capacity of water is 4200 $J kg^{-1} °C^{-1}$.
6 Calculate the calorific value of the fuel.
7 Decide what other factors contribute to what is called a good fuel. Show all the information you have found in the form of a chart.
8 Repeat these steps with each of the other fuels you have been given.

Questions

1 Why was each precaution in the instructions needed to make sure that there was no danger to you or the other students in the laboratory?
2 Which of the fuels you tested had the greatest calorific value?
3 Why do you think that your answers are lower than those in Table 9.13?
4 Which fuel(s) produced least smoke and ash?
5 If you have the costs available, you might like to calculate which fuel would provide 100 000 J (10^5 J) of heat energy at the cheapest rate.

burning airship on page 233 of Book 1 to see what happens when hydrogen burns uncontrolled. Why do you think hydrogen was used to provide the 'lift' for an airship?

If hydrogen and oxygen are mixed before ignition, the result will be an explosion. However, a lot of research is being done on using hydrogen as a fuel. One of the attractions is that only water is formed as a product of combustion.

Here are some other factors that need to be considered when deciding whether a fuel is a good one or not.

How much does it cost?
Is it natural or manufactured?
How easy is it to transport and store it?
How easy is it to light?
How quickly does it burn?
Does it produce smoke, ash or another form of pollution?

You can probably think of other factors that might need to be taken into consideration.

OXIDATION AND REDUCTION

The original meaning of the term OXIDATION was very simple. It meant that oxygen had been added to a substance. For example, if carbon was burned in oxygen, it was said that the carbon had been oxidised to make carbon dioxide.

$$C(s) + O_2(g) \rightarrow CO_2(g)$$

As the nature of chemical reactions came to be better understood, it was realised that the transfer of electrons was more important. So the name oxidation has been extended to cover a much wider range of chemical reactions, many of which do not involve oxygen.

Oxidation can be defined as follows: oxidation takes place when an atom or ion loses one or more electrons. Let us look at the burning of magnesium in oxygen.

$$2Mg(s) + O_2(g) \rightarrow 2Mg^{2+}O^{2-} \quad \text{(i.e. 2MgO)}$$

The magnesium atom has given up two electrons (becoming Mg^{2+}) while the oxygen has gained two electrons (becoming O^{2-}). It should be noted that *all* metal reactions that involve the metal forming a compound are classed as oxidations because, in every case, electrons are given up by the metal to form positive ions.

The opposite of oxidation is REDUCTION. Again, the original meaning of the word was very limited. It meant the removal of oxygen from a substance; for example, when an ore is reduced to the metal. If lead oxide is heated with carbon to a high enough temperature, then lead is produced.

$$PbO(s) + C(s) \rightarrow Pb(s) + CO(g)$$

Again, a more general definition is used today. Reduction takes place when an atom or ion gains one or more electrons. In the above reaction, the lead ions (Pb^{2+}) in the lead oxide are gaining electrons to become lead atoms.

In any reaction, the total number of electrons must remain constant. This means that if an atom or atoms lose a certain number of electrons then another atom or atoms must gain the same number of electrons. This means that oxidation and reduction always occur together. For this reason, such reactions are called REDOX (reduction–oxidation) reactions. If magnesium is burned in chlorine gas, the reaction can be represented as follows.

$$Mg(s) + Cl_2(g) \rightarrow Mg^{2+}Cl^{-}{}_2(s)$$

The magnesium atoms are being oxidised (losing electrons) and the chlorine atoms are being reduced (gaining electrons).

Oxidising and reducing agents

An oxidising agent is a substance that will oxidise another substance. (This means that the oxidising agent must be easily reduced itself.)

A reducing agent is a substance that will reduce another substance.

Table 9.14 Some oxidising and reducing agents

	Oxidising Agents	Reducing Agents
Elements	oxygen chlorine bromine iodine	reactive metals carbon hydrogen
Compounds	conc. sulphuric acid	sulphur(IV) oxide hydrogen sulphide

Redox reactions

Many of the reactions that you have studied are redox reactions. Here are a few examples.

Electrolysis
In Unit 4 you studied the electrolysis of copper sulphate solution using copper electrodes.

At the cathode: copper ions are deposited as copper atoms.

$$Cu^{2+}(aq) + 2e^- \rightarrow Cu(s)$$

The copper atom has gained two electrons, i.e. reduction.

At the anode: copper atoms go into solution.

$$Cu(s) \rightarrow Cu^{2+}(aq) + 2e^-$$

What is happening here, oxidation or reduction? In fact, all electrolysis reactions are classed as redox reactions. Can you explain why?

Production of hydrogen from a metal and an acid

$$H_2SO_4(aq) + Mg(s) \rightarrow MgSO_4(aq) + H_2(g)$$

Knowing that H_2SO_4 is $H^+{}_2SO_4{}^{2-}$ and $MgSO_4$ is $Mg^{2+}SO_4{}^{2-}$, can you say which substance is being oxidised and which reduced?

Displacement reactions
If an iron nail is placed in copper sulphate solution, the nail starts to dissolve and copper is deposited on it. Could you write the equation that describes what is happening? Which is the reducing agent?

Corrosion, respiration and combustion
These are some of the most important oxidation reactions we have considered. (Although some substances must be reduced in these reactions, the most important occurrence is the oxidation.) Try to write down a few examples and say, in each case, what is being oxidised and what the oxidising agent is.

EXOTHERMIC REACTIONS

An EXOTHERMIC chemical reaction is one that releases energy. Can you think of any such reactions? The energy that is released is usually in the form of heat energy. But, in a few instances, some of the released energy is in the form of light (for example, potassium reacting with water) or electrical energy (for example, the simple cell).

There is another 'class' of chemical reactions known as ENDOTHERMIC reactions. When these reactions take place, energy must be supplied. Two important examples are photosynthesis and electrolysis. In each case, can you say where the energy comes from?

Assignment – Endothermic reactions

When ammonium nitrate crystals are dissolved in water, heat energy is taken in.

1 How would you demonstrate this in the laboratory?
2 The heat energy required per mole of ammonium nitrate is 41.5 kJ. It is usually called the heat of solution. Describe how you would attempt to measure a value for this.
3 Write an equation to describe this endothermic reaction. Use the equation below to help you.
4 Melting is an endothermic process. How would you demonstrate this? How would you explain this?

But the reactions we are really interested in here are the exothermic ones. The reason for this is that by far the greatest part of the energy we use comes from exothermic reactions. Consider the burning of pure carbon in oxygen. The equation is

$$C(s) + O_2(g) \rightarrow CO_2(g) \quad \Delta H = -393.5\,kJ$$

The right hand part gives information about the energy involved in the reaction. The minus sign indicates that energy is released – the reaction is exothermic. What would a plus sign mean? In this case, the value of ΔH (pronounced delta H) is the energy released if one mole of carbon atoms (12 g) is burned in one mole of oxygen gas. The value is 393.5 kJ or 393 500 J. In the case of fuels, there are probably a number of substances present (as, for example, in natural gas) and the fuel has to be burned to find the energy released.

Assignment – Heat of combustion

Here are two substances that are used as fuels. How would you work out the amount of heat energy released when one mole of the fuel is burned in sufficient oxygen? This is called the *heat of combustion*. Look back to the section on burning fuels for some ideas on how to measure this heat energy.

The substances are

(a) ethanol, C_2H_5OH;

(b) butane, C_4H_{10} (supplied in a camping gas stove).

Fuels and exothermic reactions

When the exothermic reactions of fuels are used to provide more useful forms of energy, heat energy is produced as an intermediate stage.

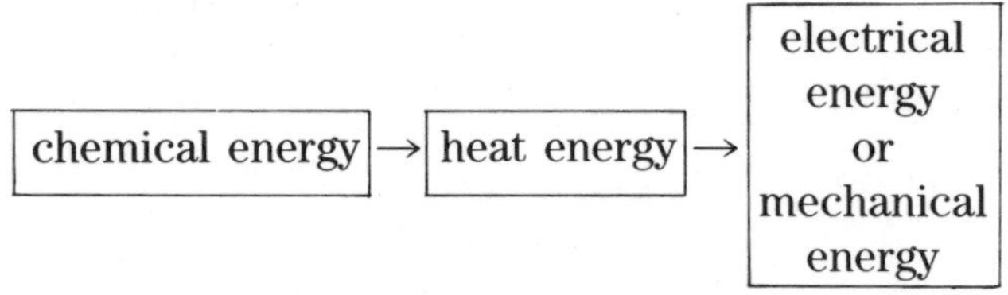

As you will probably remember from the study of energy chains in Unit 3, the more 'links' in the chain, the less the overall efficiency. All the large-scale energy conversions that are used involve heat energy being produced at one stage. For example, in power stations the fuel is burned to produce heat energy. This heat energy is used to produce steam. The steam turns the turbine attached to the generator, which produces electrical energy.

At present, and only on a small scale, there is only one type of practical device that will provide electrical energy directly from the fuel, hydrogen. In fuel cells, hydrogen and oxygen combine to produce water. The energy released due to this reaction is in the form of electrical energy. But they are expensive devices and it does not look as though fuel cells will replace the conventional power station in the near future.

Fuels and combustion

WHAT YOU SHOULD KNOW

1 The air is a mixture of gases of which oxygen is the most important.

2 Oxygen can be obtained in the laboratory using hydrogen peroxide with manganese(IV) oxide as a catalyst.

3 Oxygen is obtained industrially from the atmosphere by fractional distillation.

4 Combustion, or burning, is the general name for the reaction of a substance with oxygen.

5 Respiration and corrosion are slow forms of combustion.

6 Combustion of a hydrocarbon results in carbon dioxide and water being formed.

7 A fuel is a substance that reacts rapidly with oxygen to produce energy.

8 Coal is the most important solid fuel in this country.

9 Oil is the most important liquid fuel in this country.

10 Crude oil has to undergo fractional distillation before it is of any use.

11 Oil refineries can change the size and shape of the molecules present in the different oil fractions.

12 Chemical substances derived from oil are known as petrochemicals.

13 Natural gas is the most important gaseous fuel in this country.

14 Natural gas is mainly methane and supplies practically all our gaseous energy needs.

15 Calorific value is a measure of the energy produced from the burning of a fuel.

16 Factors other than the calorific value must be taken into account when deciding if a fuel is good or not.

17 Oxidation occurs when an atom or ion loses one or more electrons.

18 Reduction occurs when an atom or ion gains one or more electrons.

QUESTIONS

1

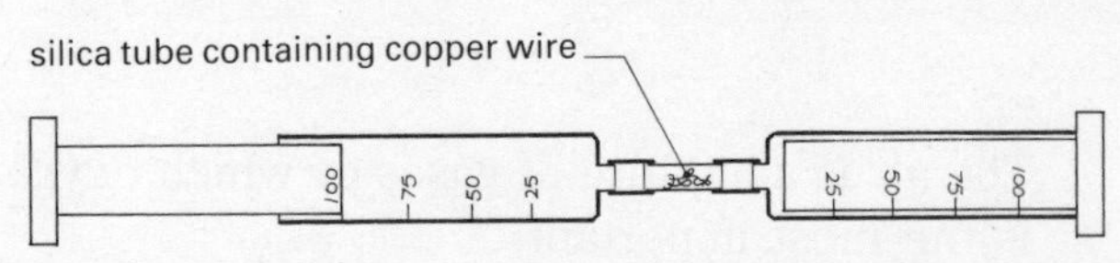

The apparatus in the diagram was used to investigate the percentage of oxygen in the air. Freshly cleaned copper wire was heated in the silica tube. At the start there was 100 cm^3 of air in one of the gas syringes and this air was passed over the heated copper until there was no further change in volume of the air. The final volume recorded was 79 cm^3. The copper had been changed into a black solid – copper (II) oxide.

(a) Why should the copper wire be cleaned before the experiment?
(b) Why must the copper be heated?
(c) Write a word equation for the reaction.
(d) From the initial and final gas volumes, what is the percentage of oxygen in the air?
(e) From weighings before and after the experiment it was found that the original mass of copper was 7.1 g and the final mass was 8.9 g. Do these figures support the statement that the product formed was copper(II) oxide?
A_r(copper) = 64; A_r(oxygen) = 16.
(f) Do you think it likely that all the copper was turned into copper(II) oxide? Explain your answer.

2 Explain the similarities and differences between each of the following: corrosion, respiration and combustion.

3 Each of the following substances could, in theory, be used as a fuel: coal, wood, bread, hydrogen and petrol. Using Table 9.13 on page 252 as a start, say whether you consider each of these substances to be a good fuel or a bad fuel.

4 The heat of combustion is the amount of energy released when one mole of a fuel is burned in air. In an experiment to measure this for ethanol, a spirit burner filled with ethanol was used to heat water. The apparatus was similar to that in the diagram in Investigation 9.14. The following results were obtained.

Mass of burner before the experiment = 26.1 g
Mass of burner after the experiment = 27.3 g
Mass of water = 100 g
Temperature rise of water = 18 °C

(Formula for ethanol is C_2H_5OH; A_r(hydrogen) = 1; A_r(carbon) = 12; A_r(oxygen) = 16; specific heat capacity of water = 4200 $J\,kg^{-1}\,°C^{-1}$)

(a) Calculate the mass of ethanol used.
(b) Calculate this as a fraction of one mole.
(c) Calculate the heat energy supplied to the water.
(d) Calculate the heat of combustion of ethanol.
(e) A data book gives the value for the heat of combustion of ethanol as 1330 $kJ\,mol^{-1}$. Comment on any difference between this figure and your answer from (d).

5

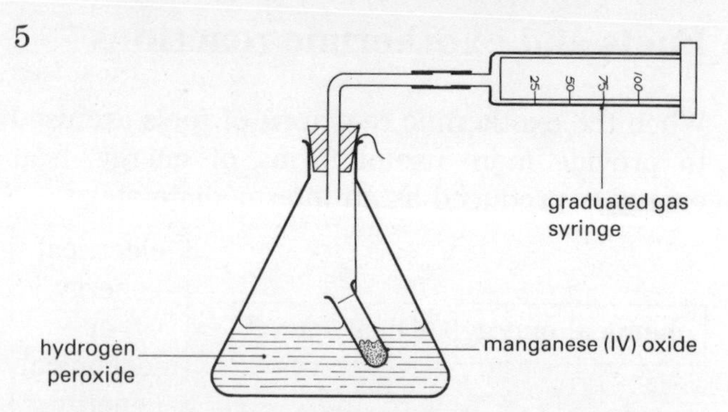

The apparatus in the diagram was used to investigate the volumes of oxygen given off when the catalyst manganese(IV) oxide was added to hydrogen peroxide solution. Below are the results obtained.

Time/min	0	1	2	3	4	5	6
Volume/cm^3	0	58	77	88	93	95	95

(a) Plot the results as a graph.
(b) How long did it take for (i) 40 cm^3 (ii) 80 cm^3 of oxygen to be produced?
(c) How does the speed of the chemical reaction change with time?
(d) What differences would you notice in the shape of the graph if no catalyst had been added?
(e) It is suggested that the enzyme catalase, which is found in the livers of mammals, is a better catalyst than manganese(IV) oxide. How would you set about comparing the two?

10. Balancing the Biosphere

WHAT IS BALANCE?

All life on earth exists in a layer of air, water and soil which is sometimes called the BIOSPHERE, Within this layer things are always changing, but after a time the change is not noticeable. In other words, the biosphere is *in equilibrium* and can return to normal after small disturbances. However, large disturbances can upset the equilibrium and things may then go wrong. In this unit we shall look at how things can go wrong.

You have learned about the cycles which help keep the biosphere in equilibrium. The water cycle, the nitrogen cycle (page 215), the carbon cycle (page 214) and the oxygen cycle are the most important of these. Table 10.1 should remind you of the ways in which the levels of each of these substances are kept constant.

The cycles listed in Table 10.1 transfer energy and matter through the biosphere. A small part of the sun's energy is 'trapped' on earth by the process of photosynthesis (see Unit 9), which uses up water and carbon dioxide. Water is the most important substance in the biosphere. It helps to control temperatures on earth, plays an important part in the rock cycle and helps in energy

Table 10.1 How the balance is maintained

	Is taken out of circulation by:	**Is put into circulation by:**
Nitrogen	micro-organisms (e.g. the nitrogen-fixing bacteria in pea plants) lightning industry (the manufacture of ammonia) mineralisation	decomposers breaking down dead organisms erosion of minerals
Oxygen	respiration oxidation and burning of organic material and minerals	photosynthesis
Carbon	photosynthesis forming bones	respiration decomposition burning fuels industrial processes chemical weathering of rocks

If you are not certain what any of the words in this table mean, consult the index and look back to the unit where the term first appears.

transfers. It evaporates from the earth's surface and some turns into clouds and falls as rain or snow, to be used by plants and animals. These, in turn, replace the water they use: plants transpire and animals 'breathe out' water vapour.

The overall equilibrium in the biosphere, or the balance of one of the four main cycles, can be disturbed by natural events or human activities. Any changes in the biosphere will affect the organisms it contains. As you learned in Unit 8, animal and plant populations are related to each other by food webs. Any change in one of the populations in a community will affect the other populations. In many ways, how humans affect their environment is most important because we use technology to change things to suit us. As the human population has grown and our technology has advanced, we have disturbed the balance in the biosphere. In the past, human technology changed slowly and any disturbances were easily absorbed. In recent years advances in technology have been more rapid and larger. Therefore we need to be aware of the effects of our ever increasing demands for more resources. Can you think why?

It would be difficult to deal with every one of the ways in which the biosphere can be thrown out of balance. It would fill several books! However we can examine some of the major causes of disturbance so far. These include:

major weather changes;
advances in agriculture;
pollution.

As you will see, it is difficult to pick out a single event as the sole cause of a disturbance. Often there are several things which work together to upset the balance – and usually human 'interference' makes things worse!

DISTURBING THE BALANCE

MAJOR WEATHER CHANGES

Changes in weather can upset the balance. Events such as earthquakes and erupting volcanoes can affect the weather, and so disturb the balance in the biosphere. As the next two assignments show, changes in the weather can have dramatic effects on human and animal life.

Assignment – El Niño and the anchovy

El Niño is a warm current that flows southwards past the coast of Peru in South America. (El Niño means 'The Christ Child' in Spanish – El Niño normally arrives near Christmas.) Its warm water, flowing by at a distance from the shore, causes plankton to rise to the surface close to the land. Anchovies are small fish that are used to make oil and meal for feeding animals. These tiny fish thrive on the rich harvest of plankton which El Niño brings to the surface. In 1971 Peru had the largest anchovy catch in the world and the anchovies teeming in the waters off the west coast of South America were much in demand.

In 1972 El Niño did not come until January and, when it arrived, flowed closer to the coast. This blocked off the plankton-rich water and deprived the young anchovies of their food. As a result of this disturbance in the environment the 1972 catch fell from the normal 10 million tonnes to 4½ million tonnes, mostly of older fish. The Peruvian fishery had to close. It opened again in 1974 and began to recover: by 1975 the catch was 3½ million tonnes. Estimates suggest that it may be possible to maintain annual catches of 5–6 million tonnes in those years that El Niño does not arrive on time.

1 Explain how the warm El Niño
   (a) caused plankton to rise to the surface closer to the coast as it passed by;
   (b) stopped the plankton reaching the surface when it flowed closer to the coast.
2 Why was the 1972 catch of 4½ million tonnes mostly of older fish?
3 Explain why the fishery was able to reopen in 1974.
4 The estimated 'safe' catch to maintain is about 5 million tonnes.
   (a) How does this figure compare with the 1971 catch?
   (b) Suggest how the size of the 1971 catch could have helped to make the 1972 catch so low.
   (c) Explain why keeping the catch at a 'safe' level should allow a catch even in years when El Niño is late.

Assignment – When the rains fail

Humans need water so badly that it has been said that the map of human population looks very like the map of world rainfall. Today the monsoon rains are vitally important: they support about half the world's population. For farmers in China, India, South East Asia and parts of Africa, the start of the monsoon rains is the most important date in the calendar. For example, in 1972 the monsoon was three weeks late and India lost almost a third of its food production.

In 1973 several countries of the Sahel of West Africa lost between 30 and 70% of their cattle as well as much of their food crops. Sahel is an Arabic word meaning 'shore', and is the name given to the southern edge, or 'shore', of the Sahara desert.

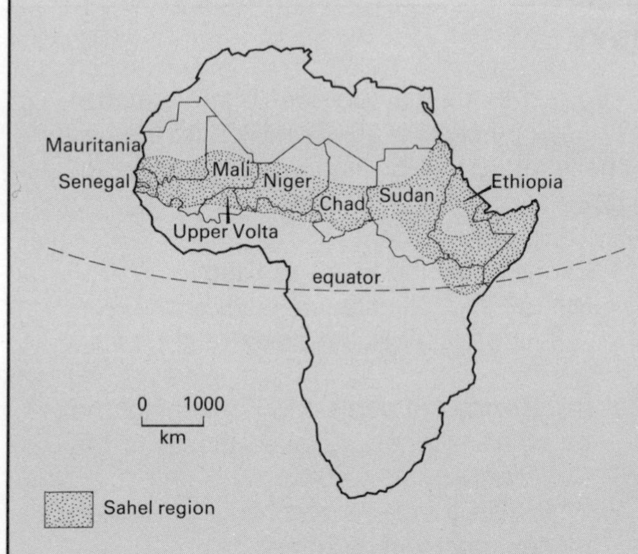

The Sahel is a zone across Africa where human life is just possible and depends on the rains. These come from the south in summer as moist ocean air passes over Africa bringing miniature monsoons. Such reliable and much-needed rain now comes less far into Africa and the Sahara desert is moving southwards (see page 263). The lessening rainfall, the advancing desert and failures of the rains brought the greatest tragedy in Northern Ethiopia, the eastern edge of the Sahel. In 1973 the government admitted that 3000–4000 people were dying each week. The immediate cause of death was often pneumonia, gastro-enteritis or cholera, but hunger was at the root of it.

Four countries lie completely within the affected zone: Mali, Upper Volta, Niger and Chad. They could become uninhabitable if the desert keeps moving southward. As you will see on page 263, the pressure of increasing population in the area is helping to speed up the creation of desert.

1 When rains fail and there is a drought, both animals and plants can die.
   (a) Why do animals (including humans) suffer when there is a shortage of water?
   (b) Why can the late arrival of monsoon rains cause whole crops to be lost?
2 The inhabitants of the Sahel who were dying at such an alarming rate in 1973 were suffering from the drought. Explain why the drought caused them to die from the diseases listed above.
3 Ten years after the worst of the Sahel drought, experts estimated that the countries involved were heading towards disaster again. Huge amounts of foreign aid were pouring in to the area, but experts felt they were going to the wrong projects. Food production was increasing by 1% a year, while population was growing at 2½% a year on average. Though more land is irrigated each year, about the same amount of irrigation projects break down from inadequate maintenance or lack of trained personnel.
   (a) Suggest why food production is increasing so slowly.
   (b) Why is such a slow increase in food production a cause for concern?
   (c) What could have been done to prevent the countries of the Sahel heading towards another disaster?
   (d) Write a brief account of the recurrence of the Sahel disaster during 1984–85.

4 Use a library, newspapers or information from organisations such as Oxfam to find out about:
(a) other areas in the world which suffer from the effects of drought;
(b) successful projects helping areas similar to the Sahel to survive.
(Your teacher may be able to give you some addresses or other information.)

THE SEARCH FOR FOOD

When human beings first evolved they lived in the same way as other animals, gathering plant foods and killing animals for meat. In the last 10 000 years we have learned to grow and harvest plants and to raise animals. In the early years farming was of the 'crop and fallow' variety. A crop was grown one year, and the next year the land was left alone ('fallow'). Different plants take different minerals from the soil: the 'rest period' allowed the soil to become fertile again.

Later, in the north of Europe, forests were cleared so that the land could be cultivated. First the larger trees were cut down, then the brush, and finally the whole area was burned. The land was planted with rye or oats, which were crops that gave a high yield for those times. After the harvest the land was allowed to go back to scrub (see succession, in Unit 8). This is described as 'slash and burn' farming and is still used today in parts of central Africa. Both of these methods of farming have the 'disadvantage' of a wasteful fallow period.

In efforts to increase food supplies, 'crop rotation' began during the Middle Ages (again in Europe). As different plants take different minerals from the soil, it was possible to grow crops each year in the same field – provided they were different ones. At first the land being cultivated was divided into three big fields, which were treated as follows.

Field 1 was planted with oats, beans or peas.
Field 2 was planted with wheat or rye.
Field 3 was left alone and ploughed three times a year.

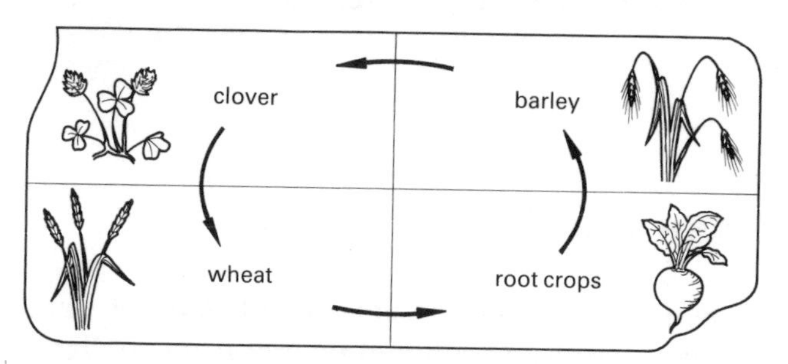

Fig. 10.1 Crop rotation in four fields

Gradually, it was discovered that using new crops meant the fallow year could be left out altogether. As you can see in Fig. 10.1, farmers began to work with four fields and four crops.

Clover and root crops, such as turnips, fed more animals as well as helping the soil fertility. The animals, in turn, produced more manure which helped to improve the crops. Crop rotation produced bigger yields and fed more people and animals.

Over the years stock farming has improved, especially in the last 100 years. This has meant that animals have improved, producing more offspring and greater yields of food in less time than before. Table 10.2 shows how dramatic the improvement has been.

Table 10.2 How crop yields have increased

1 hectare planted with wheat:	AD1200	Today
Yields	420–530 dm^3	3500–7000 dm^3 (in Europe)
Feeds	4–5 people for 1 year	20–50 people for 1 year (and provides seed for the next year)

A steady supply of good food

Over much of our history food supplies have been unpredictable and unreliable. As you have just seen, one of the main aims of agricultural science and technology has been to make supplies more reliable. We shall now see that this can be done by following two paths:

(a) working with genes to breed more productive domestic plants and animals;
(b) working with the environment to help the growth of plants and animals.

Take a field of corn . . .

A field of corn takes carbon dioxide from the air, nutrients from the soil and energy from sunlight. It releases oxygen into the atmosphere and waste organic material into the soil, whereas all we use are the edible parts that we feed to animals or eat ourselves. Humans are not the only species eating the corn. Caterpillars of moths feed on the cobs, leaves and stems. Birds and rodents attack the cobs. Wireworms and roundworms attack the crop underground. All of these animal species are in competition with us for the corn; so too are some species of bacteria and fungi. Weeds may grow up and compete with the corn plants for light, water and nutrients. The farmer has to protect his crop from all these 'dangers' and at the same time give it the best conditions for growth.

Today many crops, such as corn, are grown as a MONOCULTURE. This means that large fields are planted with the same plant species. By selective breeding we have sometimes improved the food value of plants, but in doing so the plants have become less resistant to pests and diseases. We have made life very easy for competitors by providing large areas of crops open to attack. As a result of farming in this way we have to use pesticides to kill weeds, insects and diseases. To produce better crops and more of them, we work on the soil to improve it, by ploughing and adding fertilisers (see Unit 1).

Farmers who raise animals for food suffer similar problems. Predators can kill livestock, so farmers work to reduce the numbers of these predators. However, a reduction in the population of these predators may then allow the populations of other plants or animals to grow and to cause problems. Look back to Unit 8 and the problems caused when the rabbit population in Australia was able to grow. If the livestock population is too heavy for the area, it can *overgraze* it by eating grass more quickly than it can regrow. This can allow the rapid growth of less useful plants (weeds) or can lead to the destruction of the area, as happened in the Sahel (where it was made worse by the failure of the rains).

Bigger and better yields

Food-producing *ecosystems* (see page 138) get most of their energy from the sun. Photosynthesis converts solar energy into chemical energy by 'binding' carbon dioxide and water in carbohydrates and other compounds. However, the ecosystems that produce the largest yields of food have important help from the fossil fuels. These are used in many ways:

making and applying fertilisers;
making and applying pesticides;
pumping water for livestock to drink and for irrigation;
driving machinery.

Mechanisation itself does not increase yields, but it does allow one person to manage a larger area. The machines and the energy that powers them help to provide a steady food supply. Irrigation pumps help food to be grown in dry lands and all year round in areas with short rainy seasons.

We still seek to improve the quantity and the reliability of our food supplies. However the search for 'more and better' causes some concern – and not only because of the rising cost of fossil fuels. The use of fertilisers and pesticides has proved harmful to people and animals in rural areas. If the chemicals can pass up a food chain, then they can also affect humans and wildlife far away from farms.

In addition to this, minor pests can become a great nuisance when their natural predators are killed by pesticides. Other species may become resistant to pesticides and controlling them chemically may fail.

Assignment – Desertification and intensive agriculture in the Sahel

The Sahel, in the West of Africa (see page 260) is a semi-desert which gets an average rainfall of 400 mm a year. This could fall as 800 mm in one year and none at all in the following year. The Sahel supports several species of grazing animals, mainly gazelles and antelopes. Deer and cattle prefer richer areas but nomadic farmers managed to raise cattle on the edge of the Sahel.

As the human population has increased and spread outwards from the towns, these farmers have been forced to move into the Sahel. They found that both humans and livestock could live there successfully in the wet spells, when rain fell regularly each year. Previously, they survived by staying in areas that were not more than a day's walk from a natural waterhole. This meant that the wild gazelles and antelopes could exist by keeping to the areas further away from the waterhole.

Improved technology enabled the farmers to drill bore holes and make more waterholes. This, in turn, upset the balance in this part of the biosphere. Cattle were able to graze everywhere and the wild animals lost grazing and water.

Encouraged by plentiful water and reasonable grazing the farmers allowed their herds to get bigger. These overgraze the land and loosen the soil by eating the sparse soil-binding vegetation. Each time the cycle of dry seasons returns, humans and animals suffer in greater numbers than before. The drought years (1969–73) affected 8 million km^2. It was estimated that more than half the cattle died and 150 000 people. The same cycle of events has happened again, with recent droughts causing even more land to become desert and impossible to use for domestic or wild animals. According to one forecast, the desert will have advanced another 40 miles southward by the end of the century.

1 Explain why the nomadic (wandering) farmers were forced to move into the Sahel.
2 One of the key changes that eventually led to the disasters of 1969–73 was the improvements in technology the farmers were able to use.
   (a) What did the technology enable the farmers to do?
   (b) Explain how this affected the populations of gazelles and antelopes.
3 When, in the years from 1969 to 1973, rains did not fall, animals and humans were affected.
   (a) Explain why this happened.
   (b) Why was the effect much worse than in previous dry spells?
4 Suggest why heavy grazing on the desert vegetation helped to loosen the soil.
5 Describe how the desert could 'advance 40 miles by the end of the century'.

Assignment – Another approach to the problem

A picture of life in a rural Chinese community

All manure, human excretion, crop debris and grasses are turned into compost or used in methane generators to produce 'biogas' for cooking and lighting. The sludge produced is a better compost than organic fertilisers. China's population is increasing by 15 million a year, so the Chinese also use inorganic fertilisers. These are made in ten new anhydrous ammonia and urea factories fuelled by natural gas or crude oil.

Plant protection is mainly carried out by ducklings: they are herded through the young rice plants to feed on insects and snails. Different varieties of rice are planted in alternate rows to stop the spread of disease. Insecticides were used against major pests, but the problems of poisonous DDT, and growing resistance to it, are causing increased use of alternative controls such as bacteria lethal to caterpillars.

Adapted from 'Energy Solution in China', by Vaclau Emil, in *Environment*, **19**(7) 27–31 (1977)

1 (a) Write down the chemical formula for methane.
  (b) Write an equation to show the reaction that happens when methane is burned.
2 What advantages would you expect that using ducklings to kill pests would have over using pesticides?
4 Explain why the rice crop is planted with different varieties in alternate rows.
5 Why do the Chinese still have to make inorganic fertilisers?
6 What are the 'problems of poisonous DDT'?
7 Why are the Chinese developing other methods of controlling pests?

Assignment – DDT

DDT was developed during the Second World War and proved a very effective insecticide. It has been very useful in increasing crop yields around the world. Unfortunately it has three serious disadvantages.

1. DDT is very stable, and so can spread over a wide area without breaking down. As a result there is nowhere on earth that DDT cannot be detected.
2. DDT is more soluble in fat than in water. Therefore it builds up in the fat stores of animals, and so can pass up the food chains as one animal is eaten by another. As this happens, the level of DDT in each organism is increased and it eventually reaches lethal proportions.
3. Pests become resistant to poisons such as DDT, and develop new and possibly more damaging strains.

More farming can mean less food

In the assignment on page 263 you learned how over-grazing can help valuable soil to 'disappear'. Sometimes the effects can be dramatic, as the photograph (Fig. 10.2) shows.

Figure 10.3 summarises the sequence of events which will result in the loss of top soil. As you can see, there are some other unwanted side effects such as the increased risk of flooding.

Fig. 10.2 Soil erosion in Tunisia

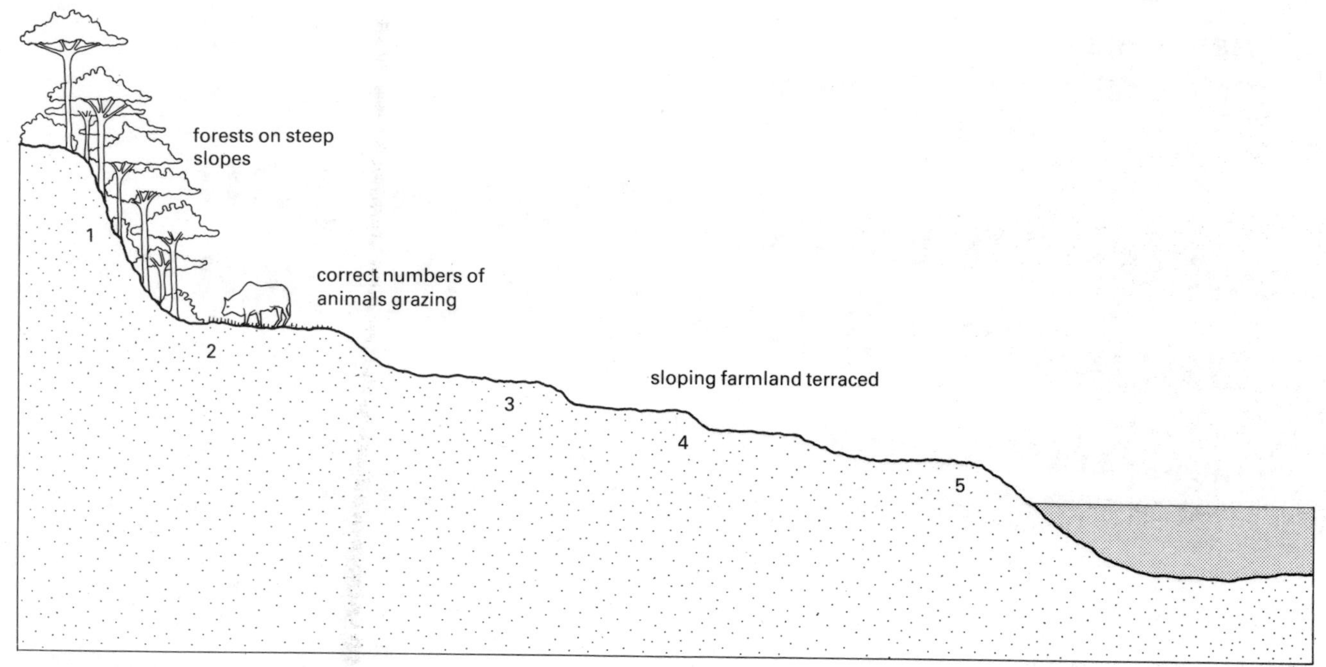

What happens

1 Trees collect rain, prevent soil being washed away
2 Soil held by grass roots
3 No 'gulley' erosion
4 Soil structure maintained
5 Streams and rivers clear

WHAT HUMANS DO

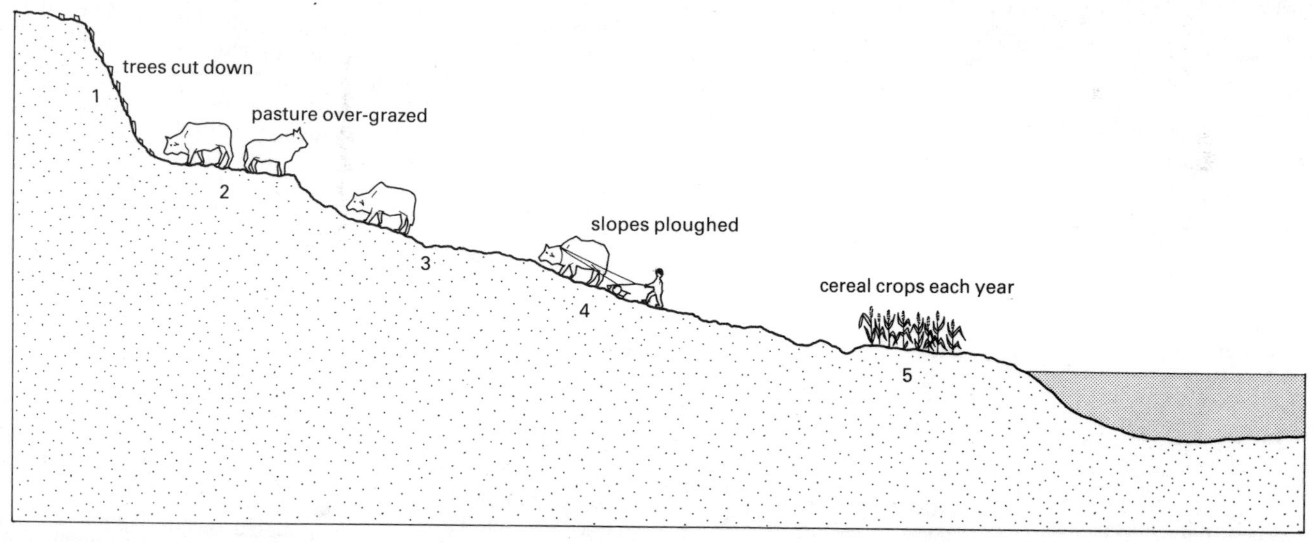

What happens

1 Thin soil washed away by rain
2 Little grass left; soil washed away
3 Water flows in furrows and causes erosion
4 Soil structure breaks down; soil blows away or is washed away
5 Streams carry away top soil; lakes and rivers choked with silt; floods likely

Fig. 10.3 The causes of erosion

Assignment –
Soil erosion in the Himalayas

Erosion sweeps Himalayan mountains down to the sea

from GEOFFREY LEAN in Lukla, near Mt Everest

THE Himalayas are disappearing into the sea. Tons of soil are being stripped from every acre of the mountains by each monsoon, destroying the land on which Nepal's people depend.

The soil is swept down the Ganges river system and is forming a new island in the Bay of Bengal.

This disaster has developed with extraordinary swiftness. In the three decades since Everest was conquered half of the forests that once coated Nepal's mountains have been cut down and the rest are expected to disappear by the end of the century.

The trees bound the soil to the mountain slopes and enabled it to soak up the fierce monsoon rains, releasing them gradually to enrich the whole region.

Now the rains are a curse, not a blessing. Each year they wash away more than 12 tons of soil from every acre of the bare hillside. In the worst areas, 80 tons of soil are ripped from each acre.

Villagers watch all night for landslides during the monsoons. Little wonder – 20 000 landslides have been recorded in a single day, sweeping away the terraces of fields carved laboriously into the mountainside and burying whole villages and their inhabitants.

As the soil goes, crops fail. Rice yields have dropped by a fifth in just five years in the hills, maize yields by a third. In many areas the once-tall maize now grows no higher than millet.

Already people are hungry. The average hill farmer can grow only enough food to feed his family for eight months of the year, and 60 per cent of the children are stunted by famine.

Pemba Sherpa, a leading member of the tribe made famous by mountaineering, told me in this village 9000 ft up the slopes of Everest: 'I am afraid that one day the land around here will be desert.' Sir Edmund Hillary calls one valley higher on the mountainside 'an ecological slum.'

Two hundred miles west, in the shadow of the four peaks of Annapurna, it is the same story. Ram Bahadur, a 40-year-old farmer in the village of Kasikot, told me that crop yields were half what they used to be, milk production was down by a third as land was over-grazed, three-quarters of the farmers had lost part of their fields, and a third of the people had emigrated to follow the soil into the plains.

But the plains are dangerous, too. The mountain soil coats the riverbeds, raising them six to 12 inches a year. Every year the rains pour more forcefully off the bare hillside, into shallower channels. Inevitably the rivers flood and the land on their banks is swept away.

The erosion of the Himalayas has greatly increased flooding in the Ganges basin, one of the world's most densely populated and important food-growing areas. In all, 300 million people are affected by what goes on in the mountains.

Some erosion is inevitable. The Himlayas are the youngest, as well as the highest and longest, mountain range on earth. They are still inching upward from the impact of India with the rest of Asia, and are naturally being weathered away.

Geological process are being vastly aggravated by man's struggle to survive. Fifteen million people are trying to wrest a living from the two-thirds of the country that can be cultivated, a population density comparable to that of Bangladesh.

The population is growing rapidly, and as it does so more and more farmers try to carve fields out of steeper and steeper slopes, exposing them to erosion, and more and more trees are cut down for the firewood that supplies 87 per cent of the country's energy.

There are as many cattle as people and they graze the land nine times more intensively than it can take, tearing up the grass, crippling trees, and eating seedlings before they can grow.

Despite growing population, the hill people managed to conserve their trees and soils until 25 years ago, voluntarily restricting felling and grazing in their forests, which had traditionally belonged to them. Then the Nepalese Government nationalised the forests. The villagers responded by indiscriminate felling and the cycle of destruction began.

Pioneering projects by the UN agencies have shown that land can be saved if it is left alone for a year or two. Trees and grasses grow rapidly, stabilising the soil and cutting erosion by 90 per cent. If the grass is then cut, it feeds four times as many cattle as could live by grazing the area. Planting fruit trees saves the soil and provides the farmers with produce.

Farmers, at first suspicious of the projects, are now anxious to give them land to rehabilitate. Aid agencies are ready to provide money to help Nepal tackle its crisis, and the top levels of its Government recognise the problem.

The country's slow-moving and labyrinthine bureaucracy has been unable so far to come up with a co-ordinated programme to save the soil.

Observer, 29/8/8

▶

▶

Read the newspaper report on soil erosion in the Himalayas, then answer these questions.

1 Explain why soil has been stripped from the Himalayas.
2 What causes the soil to tumble down the mountains in landslides?
3 Why have rice and maize crops failed?
4 What are the effects of such erosion on the health of the hill farmers?
5 How is the erosion of the mountain soils affecting life on the plains below?
6 How is the increasing population in the area affecting the soil erosion on the mountains?
7 What change happened about 25 years before the article was written that began 'the cycle of destruction'?
8 How are UN agencies attempting to halt the erosion?

More land for cultivation

Apart from improving the yield of food on available land, we can increase food supplies by making more land suitable for crops. We can 'create' more land for growing crops by irrigating dry areas, chopping down forests (deforestation) or by reclaiming land that is waterlogged. It has been said that any land without green cover is wasting the sun's energy. You should be able to see why by now! However, each of these human activities can upset the balance in the biosphere. Let us examine each in turn.

Irrigation problems
Water can be brought to dry land by digging irrigation canals to carry it. If the canals are not properly lined, water will leak out and 'swamp' the crops, killing them. Minerals dissolved in the water are deposited around the plant roots and if there is not enough water to flush them away, the soil will not allow crops to grow. This deposition of salts is called *salination*.

Irrigation canals can also be a source of disease. For example, mosquito larvae can live in the water, growing into malaria-carrying adults. Humans who come into contact with irrigation water can pick up blood flukes – organisms that cause a weakening (debilitating) disease called bilharzia, also known as schistosomiasis. This disease affects over 200 million people around the world today.

Fig. 10.4 Irrigation canal in Ghana

Deforestation problems
Tropical rain forests occur in a band around the earth, as you can see in Fig. 10.5. Each forest has different species of animal and plant life and, between them, they provide a home for about half of the species of animals and plants on earth.

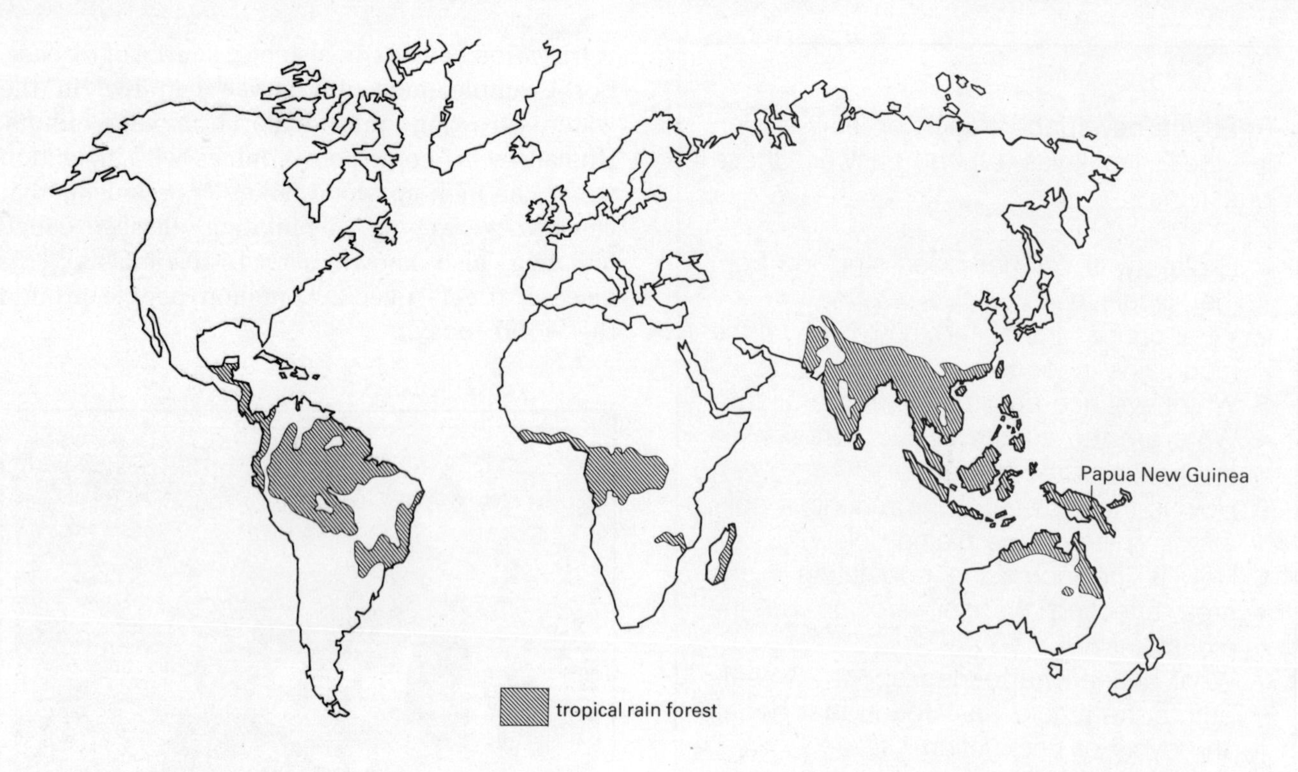

Fig. 10.5 Tropical rain forests of the world

They play a major part in maintaining the balance of oxygen in the atmosphere and are a source of many useful materials, such as bananas, avocados, cocoa, rubber, timber, and also chemicals such as derris, strychnine, quinine and curare.

The combined need for land to cultivate and for timber has increased the destruction of the rain forests. As well as providing land to cultivate, clearing the forests also enables less developed countries to earn money from the timber. However, they do not get a high price and so must go for 'cheap extraction'. This means they fell trees on a large scale, which is very destructive.

As the trees grew, many of the minerals in the soil were 'trapped' in the timber. So, exporting the trees exports the soil nutrients, leaving poor soil behind. In extreme cases constant rain changes this into a solid mass called *laterite* which will not allow plant growth.

Some organisations, such as the World Wildlife Fund, are helping to manage the rain forests in order to preserve them. One way is to establish areas of forest as 'reserves' permanently. Education also plays an important part, for local people may see forest as waste land and, especially if they are short of food, they may want to cut it down.

Fig. 10.6

Assignment – The real cost of paper?

Every month 20 000 tonnes of woodchip is shipped out of Papua, New Guinea. A huge area inland is 'clear-felled', which means that every tree, regardless of its size or value, is bulldozed to the ground. The logs are minced into woodchip, which is later converted into paper in Japan. This process earns valuable currency for the Papua New Guinea government.

The project is making a tiny contribution to the speedy disappearance of the world's forests. It is estimated that an area of forest bigger than Belgium, the Netherlands and Denmark put together disappears from the world every year.

'Clear-felling' seems like a nightmare. It reduces forest to a bare landscape. But nature in the tropics works quickly. Provided the soil is not washed away, forests will begin to grow again. However, there is debate about whether this happens fast enough to make up for the 'clear-felling'. 'Clear-felling' must have a dramatic impact on nature and humans alike. But the world needs woodchip, and some of the trees in the rain forest are useless for anything else. Further similar schemes seem unavoidable. The real question is whether 'clear-felling' is too high a price to pay.

Adapted from an article by Robert Lamb in *Earthwatch* No. 1, 1980

1 Make a list of the advantages of 'clear-felling' rain forests.
2 Make a list of the disadvantages of felling trees in the rain forests.
3 Why are the countries containing rain forests likely to be tempted to fell the trees?
4 The woodchip in the article was to be used to make paper. Find out how paper is made.
5 Compose an answer to the writer's final question: 'Is "clear-felling" too high a price to pay for paper?'

Reclamation problems

Humans have been draining wetlands for thousands of years. In recent times more and more land is being 'dried' to make crop land. The draining of 400 000 hectares of swamp in Sumatra is just one example of land reclaimed to grow rice. Big projects like this, and smaller ones like the draining of the East Anglian marshes in England, need skill and knowledge of their ecosystems if they are to be successful. Reclaimed wetlands, like the Thailand rice paddies in Fig. 10.7, can become stable ecosystems. However, they can be at risk from floods and storms and need dykes to protect them. Chemicals used on the crops on such land can kill fish and cause other damage. Organisms living there may disappear. Northern Europe lost bird species like pelicans, egrets and the ibis when its wetlands were drained.

Fig. 10.7 Rice paddies in Thailand

Despite all these problems, the increasing human population and need for food means that we shall still have to reclaim land, cut down forests and strive for bigger yields. However we shall need to do so in a way that will least upset the balance in the biosphere. In the next section we shall examine what can be done without making matters worse.

Improved harvesting and extraction

Some methods of harvesting crops and getting food from them can be very wasteful. Extracting the oil from groundnuts or soya beans, for example, wastes the food value of the residue which is fed to animals. About 90% of the energy in plants is lost to us when they are fed to animals and become animal tissue. At present the demand is for oil, so it is uneconomical to develop ways of preventing the 'loss' of the energy value in the waste. In the future our need for food may encourage us to find better ways of using *all* of these valuable plants. It may also force us to extract protein from poisonous or unpleasant plants.

The most efficient source of energy for us would be to eat plants directly. A great deal of work has gone into developing protein-rich plant food that is acceptable to humans. One such plant food is TVP which is made from soya beans and textured to resemble meat.

Fig. 10.8 Soya beans can replace meat

Other sources of protein

Fish farming

Japan, Britain and the USA are the main fish-catching nations. The sea has been a vast hunting ground for food. Over the years the size of catches has grown as technology has improved. Developed nations are now so good at catching fish that many people are concerned about the balance in this part of the biosphere. If the numbers of fish taken from the sea are so great that only a few are left to breed, the situation could be serious. As time passes the fish caught will get smaller in size, but the total catch may weigh the same – or even increase. This means that fewer young fish are living to be adults, and so the population is at risk.

There are two ways that humans can feed on fish and yet 'preserve' their populations.

(a) We can eat different species.

(b) We can make use of fish farming techniques. Fish farmers can rear young fish and release them when they are 'old enough'. They can also 'grow' fish (from eggs or by capturing youngsters) and sell them for food when they are large enough. In this way fish farming can be used to feed us directly, or to replenish stocks in the seas.

Fig. 10.9 Fish are 'farmed' to provide food

'Farming microbes'

Single-celled micro-organisms, such as yeast, are being harnessed to produce protein on a large scale. The yeasts are rich in protein and grow so fast that it is estimated that half a tonne of yeast can make 50 tonnes of protein in the time that it takes a ½ tonne bullock to make 500 g. However, at present such single-cell protein is not suitable for humans; but it can make animal feedstuffs.

Assignment – Food from a single-celled plant

A plant, *Spirulina maxima*, is being investigated to see if it could be used as a suitable crop to grow on a large scale and be processed into a concentrated animal food. The following features of the plant are now known:

(i) it grows in parts of South America and Africa, floating on the surface of ponds, supported by gas-filled spaces in its cells,

(ii) it carries out photosynthesis very efficiently, producing starch which it uses to make protein with an excellent balance of amino-acids. Table A shows the yields of traditional crops and of cultures of *Spirulina*,

(iii) the plant cell walls are easily digested by all animals, including humans,

(iv) when fed to animals it results in a very high yield of protein. Table B shows the yield of protein from animals fed on traditional crops and on cultures of *Spirulina*.

Table A Annual yield/tonne per hectare

Crop	Dry mass	Protein
Wheat	4	0.5
Maize	7	1.0
Soya bean	6	2.4
Spirulina	50	35.0

Table B Annual animal protein yield/tonne per hectare

Crop	
Wheat	0.028
Legumes	0.035
Pasture	0.08
Spirulina	2.20

1 Give a brief description of the process of photosynthesis.

2 Suggest how the gas-filled spaces might help *Spirulina* carry out photosynthesis effectively.

3 Name *one* gas that *Spirulina* could produce. Explain briefly the process of production of this gas in the plant and describe a test to identify the gas. State the results you would obtain.

4 Which *four* groups of substances, other than protein, would have to be present in reasonable amounts in the processed plant to make it suitable as a balanced diet for animals?

5 Explain the significance of the expression 'having an excellent balance of amino-acids'.

6 (a) Of what material are plant cell walls usually made?

(b) Are the cell walls of *Spirulina* made of this material? Explain your answer.

7 (a) Give *three* advantages of *Spirulina* compared with traditional crops as a source of food for animals.

(b) Outline any further *investigations* and *developments* that a food-processing company might carry out before starting large-scale production of *Spirulina* for animal food.

(c) It has been suggested that the plant should be eaten directly by humans, perhaps in the form of artificial meat, instead of feeding it to animals.

Write an account giving arguments for and against this point of view, paying attention to social, technological and economic considerations.

AEB, SCISP 'O' level, 1983

POLLUTION

A substance is called a POLLUTANT when its presence harms organisms. Here are some examples:

(a) Carbon dioxide is essential for plants and usually harmless to animals. However, *too much* can kill animals.
(b) Nitrates are needed for plants to grow, but too great a concentration in lakes can cause over rapid growth of algae.
(c) Seeds can be treated with chemicals to prevent fungus, but these chemicals can pass through the food chain when birds eat seeds from the plants.

Some pollutants can become harmless after a while. Sewage in the sea, for example, is eventually broken down by bacteria. Other pollutants, like lead and mercury, stay harmful for a long time.

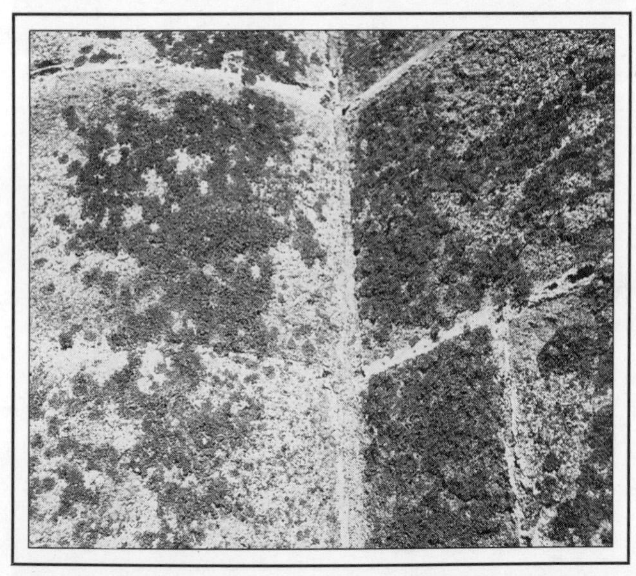

Fig. 10.10 Lichens growing on a wall show that the air is clean

Some organisms can tell us when pollutants are about. Lichen will grow only where the air is clean. Roses are free from the fungus 'black spot' only when there *is* a pollutant (sulphur dioxide) in the air. Birds, too, can act as an early-warning system. In the Netherlands, species of falcons and golden eagles began to disappear as pesticides caused them to lay thin-shelled eggs which gave no protection to the young. In Britain, the first sign that PCB (polychlorinated biphenyl pesticides) wastes were accumulating as poisons in food chains came when they were found in the eggs of birds.

Polluting the air

Human activities are usually the cause of air pollution. The air in cities and industrial areas contains gases which are not normally in the atmosphere. It may also contain a greater concentration of dust particles, which makes the atmosphere thicker and lowers visibility. The main source of air pollution is the large-scale burning of fossil fuels, producing carbon dioxide and sulphur dioxide. Sulphur dioxide is the most harmful of these, as it dissolves in water to make an acidic solution that can kill plants and damage buildings (see page 274).

In the high temperatures in furnaces and engines, oxides of nitrogen can also be formed. These are acidic gases, too, but can react in sunlight to form clouds. They are the *photochemical smogs* that damage plants and cause eye irritation.

Fig. 10.11 The city of Sydney in Australia, buried in thick smog

Unburned hydrocarbons and other particles also get into the atmosphere as soot and ash. These land on buildings and plants, or mix with water droplets to make fogs and smogs.

Motor cars are a major source of air pollution. As well as oxides of nitrogen and unburnt hydrocarbons, exhaust fumes contain carbon monoxide, lead compounds and other poisons. Carbon monoxide is a poisonous gas that can reach dangerous concentrations in cities. Once lead compounds have prevented 'knocking' (see Unit 6, page 25), they are released into the air. Lead is one of the harmful substances which are dangerous because they build up in food chains and have a long term effect.

A small but growing source of air pollution is the radioactive particles released during tests of nuclear weapons and other, more peaceful, uses of nuclear energy. Many new organic chemicals released into the air are suspected of upsetting the balance. Freons, for example, are chemicals used in fridges and aerosols. It is suspected that they may interfere with the layer of ozone gas at the top of the atmosphere. This could allow dangerous concentrations of ultraviolet light to reach us, damaging skin and other body tissue. People are also concerned that the increased amount of carbon dioxide in the atmosphere may be upsetting the biosphere's heat balance. By concentrating the heat energy in the atmosphere, as glass 'concentrates' heat energy in a greenhouse, the increased carbon dioxide could cause temperatures on earth to rise – perhaps even melting the ice at the Poles!

Polluting the water

Water goes through a cycle of evaporation, cloud-forming, rainfall, collecting (in rivers and lakes) and evaporating once again. As it does, it can leave behind many of the impurities it has picked up. However, the rate at which human and animal populations are growing makes it harder and harder for the natural water cycle to keep water pure.

Pollution in rivers and lakes

Water-living organisms need oxygen, dissolved in water, to survive. This dissolved oxygen is not affected by small amounts of organic pollution. Large amounts will reduce the oxygen concentration so much that the water organisms die. This can happen when the water is used to carry sewage. If a small amount of sewage is put on the land, soil organisms break it down. The nutrients return to the soil and nearly pure water filters through to rivers and streams. When sewage goes into water directly, it will break down there. As it does, it uses up the dissolved oxygen. In extreme cases the oxygen level may drop so low that all the living organisms die – except for bacteria that can respire anaerobically (see Unit 9, page 220). These thrive without oxygen, producing gases like hydrogen sulphide (which is also produced as eggs go bad).

The same drop in oxygen level can happen in water supplies rich in substances that encourage algae to grow. Where excess fertilisers (nitrates and phosphates) are washed off fields, or detergents are emptied into the water, algae plants grow rapidly. As they grow, they use up the oxygen in the water. The other organisms will be deprived of oxygen and die. Eventually the algae population may be so great that there is too little oxygen to support it. The algae die and, as they decompose, even more oxygen is used up.

Oxygen is less soluble in hot water, so the oxygen level will drop if the water temperature rises. The electricity generating industry, especially, uses large amounts of water for cooling. The heated water is released into the waterways and upsets their balance. Some species die, but others thrive as they prefer the warmth.

In recent years the ecosystems in rivers and lakes have been polluted with other substances. Most of these are pesticides, washed off fields, or waste from industrial processes. With many of these pollutants, the danger lies in the way they get into food chains, as you have seen (page 264).

Pollution at sea

Human activity pollutes the oceans of the world in many ways. Ships carry waste out to sea and dump it. Passenger ships throw out their waste as they go. Rivers bring down effluents and silt and dump them into the sea. Chemicals polluting the air fall into the sea as rain.

Some seas, like the Mediterranean and the Baltic, are almost 'closed systems' (look at them in an atlas, or in the world map in Fig. 10.5). In the Mediterranean water movement is so slow that it takes over 100 years to be replaced. As a result, pollution has built up:

pollutants from poisoned rivers collect in deep water, killing life;
tanker waste puts a coating of oil on the surface which stops light reaching plankton, so that they die;
the Aswan Dam, in Eygypt, has reduced the amount of fresh water reaching the Mediterranean;
the Suez Canal allows in more salt water from the Red Sea.

The Mediterranean is already showing signs of serious damage to its balance of life. In some parts of it, people are at risk if they swim in it.

Assignment – Acid rain

Acidic rain has been damaging buildings, forests and streams in Scandinavia. Trees are dying, fish and other organisms, including people, are suffering.

The rain is made acidic when sulphur dioxide gets into the air and dissolves in the water vapour. It is thought that the large amounts of sulphur dioxide causing the damage come from the industrial areas of Northern Europe. Winds blow the gas, then the rain, north-westwards. Recently the trail of damage has spread to the north of England and parts of Scotland. By 1983 news of the damage was beginning to appear in newspapers. Read the newspaper article on the opposite page, then answer the questions.

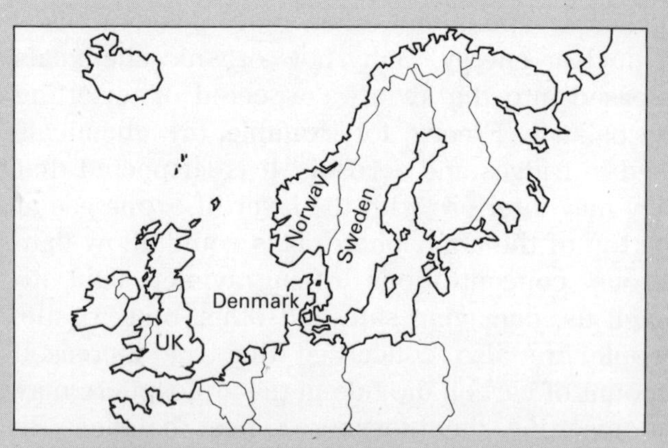

1 Write an equation to describe what happens when sulphur dioxide dissolves in water.

2 (a) Write an equation to show what will happen when acid rain falls on buildings made of limestone (calcium carbonate).

(b) Why will this reaction help to break down the structure of the buildings?

3 When the pH of lakes and rivers drops below 6.5, fish begin to die. The only creatures to survive are insensitive insects and some types of plankton. As the pH drops further, white moss will grow more readily.

(a) What will cause the pH of the water to drop?

(b) What will be the long term effect, in a lake, if the pH continues to drop?

4 How should the dangers of acid rain be prevented?

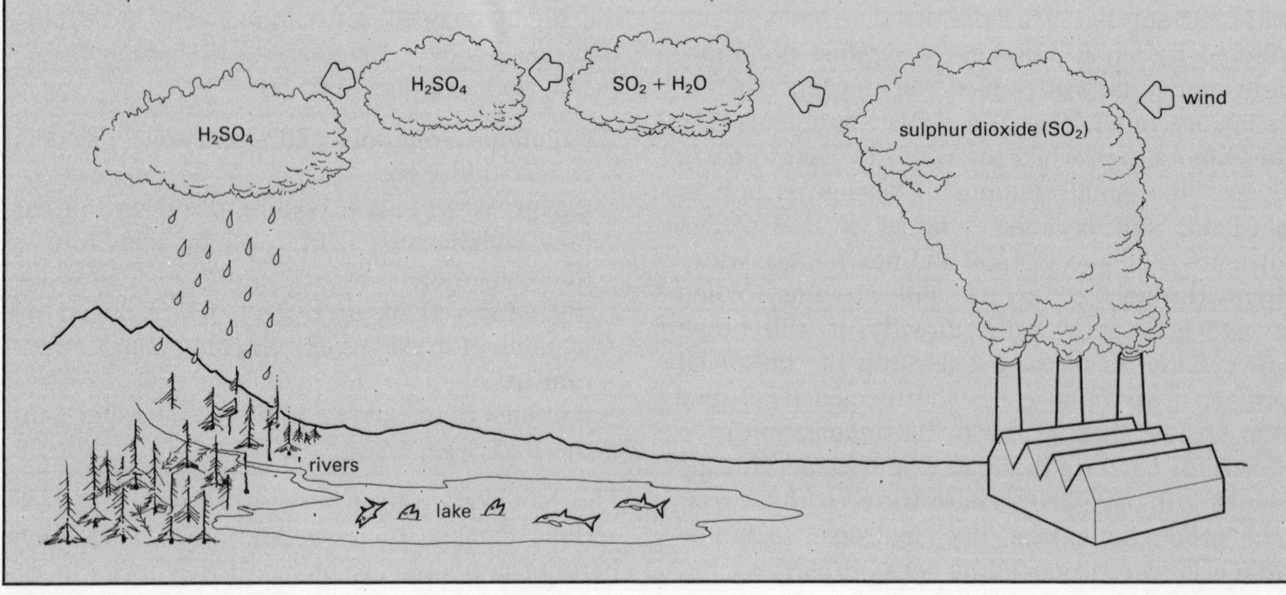

Fish stocks 'hit by acid rain'

ACID rain has killed all the fish in 26 lochs on 200 square miles of southern Scotland, according to local fish farmers, and 72 other lochs are rapidly losing their fish.

The owner of one East Galloway hatchery said yesterday that he had lost virtually his whole stock and was going out of business. Another complained that the loss of 20 000 rainbow trout had cost him £10 000 during the year. The raising of rainbow trout has become a substantial Scottish industry.

'We know that fish and virtually all aquatic life have vanished from Loch Grennock, Loch Fleet, Loch Enoch and Loch Doon and there is very little left in Loch Katrine in the Trossachs,' said a fish farmer in Newton Stewart, Galloway.

The only assessment of the situation had come from the Norwegians. He said: 'They invited some of us to Norway so that we could see for ourselves how acid rain had denuded their rivers – some of the finest game fishing rivers in the world – of fish.'

The fish farmer added: 'After surveying Scottish lochs, Norwegian scientists prepared a detailed report of their fish content. The Norwegians blame our industrial chimneys for what has happened to their rivers and have been making tests all over the Northern hemisphere.

'But our Government will go no further than to say that, even though so many lochs and burns are now fishless, the experts from the Pitlochry Fresh Water Research Laboratory have not yet been able to find proof that acid rain is the responsible pollutant.'

A spokesman for the Scottish Office said yesterday: 'The Department heard of one substantial loss of fish on a farm in Galloway because of the quality of the water. It could have been caused by the melt of early snow. It was difficult to determine the extent of acid rain.'

Research continues at the Scottish Office laboratory at Pitlochry but it is described as a long-term industrial problem. Recent tests have shown increased acidity in lochs, rivers and burns.

By AUSTIN HATTON
Nature Correspondent

Dr David Fowler of the Institute of Terrestrial Ecology in Edinburgh, said there was reason for concern. It was found after research in Europe and Scandinavia, as well as Scotland, that the acid rain affecting fish particularly in Scandanavia, was not chemically the same as that which is killing thousands of acres of trees in Germany, Poland, and Czechoslovakia.

A Pitlochry laboratory spokesman said: 'There are now higher degrees of acid with additional sulphur in the water. Where the increase has reached over 50 per cent, fish are unable to survive.'

Urgent international efforts are being made to minimise the effects of acid rain. A government spokesman said that Britain's contribution was the setting up recently of a five-year research project by the Royal Society to study 'the causes of acidification of surface waters in affected areas of Norway and Sweden'.

The project was being financed with £5 million from the Central Electricity Generating Board and the National Coal Board.

Officials added that Scotland's problem would be among the subjects discussed when Mr William Waldegrave, Under Secretary of State in the Department of the Environment, meets Norwegian experts in Oslo this week to learn more about the five-year research programme in which Canadians and Norwegians are jointly to study atmospheric pollution.

Dr Gwyneth Howells, of the Central Electricity Research Laboratory at Leatherhead, has already been studying the effects of acid rain in Scotland.

She explained that fish farmers in southern Scotland were losing their stocks partly because imported rainbow trout, large quick growing fish, are more sensitive than our native brown trout and cannot survive so well in acid water

Sunday Telegraph, 30 Oct. 1983

Toxic waste Heavy metals like lead, cadmium and mercury remain poisonous (*toxic*) permanently. Some sea organisms can make them even more toxic as they pass these elements up the food chains. In Minamata Bay, Japan, mercury waste was discharged during the 1950s and 1960s. It was thought to be in small enough quantities to be safe. However, it was converted to methyl mercury by organisms in the water. This very toxic substance built up in fish and shellfish. As the Japanese eat fish as a major part of their diet, many people were harmed at the end of the food chain.

Assignment – Polluting the sea

Whiteners should be blacked

Said one of the last of the Humber's inshore fishermen, 'That factory has killed almost everything in the river. Until that plant started, the Humber was an ideal breeding ground for all types of fish.'

The factory, opened in 1948, makes titanium dioxide. This is used as a whitener in paints, plastics and toothpaste. During production the plant discharges up to the legal limit of 26 000 tons of diluted waste every day. This passes into the Humber through a pipeline, just upstream from Grimsby. On a sunny day the bright orange waste can be seen flowing downstream to the sea. It contains:

diluted sulphuric acid;
iron sulphate;
small quantities of heavy metals such as arsenic, cadmium and mercury.

Opposers to the dumping say that the acid attacks sea organisms like plankton and fish and that the iron sulphate turns into flakes of iron oxide which carpet the sea bed and stifle the growth of plants, worms and shellfish.

The representatives of the factory say that the purity of the water is carefully monitored and there is no harmful pollution. The local Water Authority agrees with this view.

Adapted from the *Sunday Times*, 21 August 1983

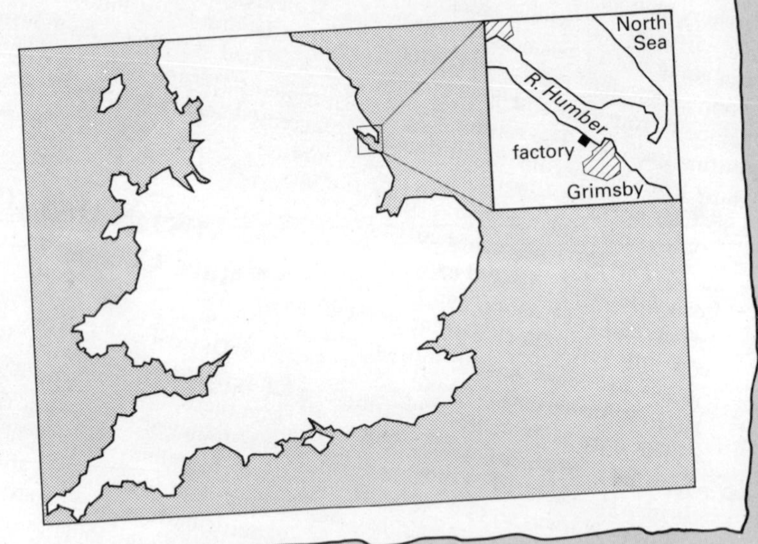

1. What are the harmful effects of each of the substances in the orange waste from the factory?
2. Draw a food web to help you explain the effect of killing plankton in the sea.
3. What effect would you expect there to be on the bird population in the area?
4. Do you feel that the factory representatives are correct to seem confident that all is well?
5. (a) Write a letter to the factory managers. Explain to them why people are upset, and what *you* feel ought to be done.
   (b) Write a letter to the local paper in Grimsby, expressing your view. Outline what you suggest could be done to make local residents happier. At the same time, is it possible to satisfy the people who buy toothpaste, plastics and paint?

Oil pollution Each year millions of tonnes of oil spill into the seas. As oil tankers get bigger, so does the amount of pollution they cause as they rinse out their tanks, or break up after accidents.

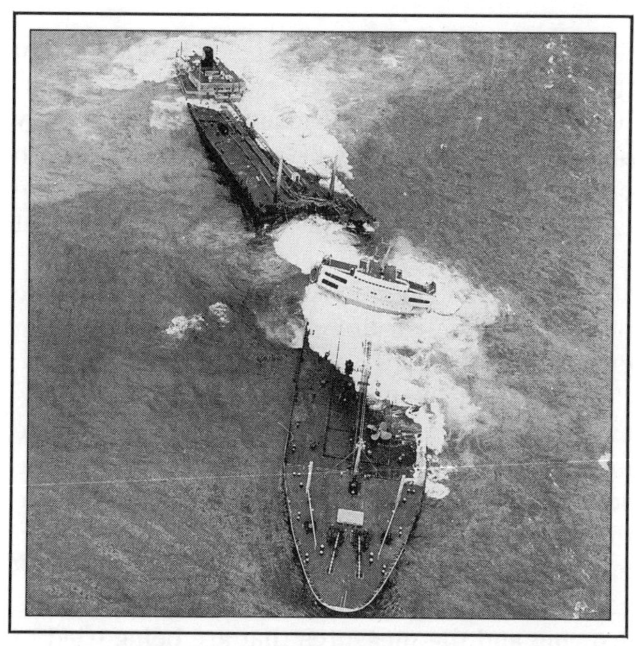

Fig. 10.12 The first serious oil spillage disaster occurred when the tanker Torrey Canyon *went aground in 1967, covering the sea around it with oil*

Tankers are not the only source of oil pollution. One of the worst oil spills came from an oil drilling platform off the coast of California. However, the largest source of oil at sea comes from the land. Waste oil from industries and motor vehicles is dumped into rivers, which then take it to the sea. Oil covers the sea-surface with thick sludge. This kills many animals, especially those birds that feed from the surface.

Oil slicks can be cleared by detergents, but these can often cause more harm than good. Because oil is organic, it can be broken down by sea organisms. After about a year bird colonies can be back to normal. Detergents stop the balance being restored so quickly, because they kill organisms. Chalk can be used to sink the oil to the sea bed where it will be decomposed by bacteria. Various methods of skimming the oil off the surface have also been tried.

Balancing the Biosphere

Fig. 10.13 *Health education can make a big difference. These health visitors in an Indian village are giving advice on the care of a new-born baby*

WHAT DOES THE FUTURE HOLD?

So far in this unit, we have considered some of the events that can cause large-scale disturbances. Often these are caused by human activity and at the time the results could not have been predicted. Can we use our understanding of the delicate balance that exists to help us prevent further upsets?

In this final section we shall look at some present-day human activities which could bring dramatic changes to our world. As time goes by, the increasing human population on earth puts ever greater demands on resources. More people means more food, more fuel, more goods. Resources are limited, so any changes which use them up even faster will certainly upset the balance in the biosphere somewhere. In Unit 8 you learned about how the human population is growing and the measures that are being tried to control this growth. It may be helpful to remind yourself of those before going on.

ADVANCING MEDICINE

Medical knowledge and skills have improved steadily. In the developed world, epidemics of infectious diseases are very rare. Doctors are able to speed recovery from most illnesses. However, treatments are usually helping us to get better more quickly. Many diseases are still not curable, and there is yet much to be done in preventing us becoming ill in the first place!

Where living conditions and diet have 'advanced' more people are suffering from conditions such as heart trouble, strokes, arthritis and mental illness. The real reason for this is that more people live until old age, when these ailments normally occur. The link between healthiness and such things as proper diet and improved living conditions is strong. As a result a great deal of effort is going into helping to speed improvements in developing countries. As you will see in the next assignment, sometimes a simple action can bring tremendous results.

Assignment – The fight to save 10 million lives

In 1983 the director of UNICEF put forward a plan which could save 10 million lives a year. It included two straightforward actions:

(a) immunising all children;

(b) providing UNICEF salt packets around the world. At the time 40 000 children were dying *every day*. They died from lack of food (malnutrition) or diseases such as measles, diptheria, tetanus, whooping cough and tuberculosis. Many more were crippled by polio. Underfed children are more likely to get diseases.

UNICEF found that a high birth rate goes together with a high incidence of such diseases. As people become more confident that their children will live to be adults, the birth rate drops. It is now possible to control, if not eliminate, serious children's diseases. New vaccine technology makes it easier to do. It is possible to combine doses of different vaccines, which themselves need fewer 'shots'. Therefore, children in developing countries can be protected against the serious diseases with just two injections, about 6 months apart. What really made the task possible was improved technology. It became possible to produce the vaccines on a large scale efficiently and cheaply. The vaccine could be produced in large tanks containing cells growing on microscopic beads.

The UNICEF salt packets were to tackle the other cause of 5 million deaths a year among children in developing nations. They are small packets, containing a mixture of glucose, sodium chloride, sodium hydrogencarbonate and potassium chloride, costing just a few pence. 500 million children suffer from acute diarrhoea every year. This makes them dehydrated and they will die if they are not treated. The salt packets have proved an effective treatment for dehydration and could save the 5 million who die.

UNICEF say that the lives of tens of thousands of children a day can be saved, *if the world wants it.* To answer the question, 'Of course, we must save these children's lives; but won't it simply make matters worse? More mouths to feed, more jobs, more education', UNICEF points out that, in country after country, a decline in the death rate has been accompanied by an even greater drop in the birth rate.

Adapted from the *Sunday Times*, 6th and 20th Feb. 1983

1 Explain why so many children are dying each year.
2 Suggest three reasons why most of these deaths are in the developing countries.
3 Why has vaccine become cheaper?
4 Explain why the immunisation scheme described is more effective than previously.
5 (a) What is the pattern linking birth rate and childhood illness?
  (b) Suggest why the birth rate goes down as the treatments take effect.
6 What do *you* feel? Do you agree with UNICEF or the questioner? Write a short account of your point of view. Use the information in the text, and any other source, to illustrate your ideas.

At the other end of the scale, many researchers have been trying to find out why cancers happen. Perhaps then we shall be able to conquer one of the world's most feared diseases.

Assignment –

How they cracked the cancer code

The breakthrough was simple and dramatic. Peter Stockwell found a close likeness between a virus that causes cancer in monkeys and PDGF. PDGF is a substance found in human blood. It is produced in wounds by clotting blood, and makes the cells around the cut grow again. Dr Mike Waterfield, leader of the team of scientists, set out to find out how PDGF made normal cells grow. He was joined in the search by other researchers. They discovered that PDGF was a very unusual protein. Most of the human body is made from proteins – but PDGF was different. It could be boiled in strong acid for 10 minutes without changing. Once they worked out the structure of PDGF, the next problem was to find out if it was related to any other protein which had ever been analysed. It turned out to be very like a gene of the monkey cancer virus. How could this be?

The cycle which forms a tumour

Normal body cells contain a complete genetic 'masterplan' for every cell in the body. They contain the gene for PDGF, which is normally switched off. However it can be switched on by chance mutation, so that it works when it should not. Radiation or chemicals such as those in cigarette smoke can cause such mutations.

It is now possible to explain how viruses can cause cancer. Viruses infect body cells, and usually kill the cells as they grow. But, sometimes a virus will stay in the cell until something disturbs it. It may then break away taking a 'rogue' gene like the one for PDGF, with it.

Rogue genes such as the one for PDGF have been given the name *oncogenes*. At least 15 different oncogenes have been identified. It is believed that they must serve some very important function, otherwise evolution and natural selection would have 'lost' them. Some of them probably have vital roles to play as an embryo develops, but are switched off in a normal adult. When the gene is switched on by accident, it bursts into action. Unless it is destroyed by the body's defences a tumour will develop.

Adapted from *Sunday Times*, July 10 1983

1 Why does a normal body need the PDGF gene?
2 How is cell growth related to cancer? (See page 121)
3 (a) What is meant by a 'mutation' in a gene?
  (b) What can trigger a gene mutation?
4 What do the researchers think is the role of oncogenes in causing tumours?
5 What is a tumour?
6 (a) Explain what is meant by the phrase 'evolution and natural selection'.
  (b) How could they 'lose' the oncogenes?

NON-RENEWABLE RESOURCES

Minerals, energy and food are all resources that are having difficulty keeping pace with the rate at which humans are using them up. Figure 10.14 shows one estimate of the supplies of minerals that are left.

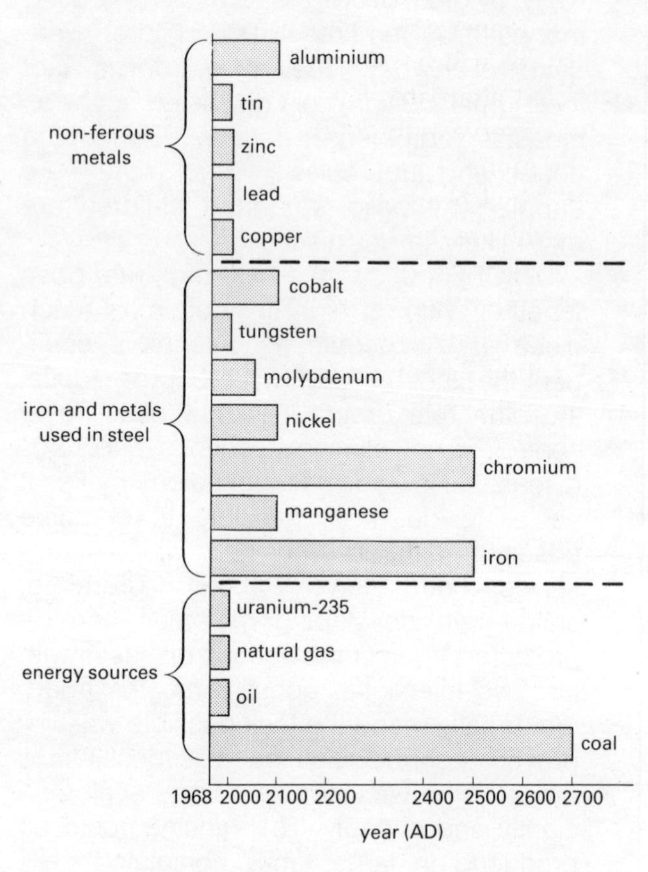

Fig. 10.14 One estimate of how long minerals will last

In both industry and agriculture increasing production depends on cheap and plentiful supplies of raw materials and fuels. As a result, industrial nations import large amounts of raw materials. These are turned into goods – which are thrown away after a few years. Opinions vary on the supplies of raw materials that are left in the biosphere. Some experts suggest that they will last longer than Fig. 10.14 shows. This is because we will be forced:
(a) to find alternatives as they get scarcer;
(b) to conserve and recycle the materials already in circulation.

Unlike hydrocarbons, metals do not disappear when they are burned. Metals are spread and 'diluted' as we use them and dump them. Conserving and recycling them will also have the added advantage of reducing the amounts of toxic metals in the environment. The next two assignments explore two ways of increasing resources.

Fig. 10.15 Returnable aluminium cans are saved for recycling

Assignment – Food or fuel?

The coming shift of farmland into energy crops, from which fuel can be distilled, poses a major threat to the world's food supplies. The rise in oil prices is making it economical to distil crops to make alcohol. The alcohol can then be used to supplement, or replace, petrol. Several crops look promising, amongst them sugar, maize, temperate sorghum and cassava.

Plant alcohol is renewable and burns cleanly. It does not increase carbon dioxide levels in the atmosphere because the plants absorb as much gas while growing as they release when their alcohol is burned. The production of alcohol creates more jobs than the oil industry, many of these in rural areas. It also helps countries become more self-sufficient in energy.

The disadvantages of production of such plants on a very large scale arise from the large amounts of land which would be needed. The land normally used to grow food crops could be taken to grow alcohol-producing crops.

Lester Brown in *Earthwatch* No. 4, 1980

1 Why is the distillation of plants to make alcohol becoming economical?
2 The crops used to produce alcohol must be rich in carbohydrates.
   (a) What are carbohydrates?
   (b) Why are they necessary to produce alcohol?
   (c) Explain what Mr Brown *really* means when he says that 'the plants are distilled' to make alcohol.
3 (a) Write down the formula for ethanol
   (b) Write an equation to summarise what happens when ethanol burns
   (c) Why does the alcohol 'burn cleanly'?
4 Explain why using alcohol as a fuel does not increase the level of carbon dioxide in the atmosphere.
5 Describe carefully the disadvantages of large-scale alcohol production.

Assignment –

Genetic engineers to feed the growing world

'Within the next few years the world will have its first genetically engineered plants capable of making their own food.

The crucial natural food element is nitrogen; which is why about a third of the world's people depend on chemical fertilisers for their survival. Genetic engineering could change this picture dramatically but the change may come only just in time to feed a growing population, Professor John Postgate told the British Association's agricultural section. He warned that there is no easy way to meet the fertiliser needs of the world in 20 years' time, when the population will have grown by at least 2½ billion. Chemical fertilisers are inefficient and demand resources which may be needed for other purposes.

Through bacteria in their root systems, some plants are able to use nitrogen from the atmosphere as their own fertiliser. One of the main lines of research has been to identify the genes which are responsible for this 'talent'.

By direct breeding techniques and genetic engineering it appears possible to create a range of plants which need no fertilisers.'

Adapted from the *Observer*, August 1983

1 Explain why nitrogen is 'the crucial natural food element'.
2 Why do so many people depend on chemical fertilisers?
3 (a) Find out the chemical formula of some fertilisers.
(b) Explain why chemical fertilisers (i) are inefficient; (ii) demand resources which may be needed elsewhere.
4 What are the advantages of a crop that can use nitrogen from the air directly?
5 Suggest how the researchers may be able use their findings to develop plants which need no fertilisers.
6 Give your opinion of the idea of reducing the world's dependence on chemical fertilisers by developing the plant types described in the article.

AND FINALLY . . . CHIPS WITH EVERYTHING

Not long ago most teenagers did not know what existed beyond their own village. Today most of them have seen the earth from space – courtesy of space probes and TV. The space programme is guided by microprocessors ('chips'), so too are ships, planes, high speed trains. 'Chips' have made travel easier, faster and cheaper. They have also improved communications around the world. As a result, people move about the globe more, industries are larger and countries join in groups such as the EEC (Common Market). Similar improvements have been possible in the power and accuracy of weapons.

Many people are concerned that the stresses facing the earth in the years ahead will cause even more wars. Optimists say that this is unlikely. In the past, when a society needed more food and resources, the only way these could be obtained was by getting more land. That meant a war to take the land. Nowadays we have the ability to increase our supplies by developing new crops, or better machines. This is a much safer and cheaper way of operating than fighting with another nation for resources.

It is not yet possible to be so optimistic in developing countries. They have yet to reach the same stage as the industrialised nations. From our understanding of science and the delicate balance in our biosphere, it should be possible to

correct many of the mistakes we made in the past;
avoid repeating the same mistakes;
prevent the sorts of activities that will upset the balance.

THE PERIODIC TABLE

I	II											III	IV	V	VI	VII	0
1 H 1																	2 He 4
3 Li 7	4 Be 9											5 B 11	6 C 12	7 N 14	8 O 16	9 F 19	10 Ne 20
11 Na 23	12 Mg 24											13 Al 27	14 Si 28	15 P 31	16 S 32	17 Cl 35.5	18 Ar 40
19 K 39	20 Ca 40	21 Sc	22 Ti	23 V	24 Cr	25 Mn	26 Fe 56	27 Co	28 Ni	29 Cu 63.5	30 Zn 65	31 Ga	32 Ge	33 As	34 Se	35 Br 80	36 Kr
37 Rb	38 Sr	39 Y	40 Zr	41 Nb	42 Mo	43 Tc	44 Ru	45 Rh	46 Pd	47 Ag 108	48 Cd	49 In	50 Sn	51 Sb	52 Te	53 I 127	54 Xe
55 Cs	56 Ba	57 La	72 Hf	73 Ta	74 W	75 Re	76 Os	77 Ir	78 Pt	79 Au	80 Hg 200.5	81 Tl	82 Pb 207	83 Bi	84 Po	85 At	86 Rn
87 Fr	88 Ra	89 Ac	104 Rf*	105 Ha	106												

58 Ce	59 Pr	60 Nd	61 Pm	62 Sm	63 Eu	64 Gd	65 Tb	66 Dy	67 Ho	68 Er	69 Tm	70 Yb	71 Lu
90 Th	91 Pa	92 U	93 Np	94 Pu	95 Am	96 Cm	97 Bk	98 Cf	99 Es	100 Fm	101 Md	102 No	103 Lr

*Or Ku.

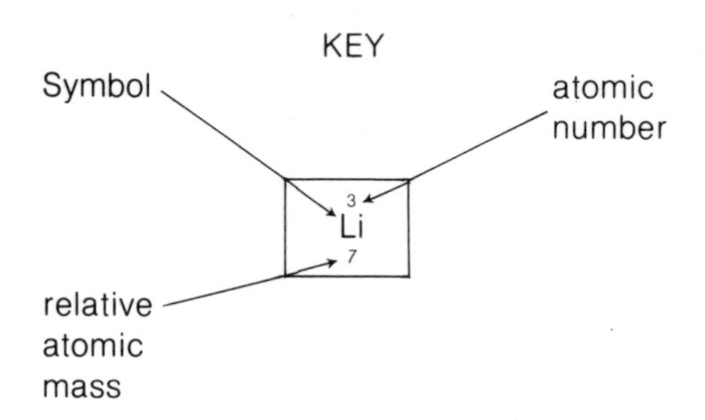

INDEX

Page numbers printed in **bold type** show the pages where the word is explained.